Nitric Oxide in Plants: A Molecule with Dual Roles

Nitric Oxide in Plants

A Molecule with Dual Roles

Edited by

Mohammad Abass Ahanger
Northwest A & F University
Xianyang, China

Parvaiz Ahmad
Government Degree College Pulwama
Jammu and Kashmir, India

WILEY

This edition first published 2022

Registered Office(s)
John Wiley & Sons, Inc., 111 River Street, Hoboken, NJ 07030, USA
John Wiley & Sons Ltd, The Atrium, Southern Gate, Chichester, West Sussex, PO19 8SQ, UK

Editorial Office
The Atrium, Southern Gate, Chichester, West Sussex, PO19 8SQ, UK

For details of our global editorial offices, customer services, and more information about Wiley products visit us at www.wiley.com.

Wiley also publishes its books in a variety of electronic formats and by print-on-demand. Some content that appears in standard print versions of this book may not be available in other formats.

Library of Congress Cataloging-in-Publication Data
Names: Ahanger, Mohammad Abass, editor. | Ahmad, Parvaiz, editor.
Title: Nitric oxide in plants : a molecule with dual roles / edited by
  Mohammad Abass Ahanger, Northwest A & F University, Xianyang,
  China, Parvaiz Ahmad, S.P. College, Jammu and Kashmir, India.
Description: Hoboken, NJ : John Wiley & Sons, 2022. | Includes
  bibliographical references and index.
Identifiers: LCCN 2021060389 (print) | ISBN 9781119800125 (hardback) |
Subjects: LCSH: Nitric oxide. | Botanical chemistry. | Nitric
  oxide--Metabolism.
Classification: LCC QK898.N58 N587 2022 (print) | LCC QK898.N58 (ebook) |
  DDC 572/.544--dc23/eng/20220103
LC record available at https://lccn.loc.gov/2021060389
LC ebook record available at https://lccn.loc.gov/2021060390

Cover image: © Vasilenko Natalya/Shutterstock
Cover design by Wiley

Set in 9.5/12.5pt STIXTwoText by Integra Software Services Pvt. Ltd, Pondicherry, India

10 9 8 7 6 5 4 3 2 1

Contents

Preface

Plant growth and development is significantly influenced by environmental factors. Due to rapid global climate change both biotic and abiotic stresses have intensified and so have the deleterious effects on normal plant growth and productivity. Environmental stress-mediated decline in plant productivity imparts significant pressure on global food security, and therefore threatens the likelihood of serious food crises in the near future for the increasing world population. Stresses result in oxidative damage through excessive generation of reactive oxygen species (ROS), thereby inducing the oxidation of lipids, proteins, and nucleic acids. Plants have evolved key mechanisms to counteract the damaging effects of stresses. Osmolyte and secondary metabolite accumulation, the antioxidant system, phytochelatin production, ion compartmentation and exclusion, etc. are some of the key mechanisms to counteract the stress factors. All these tolerance mechanisms are regulated at the gene and protein level. Plants exhibiting upregulation of the tolerance mechanisms show better performance in terms of photosynthesis, mineral uptake and assimilation, enzyme functioning, and hence yield productivity. However, it should be mentioned here that enhancing the tolerance potential by employing efficient management practices can be very effective in protecting the yield potential of plants. Every tolerance pathway mentioned above is a combination of many components, which could be organic molecules or enzymes or metabolites. From the past decade onward there has been increasing advocacy for exploiting these individual beneficial molecules to improve the tolerance pathways. In this respect exogenous usage of the key components of tolerance pathways means either foliar or through the roots. Nitric oxide (NO) is a gaseous signaling molecule that was considered toxic for plant metabolism; however, advances in research have confirmed its beneficial role in stress tolerance through its role in signaling and regulation of key developmental events including germination and programmed cell death. Fine-tuned mechanisms elicited by NO have been confirmed through metabolomic, transcriptomic, and proteomic studies. Identification of genes and other key molecules interacting in NO-mediated growth and developmental regulation under stressful conditions is being investigated. Modulations in the endogenous NO concentrations, either through stress exposure or by exogenous application of protectants, confirm the role of NO in plant stress management. In addition, physiological and biochemical studies have confirmed the vital role of optimal NO concentrations in regulation of photosynthesis, carbon assimilation, osmolyte synthesis, antioxidant and secondary metabolite metabolism, and nutrient uptake and assimilation. Interactions with phytohormones like abscisic acid, ethylene, salicylic acid, jasmonic acid, and cross-talk with other signaling molecules have been reported to play pivotal roles in

NO-mediated stress tolerance in plants. Keeping in view the above mentioned facts this book is compiled with the aim of providing the scientific community with the latest updates and future goals of NO research. *Nitric Oxide in Plants: A Molecule with Dual Roles* has 13 chapters, with every chapter having updated information about the relevant topic. The book aims to fill the existing knowledge gap in NO and plant metabolism regulation.

Chapter 1 provides an overview of the biosynthesis and regulation of NO synthesis in plants vis-à-vis its regulatory role in plant growth regulation. In addition the interaction of NO with certain key stress molecules is also discussed. Chapter 2 examines the enzymatic and nonenzymatic biosynthesis of NO, and the influence of different stress factors on NO synthesis. Moreover, modulations in NO synthesis due to osmolytes are considered, and a brief discussion about signaling components including transcription factors and phytohormones is also included. Chapter 3 discusses the reductive and oxidative pathways of NO synthesis, and the physiological, biochemical, and molecular modulations resulting from interactions of NO with auxin, gibberellic acid, cytokinin, ethylene, and abscisic acid. NO-mediated regulation of growth, photosynthesis, and tolerance mechanisms under different environmental stresses are also discussed. Chapter 4 reviews the molecular interventions for enhancing NO synthesis for its optimum exploitation for plant growth enhancement. NO-mediated regulation of phytohormones and fatty acids vis-à-vis signaling mechanisms is also the subject of this chapter. In addition, gene expression modulation and stress tolerance are examined. The role of nitrogen in production of NO and the subsequent regulation of major plant cellular pathways are described in Chapter 5. The role of beneficial microbes in NO production under normal and changing environmental conditions are dealt in Chapter 6, with a focus on physiological-, biochemical-, and molecular-level tolerance mechanisms. In addition, the interactive effects of ROS and NO in tolerance to biotic and abiotic stresses are mentioned. There is also a focus on the contribution of nitrifying and denitrifying bacteria in NO production and subsequent alterations in gene expression, as well as the role of polyamines. Chapter 7 discusses the synthesis, metabolism, and transport of ROS and reactive nitrogen species (RNS). The role of reactive oxygen, sulfur, and nitrogen species in stress signaling and the underlying cross-talk for tolerance against abiotic and biotic stresses are explained. Chapter 8 deals with understanding the role of endogenous and exogenous NO in modifying the key antioxidant pathway – the ascorbate–glutathione cycle. The role of NO-induced changes in enzymatic and nonenzymatic components of the ascorbate–glutathione cycle in plant growth regulation is examined. The authors also explain the posttranslational modifications of NO and their role in signaling and stress tolerance. In Chapter 9 the authors take the opportunity to explain the cross-talk mechanisms underlying environmental stress tolerance mediated by NO and phytohormones like auxins, cytokinins, gibberellins, abscisic acid, ethylene, jasmonic acid, salicylic acid, and brassinosteroids. The regulation (positive and negative) of several signaling pathways and the elicitation of tolerance mechanisms for improved plant performance are also discussed. Phytohormone–NO cross-talk for improved nutrient acquisition and photosynthetic regulation is the topic of Chapter 10. In addition the authors also examine the antagonistic and synergistic influence of NO on phytohormone synthesis, photosynthate partitioning, gene expression, and programmed cell death regulation. Chapter 11 covers the role of polyamines in NO synthesis and their interactive effects in alleviating the damaging effects of stresses. Polyamine metabolism-induced ROS production and subsequent triggering of NO accumulation are also discussed. In Chapter 12 the authors consider the NO-mediated modulation of the synthesis and transport of auxins, abscisic acid, and brassinosteroids under normal and adverse environmental conditions. In addition the molecular mechanisms of the signaling events resulting from their interactions are discussed. Finally, Chapter 13 looks at the interactive effects of NO and salicylic acid, jasmonic acid, and ethylene, with a focus on the antioxidant functioning of plants with regard to stress tolerance. Posttranslational modifications by NO and its cross-talk with salicylic acid, jasmonic acid, and ethylene for regulation of several downstream targets are discussed.

This volume presents a wealth of knowledge on the current understandings of NO and growth regulation in plants, and provides insights into the abiotic and biotic stress tolerance mechanisms regulated by NO. The chapters included in this volume are the authors' own work and necessary editorial modifications were incorporated wherever the need was felt. We have tried our best to gather the most up-to-date information about NO research; however, some areas are inevitably missing and there is always the possibility of errors creeping into the book. Therefore, we seek the readers' indulgence, and suggestions are welcome for improvements in future editions. We are very thankful to the well-versed contributors who accepted our invitation and contributed chapters despite the hardships created by the pandemic. Moreover, we extend our sincere thanks to the entire team at Wiley, especially Kerry Powell, Rebecca Ralf, and Camille Bramall, for their invaluable suggestions and constant help in accomplishing this project and its publication.

Dr. Mohammad Abass Ahanger
Dr. Parvaiz Ahmad

List of Contributors

Altaf Ahmad
Department of Botany
Aligarh Muslim University
Aligarh
Uttar Pradesh
India

Muhammad Ashraf
The University of Agriculture
Faisalabad
Pakistan

Mehtab Muhammad Aslam
Center for Plant Water Use and Nutrition Regulation
College of Life Sciences
Joint International Research Laboratory of Water and Nutrient in Crops
Fujian Agriculture and Forestry University
Fuzhou
Fujian
China

Habib-ur-Rehman Athar
Institute of Pure and Applied Biology
Bhauddin Zakria University
Multan
Pakistan

Ester Badiani
Dipartimento per la Innovazione nei Sistemi Biologici
Agroalimentari e Forestali
Università della Tuscia
Viterbo
Italy

Maurizio Badiani
Dipartimento di Agraria
Università Mediterraneadi Reggio Calabria
Loc. Feo di Vito
Reggio Calabria
Italy

Sagar Bag
Laboratory of Microbial Interaction
School of Biotechnology
Presidency University
Kolkata
West Bengal
India

Avishek Banik
Laboratory of Microbial Interaction
School of Biotechnology
Presidency University
Kolkata
West Bengal
India

Renu Bhardwaj
Plant Stress Physiology Lab
Department of Botanical and Environment Sciences
Guru Nanak Dev University
Amritsar
Punjab
India

Priti Chauhan
Department of Bioscience and Biotechnology
Banasthali Vidyapith
Rajasthan
India

Mario Ciaffi
Dipartimento per la Innovazione nei Sistemi Biologici
Agroalimentari e Forestali
Università della Tuscia
Viterbo
Italy

Loitongbam Lorinda Devi
National Institute of Plant Genome Research
Aruna Asaf Ali Marg
New Delhi
India

Debjani Dutta
Department of Plant Physiology
Faculty of Agriculture
Bidhan Chandra KrishiViswavidyalaya
Mohanpur
Nadia
West Bengal
India

Padmanabh Dwivedi
Department of Plant Physiology
Institute of Agricultural Sciences
Banaras Hindu University
Varanasi
Uttar Pradesh
India

Kaneez Fatima
Department of Life Sciences
University of Management and Technology (UMT)
Johar Town
Lahore
Pakistan

Abah Felix
Crop Protection College
Fujian Agriculture and Forestry University
Fuzhou
Fujian
China

Arajmand Frukh
Department of Botany
School of Chemical and Life Sciences
Jamia Hamdard
New Delhi
India

Shreya Gupta
National Institute of Plant Genome Research
Aruna Asaf Ali Marg
New Delhi
India

Asma Imran
National Institute for Biotechnology and Genetic Engineering – Campus Pakistan Institute of Engineering and Applied Sciences (NIBGE-C-PIEAS)
Faisalabad
Pakistan

Yachana Jha
N.V. Patel College of Pure and Applied Sciences
CVM University
Vallabh Vidyanagar Nagar
Anand
Gujarat
India

Snehashis Karmakar
Department of Plant Physiology
Faculty of Agriculture
Bidhan Chandra Krishi Viswavidyalaya
Mohanpur
Nadia
West Bengal
India

Harsimran Kaur
Plant Protection Division
PG Department of Agriculture
Khalsa College
Amritsar
Punjab
India

Mushtaq Ahmad Khah
Department of Botany
Shri JJT University
Jhunjhunu
Rajasthan
India

Sukhmeen Kaur Kohli
Plant Protection Division
PG Department of Agriculture
Khalsa College
Amritsar
Punjab
India

Plant Stress Physiology Lab
Department of Botanical and Environment Sciences
Guru Nanak Dev University
Amritsar
Punjab
India

Asha Kumari
ICAR-Vivekananda Parvatiya Krishi Anusandhan Sansthan
Almora
Uttarakhand
India

Saima Liaqat
Department of Botany
School of Chemical and Life Sciences
Jamia Hamdard
New Delhi
India

Rayees Ahmad Mir
School of Studies in Botany
Jiwaji University
Gwalior
India

Anupam Mondal
Laboratory of Microbial Interaction
School of Biotechnology
Presidency University
Kolkata
West Bengal
India

Sananda Mondal
Department of Crop Physiology
Institute of Agriculture
Visva-Bharati
Sriniketan
Birbhum
West Bengal
India

Anna Rita Paolacci
Dipartimento per la Innovazione nei Sistemi Biologici
Agroalimentari e Forestali
Università della Tuscia
Viterbo
Italy

Stefania Pasqualini
Dipartimento di Chimica
Biologia e Biotecnologie
Università di Perugia
Perugia
Italy

Fozia Sardar
National Institute for Biotechnology and Genetic Engineering – Campus Pakistan Institute of Engineering and Applied Sciences (NIBGE-C-PIEAS)
Faisalabad
Pakistan

Iffat Shaheen
College of Agriculture
Bahauddin Zakariya University
Multan
Pakistan

Binny Sharma
Department of Plant Physiology
Institute of Agricultural Sciences
Banaras Hindu University
Varanasi
Uttar Pradesh
India

Amar Pal Singh
National Institute of Plant Genome Research
Aruna Asaf Ali Marg
New Delhi
India

Bansh Narayan Singh
Department of Plant Physiology
Institute of Agricultural Sciences
Banaras Hindu University
Varanasi
Uttar Pradesh
India

Agostino Sorgonà
Dipartimento di Agraria
Università Mediterranea di Reggio Calabria
Loc. Feo di Vito
Reggio Calabria
Italy

Muhammad Waseem
Horticulture
South China Agricultural University
Guangzhou
China

Rinukshi Wimalasekera
Department of Botany
Faculty of Applied Sciences
University of Sri Jayewardenepura
Gangodawila
Nugegoda
Sri Lanka

Mobina Ulfat
Department of Botany
Lahore College for Women University
Lahore
Pakistan

Zafar Ullah Zafar
Institute of Pure and Applied Biology
Bhauddin Zakria University
Multan
Pakistan

1

Nitric Oxide: A Dynamic Signaling Molecule Under Plant Stress

Asha Kumari[1], Binny Sharma[2], Bansh Narayan Singh[2], and Padmanabh Dwivedi[2]

[1]*ICAR-Vivekananda Parvatiya Krishi Anusandhan Sansthan, Almora, Uttarakhand, India*
[2]*Department of Plant Physiology, Institute of Agricultural Sciences, Banaras Hindu University, Varanasi, Uttar Pradesh, India*

1.1 Introduction

Nitrogen monoxide, often known as nitric oxide, is a physiologically active chemical that is widely used in animal and plant signaling mechanisms. In plants and animals, it is an intracellular and intercellular signaling molecule with a variety of regulatory roles. Its functions in the central nervous, cardiovascular, and immunological systems, platelet inhibition, programmed cell death, and host responses to infection, among other things, have been widely studied in animals.

Because of the presence of an unpaired electron, it is a highly reactive gaseous molecule that occurs with oxygen in a variety of reduced states such as nitroxyl ion (NO^-), nitric oxide free radical ($NO^•$), and nitrosonium (NO^+). These NO-derived molecules are referred to as reactive nitrogen species (RNS). NO influences signaling in biological systems through a variety of mechanisms. The interaction of $NO^•$ with O_2 results in the formation of several redox compounds (including $NO_2^•$, N_2O_3, and N_2O_4), which may react with cellular amines and thiols or simply change to form the metabolites nitrogen dioxide radical (NO_2^-) and nitrate (NO_3^-) (Wendehenne et al. 2001). NO combines with dioxygen to form NO_2 or with reactive oxygen species (ROS) to form peroxynitrite ($ONOO^-$), which triggers cellular damage. $NO^•$ facilitates electrophilic assault on reactive sulfur, oxygen, nitrogen, and aromatic carbon centers, with thiols being the most reactive of the reactive teams. Nitrosation is the name given to this natural process. Nitrosation of numerous enzymes or proteins results in chemical change, which may affect the function of those entities. These alterations are reversible, and supermolecule nitrosation–denitrosation might be a crucial mechanism for controlling signal transduction (Hayat et al. 2010).

In contrast to the mammalian system, the cellular/subcellular localization of NO production in plants is exceedingly diverse and contentious. The production of NO in plants is determined by the plant's physiological condition. This includes NO production during root development, stomatal movement control, blooming, plant component expansion, and leaf senescence (Neill et al. 2002; Mishina et al. 2007). NO is produced in plants through nonenzymatic and accelerator systems, depending on the plant species, organ, or tissue, as well as the plant's state and ever-changing environmental circumstances. The most effective recognized NO sources in plants are as a substrate by cytosolic (cNR) and membrane-specific nitrate enzyme (PM-NR), and NO synthesis by many arginine-dependent gas synthase-like activities (NOS).

According to studies, mitochondria are a major source of arginine- and nitrite-dependent NO synthesis in plants. Tischner et al. acquired the first evidence for mitochondrial NO synthesis in plants when they assessed NO production under anoxic conditions from the unicellular blue green alga *Chlorella sorokiniana* (Tischner et al. 2004). This green alga does not create NO when exposed to nitrate (NO_3), but it does create NO when exposed to nitrite (NO_2). NO generation was also inhibited by mitochondrial electron transport inhibitors. Shortly after, mitochondrial NO synthesis in higher plants was discovered. Gupta et al. discovered mitochondrial NO production in barley plants grown in anoxic conditions (Gupta and Kaiser 2010). Under anoxic circumstances, a tobacco Nia 1, 2 (nitrate reductase-deficient) cell suspension was able to manufacture NO from exogenous nitrite, despite the absence of nitrate reductase (which can also manufacture NO from nitrite) (Gupta et al. 2011). Other putative NO producers in plants include xanthine oxido-reductase, peroxidase, and cytochrome P450. NO is a ubiquitous chemical that is found in all eukaryotes. The NR system is by far the most effective and well-characterized mechanism for NO generation in plants. In this case, the cytosolic NR mostly catalyzes the reduction of nitrate to NADH as the predominant negatron donor. NR's NAD(P)H-dependent NO production has been demonstrated in vitro and in vivo (Rockel et al. 2002). The biological significance of NR activity as a source of NO was first shown in *Arabidopsis* guard cells by Desikan et al. (2002). The peroxisomal catalyst organic compound enzyme can also catalyze group reduction to NO (XOR). XOR activity in pea (*Pisum sativum*) leaves is linked to peroxisomes, and as a result, the possibility of interaction between the construction of reactive chemical elements and reactive gas species (ROS and RNS, respectively) has been suggested (del Río et al. 2004). NO production in animals was demonstrated by a chemical reaction of arginine transforming into citrulline mediated by the enzyme NO synthase (Palmer et al. 1987). Following the discovery of a purpose for NO in plants in 1998 (Delledonne et al. 1998; Durner et al. 1998), several researchers began to look for NOS activity in plants, despite the fact that the *Arabidopsis thaliana* ordering failed to reveal any factor with significant similarity to animal NOS (Moreau et al. 2010).

1.1.1 Historical Evidence and Biosynthesis of Nitric Oxide

Several studies have been conducted over the past several decades to investigate the presence and characteristics of NO gas in living beings. Gas, as a versatile molecule, has piqued curiosity and opened up new opportunities for research. NO gas is an atom gas with well-defined communication roles in mammalian systems, serving as a second messenger during vasorelaxation, neurotransmission, immunity, and toxicity. It is now clear that NO performs a critical role in animal physiology. Because of its extensive biological relevance, NO was designated "Molecule of the Year" in 1992 by *Science*, and Furchgott, Murad, and Ignarro were given the Nobel Prize in Physiology and Medicine in 1998. Furchgott discovered in 1980 that an unknown molecule found in animal tissue could relax smooth muscle cells, and he dubbed it EDRF (endothelium-derived reposeful factor). Murad discovered years ago that vasodilators activate guanylate cyclase (GC), which creates cyclic guanosine monophosphate and relaxes muscle fibers. This discovery begged the question of how a vasodilator outside the cell could influence a catalyst inside the cell. The solution was that the vasodilator was tainted with NO residues. Murad then bubbled NO gas across swish muscle cells, activating gigacycles per second. Thus, even before eukaryotes were thought to produce NO, he postulated that hormones may regulate swish muscles via NO. Years later, Ignarro demonstrated that NO has comparable chemical behavior to EDRF and is, in fact, a twin of EDRF. NO release from plants was initially suspected by Klepper in soybeans in 1975, much earlier than in mammals (Klepper 1979). The vast biological importance of gas in plants was established in the 1990s (Gouvea et al. 1997; Leshem et al. 1998).

In animals, NO has since been recognized as an important signaling molecule in maintaining blood pressure within the circulatory system, stimulating host defenses within the system, controlling neural transmission within the brain, controlling organic phenomena, protoplasm aggregation, learning and memory, male sexual function, toxicity and cytoprotection, the development of artery hardening, and a variety of other functions. It functions as a secondary transmitter in mammalian systems during vasorelaxation, neurotransmission, immunity, and toxicity. As a result, they play critical roles in animal physiology.

However, unlike animal physiology, the physiology and chemical chemistry of NO in plants is less well known. NO has the potential to be a dynamic bioactive molecule that plays an important physiological role in plants and animals.

1.1.2 NO Biosynthesis in Plants

The process of NO production has been explored in a variety of organisms, including microorganisms, alga, lichens, gymnosperms, and angiosperms (Rőszer et al. 2014). NO synthesis utilizes both accelerator and nonaccelerator mechanisms. Body parts, plastids, mitochondria, and peroxisomes are important sites for NO production (Rőszer 2012a, 2012b). Furthermore, multiple organelles, including protoplasm, cell wall, endoplasmic reticulum, and apoplast, generate NO in higher plants (Fröhlich and Durner 2011). Chakraborty and Acharya (2017) distinguish between subtractive and aerobic NO production mechanisms. The protoplasm, mitochondria, plastid, peroxisomes, and apoplast are the primary sites of subtractive NO_X production, which is mediated by the nitrate enzyme or mitochondrial negatron transport chain and deoxygenated proteins containing heme. The aerobic route of NO production begins with L-arginine, which appears to include the enzyme NO synthase. Despite the fact that numerous genes and proteins coding for NOS enzymes are known in the class system, prokaryotes, and eukaryotes, the kingdom Plantae is still little characterized (Figure 1.1).

1.2 The Function of Nitric Oxide in Plants

Nitric oxide is a common chemical molecule that is important in plant physiological activities. NO sources are useful in breaking dormancy and boosting seed germination in a variety of crops (Bethke et al. 2007; Giba et al. 2007; Prado et al. 2008; Albertos et al. 2015; Sanz et al. 2015). Similar research has been done to show the impact of organic nitrates in enhancing light-dependent and phytochrome-regulated germination in *Pauwlonia tomentosa* and *Stellaria* medium (Grubisic et al. 1992; Jovanovic et al. 2005). NO is involved in the

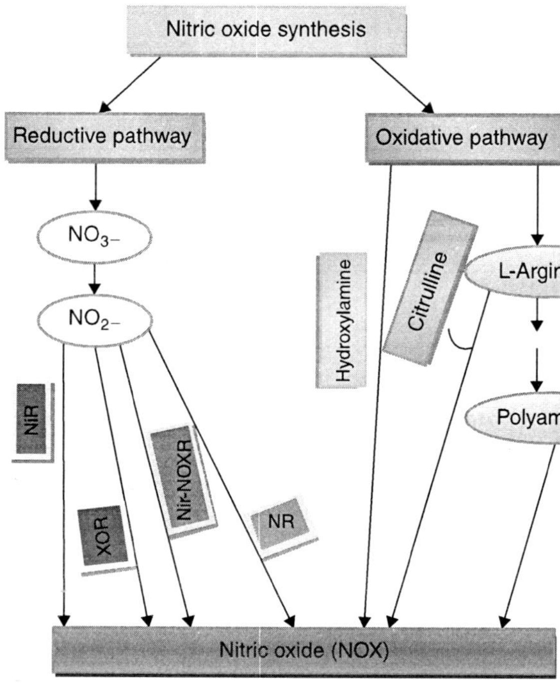

Figure 1.1 Diagram representing NO synthesis. NO_3^-, nitrate; NiR, plastidial nitrate reductase; XOR, xanthine oxido-reductase; NiR-NOR, nitrite reductase; NR, nitrate reductase; PA, polyamines.

regulation of catalase, ascorbate peroxidase, and aconitase activities (Clarke et al. 2000; Navarre et al. 2000), in cell wall lignification (Ferrer and Ros Barcelo 1999), the regulation of ion channels of guard cells(Garcia et al. 2003), mitochondrial and chloroplastic functionality (Yamaski et al. 2001), cell death (Pedroso et al. 2000), senescence (Hung and Kao 2003), accumulation of ferritin (Murgia et al. 2002), wound signaling (Orozco-Cardenas and Ryan 2002), cytokinin-induced programmed cell death (Neill et al. 2003), and abscisic acid (ABA)-induced stomatal closure (Neill et al. 2002). NO mediates maturation and senescence, operates on ethylene antagonism (Lamattina et al. 2003), induces increase of flavonoid production in *Camellia sinensis* L., and endogenous NO stimulates brassinosteroid (Li et al. 2017). Nitric oxide has a visible part to play in the formation of plant roots and shoots. It promotes root and plant development in many plants (Corpas et al. 2006, 2015). Controlling morphogenesis, growth, and development in plants requires targeted NO control (Hebelstrup et al. 2013). NO is primarily necessary for the establishment of plant-microbe interactions, which regulate N_2-fixing symbiotic relationships and nodule senescence (Hichri et al. 2015). NO influences senescence in several plant species, including *Arabidopsis*, tobacco, pea, wheat, and others (Procházková et al. 2011), as well as flower formation in olive and *Arabidopsis* (Seligman et al. 2008; Zafra et al. 2010; Procházková et al. 2011). Nitric oxide is a crucial chemical in agriculture. It promotes seed germination, reduces postharvest losses by delaying fruit ripening, and improves the shelf life and quality of cut and detached flowers. NO also increases the activity of antioxidant enzymes in plants, which influences RNS and ROS metabolism (Corpas and Palma 2018).

1.3 NO's Role in Biotic Stress

Several studies conducted within the past decade have revealed that NO is engaged in communicating defense responses during plant–pathogen interactions. Pathogen challenges typically result in hypersensitization (HR). There is evidence that NO, in addition to forming reactive oxygen intermediates (ROIs) and salicylic acid (SA), plays a vital communication function throughout life (Delledonne et al. 1998; Durner and Klessig 1999).

Hypersensitized necrobiosis via NO is a prominent example of programmed cell death (PCD). NO-donor treatment of plant structure has been found to trigger chromatin granule condensation and dioxyribonucleic acid fragmentation (Clarke et al. 2000; Pedroso et al. 2000). Furthermore, NO-induced necrobiosis may be silenced by an animal caspase-1 molecule (Clarke et al. 2000). Despite studies demonstrating proteolytic enzyme activity in plants (D'Silva and Poirier 1998; Hatsugai et al. 2004; Rojo et al. 2004), associated transgenic plants with an overexpression of a proteolytic enzyme inhibitor – protein p35 and Op-IAP – show homologous recombination (HR) inhibition (Dickmann et al. 2001; Del Pozo and Lam 2003).

Plants have practical homologs of animal caspases, known as metacaspases (Bozhkov et al. 2005). Belenghi et al. (2007) recently demonstrated that *A. thaliana* metacaspase nine (AtMC9) is rendered inactive by *S*-nitrosylation of a cysteine residue in AtMC9. In turn, the mature form of this necrobiosis fiduciary is resistant to *S*-nitrosylation by NO. Exogenous NO promoted necrobiosis in *Arabidopsis* suspension cells in amounts comparable with those produced by cells challenged with an avirulent bacterium (Clarke et al. 2000). However, reactions between NO and H_2O_2 produce either singlet oxygen or free radicals (Noronha-Dutra et al. 1993), which might cause necrobiosis. A simultaneous rise in NO and H_2O_2 triggered necrobiosis in soybean and tobacco cell suspensions, but a rise in only one of the preceding variables promoted necrobiosis very modestly (Delledonne et al. 1998; de Pinto et al. 2002). Furthermore, microscopic anatomy studies revealed that either injection of NO donors or a change in H_2O_2 level has no effect on HR in infected oat cells, despite the fact that each molecule was required for the initiation of death in neighboring cells (Tada et al. 2004). The mechanism by which NO and H_2O_2 kill remains mostly unclear. The reaction of NO with O_2 creates peroxynitrite, an exceedingly poisonous chemical for animal cells that mediates necrobiosis. $ONOO^-$ is relatively nontoxic to plants (Delledonne et al. 2001). However, it has been shown that $ONOO^-$ stimulates pathogenesis-related protein (PR-1) accumulation in tobacco leaves (Durner and Klessig 1999) and supermolecule nitration modulates cell oxido-reduction status (Delledonne et al. 2001). Furthermore, it was demonstrated that whereas peroxynitrite was responsible for the death of most *Arabidopsis* cells in response to avirulent *Pseudomonas syringae*, scavenging of this ion did not result in efficient defense against avirulent bacterium (Alamillo and Garcia-Olmedo 2001).

NO, when rebuilt into a peroxynitrite particle, may join forces in killing microorganisms (Durner and Klessig 1999; Romero-Puertas et al. 2004), though it has not been determined if NO and its derivatives are directly hazardous to plant diseases (Garcia-Olmedo et al. 2001). In vitro, it was undeniable that the growth of virulent and avirulent *Pseudomonas* bacterium was stifled by both NO and the plant system producing peroxynitrite (sodium nitroprusside + hypoxanthine/xanthine oxidase) (Noronha-Dutra et al. 1993; Garcia-Mata and Lamattina 2002). Romero-Puertas et al. (2004) proposed that $ONOO^-$ might be continuously synthesized in healthy cells, implying that plants may evolve certain detoxifying mechanisms. Ascorbates may have a significant role in the inactivation of $ONOO^-$ in animal cells (Arteel et al. 1999). Given that vitamin C (AsA) is a quantitatively dominant inhibitor in plant cells (Smirnoff 2000), it is possible that AsA also contributes to $ONOO^-$ breakdown in plant cells. The genetic composition of the plant (R genes) and the microbe (avr) seems to be strongly related to early NO production, referred to as NO burst (Mur et al. 2005; Bennett et al. 2005). Prompt gas generation (30–45 min after inoculation) was seen in noncompatible systems of *P. s.* pv. *phaseolicola*–tobacco and *P. s.* pv. *tomato–Arabidopsis*. This early NO burst happened six hours before the appearance of obvious HR-type death signs and directly preceded H_2O_2 production. As proof, plants injected with a mutant of the avirulent bacterium (hrp) incapable of supplying the avr supermolecule to the plant showed a lack of NO emission (Mur et al. 2005).

The use of fluorescent dyes in cytochemical methods has made it possible to present the mechanics of NO generation inside the stratum of tobacco leaves treated with cryptogein. After tissue treatment, a supermolecule elicitor derived from the morbific plant life *Phytophthora cryptogea* stimulated NO buildup at several minute intervals (Foissner et al. 2000). In turn, Prats et al. (2005) discovered a significant, transient rise in NO level before programmed death of barley dermal cells infected with *Blumeria graminis* f. sp. *Hordei* using DAF-2DA (5,6-diaminofluorescein diacetat) dye. Furthermore, Zeier et al. (2004) isolated an *Arabidopsis* transgenic line with overexpression of nitric oxide dioxygenase (NOD) gas, an accelerator catalyzing the twofold reaction of NO to nitrates. Transgenic plants treated with an avirulent strain of *P.s.* pv. *tomato* avr B showed decreased NO production and a significantly lower mortality rate, confirming that NO is required for HR stimulation. According to Modolo et al. (2006), *P. syringae* is diminished concurrently in NR-deficient double mutants (Nia1 and Nia2) of *Arabidopsis* because these plants lack L-arginine and NO_2, endogenous precursors for NO production.

Until recently, it was thought that HR is only seen in incompatible relationships, where the plant has a resistance gene encoding R and the microbe has a virulence gene, avr (Levine et al. 1994). However, it has recently been demonstrated that HR of host cells may also arise in plants to partially protect against a specific microbe and only in the situation of non-host-kind resistance (Vleehouwers et al. 2000).

The majority of the evidence demonstrating that NO works as a messenger in gene-for-gene defensive responses was gained by studying completely separate plant–biotrophic microbe systems (Delledonne 2005). It has yet to be confirmed what part NO plays inside the plant and the necrotrophic pathogen. In a review in 2004 Van Baarlen noted that the formation of endogenous NO and H_2O_2 was recorded in distinction to the compatible interaction, i.e. during disease development of liliaceous plant and *Botrytis elliptica*. Another event associated with plant resistance in which NO appears to be involved is phytoalexin buildup (Able 2003). Exogenous NO stimulated the accumulation of rishitin in potato tubers. Furthermore, the outcome of this compound's inhibited production was identified after the use of a NO scavenger (Noritake et al. 1996). Several times after NO treatment of soybean cotyledons, the production of particular phytoalexins was identified (Modolo et al. 2002). NO could also play a role in the emergence of general systemic acquired resistance (SAR). In tobacco, exogenous NO increases the accumulation of salicylic acid, which plays a key role in SAR (Durner et al. 1998). Activation of the PR-1 macromolecule, produced by

NO, occurs with the participation of SA, because an identical result was not seen in transgenic plants unable to accumulate SA (NahG). Furthermore, disease spots formed by tobacco mosaic virus (TMV) on leaves pretreated with NO were dramatically decreased as compared with transgenic plants. SAR was lowered by using inhibitors specific for animal NOS or NO scavengers (Song and Goodman 2001). As a result, our findings suggest that NO plays an important role in the development of a distal signal network, resulting in increased SAR in tobacco.

NO may be transferred to plants in the form of nitrosoglutathione (GSNO), much as in mammals' vascular systems (Durner and Klessig 1999). It is hypothesized that GSNO may act as both an intracellular and organismal NO carrier, and that it is distributed throughout the plant via vascular tissue bundles. Glutathione-dependent formaldehyde dehydrogenase (GS-FDH)/I-nitrosoglutathione reductase (GSNOR) may play an important role in turning off/on the NO or GSNO signal, as well as modifying the amount of intracellular thiols, which may cause nitrosative stress (Diaz et al. 2003). In *Arabidopsis*, Feechan et al. (2005) discovered that the deletion of AtGSNOR1, an *S*-nitrosoglutathione enzyme, resulted in an increase in cellular *S*-nitrosothiols, which was correlated with a decrease in resistance to microbial infection. Throughout the prevalence of avirulent microorganisms and the consequent hypersensitized response, NO demonstrates an unusually wide range of affinities to a variety of signaling chemicals. The popularity of an avirulent microbe is associated with a robust aerobic burst, during which there is a redoubled creation of ROS/RNS, primarily superoxide ($O_2^{\bullet}$), peroxide (H_2O_2), or gas (NO), and the commencement of an avirulent microbe. In soybean cell suspension infected with avirulent *P. syringae* pv. *glycinea*, treatment with NO donor, sodium nitroprusside, significantly increased the induction of death by exogenous H_2O_2 or ROS (Delledonne et al. 1998). This response was significantly inhibited not solely by the NO scavenger 2-4-carboxyphenyl-4,4,5,5-tetramethylimidazoline-1-oxyl-3-oxide (CPTIO), but additionally by diphenyleneiodonium, a substance in the leukocyte NADPH enzyme that inhibits the plant oxidative burst, and by an enzyme that destroys H_2O_2; NO reacted synergistically, resulting in a 10-fold increase in ROI-induced death (Delledonne et al. 1998).

NO production, along with H_2O_2 and O_2, was observed to be associated with elicitor-elicited mortality in Mexican cypress. The co-accumulation of ROS and NO, as well as the interplay between H_2O_2, O_2, and NO, mediated the death response. The use of protein scavengers/inhibitors was used to investigate the role of NO and O_2 in death. NO and H_2O_2 reciprocally promoted each other's assembly, whereas NO and O_2 reciprocally suppressed each other's production. The interaction between NO and O_2, but not between NO and H_2O_2, triggered PCD via peroxynitrite ($ONOO^-$) (Neill et al. 2002; Zhao 2005). The stress response in bacterially triggered PCD in soybean and *Arabidopsis* concerned reactions to NO and H_2O_2, with interactions between NO and H_2O_2 being synergistic and additive in different ways, which is consistent with previous findings. After exposing tobacco leaves to a high intensity of stress, NO-mediated death was seen in enzyme-deficient (CAT1AS) plants but not in wild types, revealing the interaction of NO with H_2O_2 throughout the NO-mediated death response (Zago et al. 2006).

Plants protect themselves from microorganisms by triggering elaborate defenses. Similarly, it was recently demonstrated that NO works synergistically with ROS to prolong host mortality in soybean suspension cells, and that NOS inhibitors impair the hypersensitized resistance response in *Arabidopsis* and tobacco (Delledonne et al. 1998). It was discovered that the host plant has intracellular defense mechanisms to combat the oxidative stress caused by pathogens. Increased activity of the catalyst inhibitor system, total phenol content, and phenyl ammonia lyase activity slowed oxidative burst, which was presumably mediated by excessive chemical group content and hence involved gas signaling; this explained the protection against pathogen stress (Dwivedi et al. 2016). Following that, extensive studies revealed that NO could influence the expression of genes encoding a variety of plant effector and restriction proteins (Polverari et al. 2003; Huang et al. 2004; Parani et al. 2004; Zago et al. 2006). Furthermore, recent research on catalase-deficient tobacco plants discovered a small number of genes that are particularly controlled by NO or H_2O_2 (Zago et al. 2006). Interestingly, most of the notable genes were affected by both NO and H_2O_2, showing a significant overlap in these two separate chemical process signaling agents (Figure 1.2).

1.3.1 Interaction of NO with Other Molecules to Confer Biotic Stress Responses in Plants

Plants defend themselves against microorganism attack by activating elaborate defenses. Similarly, it has been recently demonstrated first that NO acts synergistically with ROS to extend host death in soybean suspension cells, and second that NOS inhibitors compromise the hypersensitized resistance response in *Arabidopsis* and tobacco (Delledonne et al. 1998). NO and ROS induce SA synthesis, and elevated SA may subsequently result in improved NO and ROS levels. SA additionally induces expression of genes that may be NO targets/sensors (e.g. pathogen-induced oxido-reductase [PIOX]). What is more, SA may enhance the results of NO activity through interaction with many NO_X-regulated enzymes (e.g. catalase and aconitase). On the other hand, SA, which is a very important antioxidant in mammals, may counter the effects of NO. As an example, NO blocks respiration via inhibition of cytochrome enzyme, while SA induces production of various NO-resistant enzymes and sensors/targets in plants (e.g. pathogen-induced oxygenase, PIOX). NO and SA have an effect on common targets (e.g. catalase, aconitase). SA counteracts NO action (e.g. NO inhibits cytochrome enzyme, whereas SA induces a different oxidase). The induction of necrobiosis and/or defense factor activation occurs via the action of NO, SA, and ROS. NO is not only thought to perform throughout the event of hypersensitive cell death but is also needed in the establishment of disease resistance. Melatonin also enhances pathogen resistance in plants by increasing the levels of SA and ethylene. Lee et al. (2015) reported that decreased level of melatonin content in *Arabidopsis* mutant lowers the SA content thereby reducing resistance to pathogen infection. It also stimulates elevation of endogenous NO levels and increased expression of defense-related genes, along with disease resistance against Pst DC3000 infection.

Melatonin confers plants with innate immunity via the NO- and H_2O_2-dependent pathway by stimulating mitogen-activated protein kinase kinase kinase 3 (MAPKKK3) and oxidative signal-inducible1 (OXI1). Melatonin stimulates accumulation of NO and SA and improved tomato TMV resistance (Zhao et al. 2019).

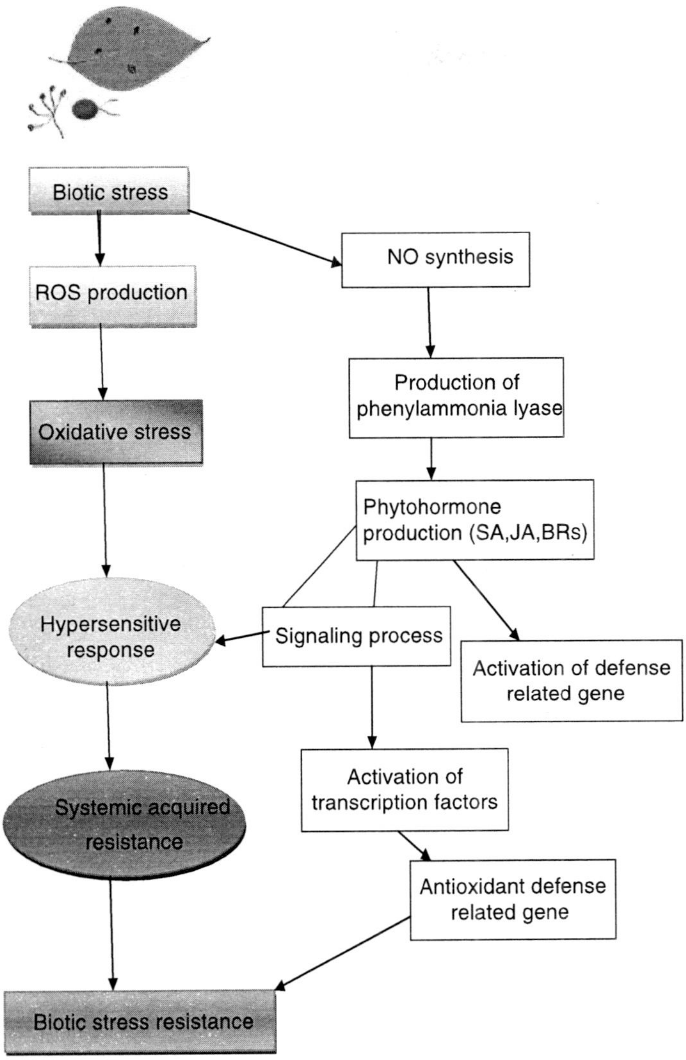

Figure 1.2 Schematic representation of NO signaling against biotic stress in plants.

Interaction between NO and glutathione is an important component to the NPR1-dependent defense signaling pathway in *A. thaliana* (Kovacs et al. 2015). Cao et al. (2019) found that brassinosteroids aid in stimulating the susceptibility of maize to maize chlorotic mottle virus infection in a NO-dependent manner. Nitric oxide further acts as a downstream signaling molecule in brassinosteroid-mediated virus susceptibility to maize chlorotic mottle virus in maize. The brassinosteroids provide systemic resistance against viruses along with H_2O_2 and NO in *Nicotiana benthamiana* (Deng et al. 2016). However, several studies have demonstrated the integrative role of brassinosteroids with NO in response to pathogen attack in various species (Hayat et al. 2010; Shi et al. 2015; Zou et al. 2018; Kohli et al. 2019).

1.4 Nitric Oxide's Role in Abiotic Stress

Existing climatic aberrations cause plants to be exposed to a variety of abiotic stresses such as drought, temperature increases, salinity, and oxidative damage. Because plants are sessile, they must deal with changing environmental conditions throughout their lives. Though plants develop various strategies to cope with adverse conditions, plant scientists and researchers are discovering new methods to learn about plants in stressful environments. In the presence of salt, substantial metals, drought, and oxidative stress, NO plays a critical role in abiotic stress management (Zhao et al. 2001, 2004; Uhida et al. 2002; Kopyra and Gwóźdź 2003). Abiotic stressors are a key stumbling block to agricultural growth while also jeopardizing food security (Figure 1.3). They also jointly

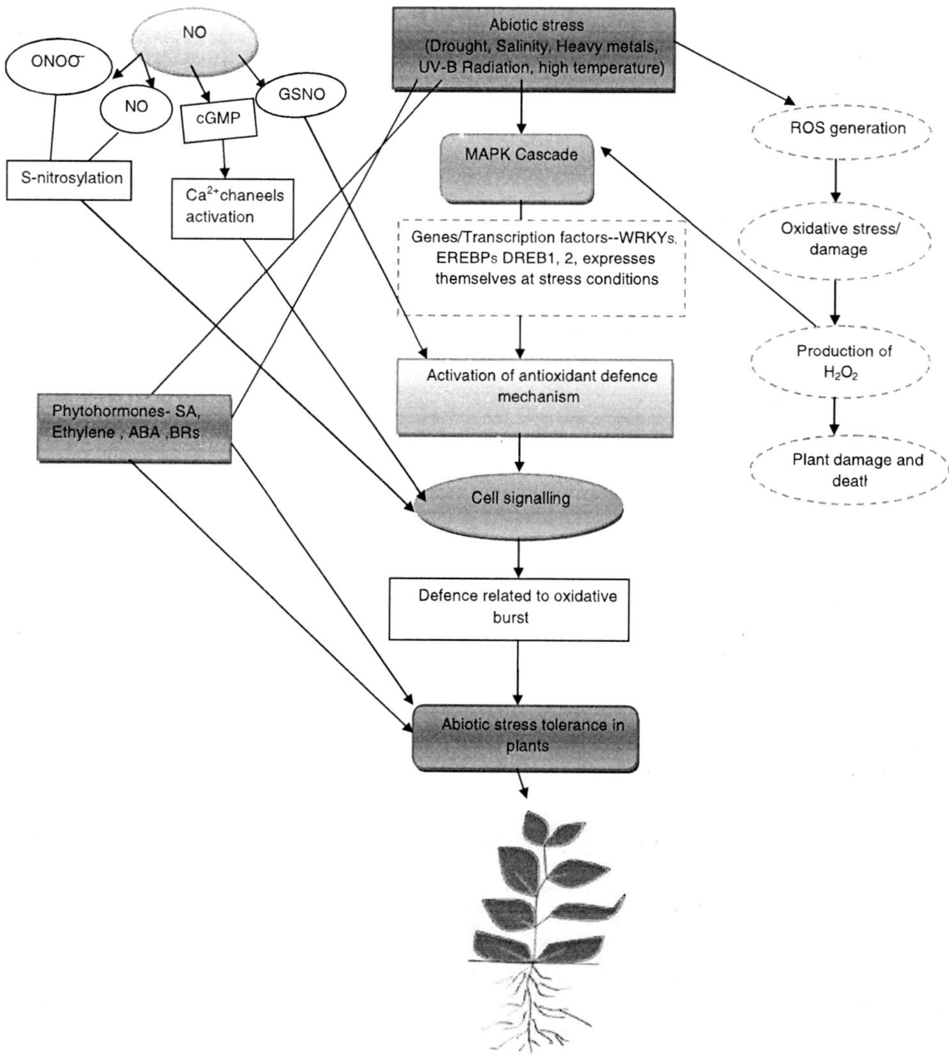

Figure 1.3 Schematic illustration of NO signaling in abiotic stress.

influence the metabolism of plants and the defense of inhibitors. Recent NO studies show that they play a serious role in plant growth and mitigate environmental stress (Aroca et al. 2015; Corpas et al. 2019). Drought tolerance of cut leaves and seedlings of wheat was improved by exogenous NO application (Garcia-Mata et al. 2001). NO also mediates drought tolerance by the activation of numerous enzymes and plant metabolism inhibitors (Filippou et al. 2014; Shi et al. 2014). In many species, NO mediates stomatal rotation and defense methods in water stress (Greco et al. 2012; Garcia-Mata and Lamattina 2013; Chen et al. 2016). ROS and NO work synergistically to address the stress in water that causes ABA synthesis in wheat roots (Zhao et al. 2001). As proof of this, ABA accumulation under stress was prevented by the administration of ROS scavengers. The accumulation of NO tested jointly is needed in broad bean throughout ABA-induced stomata closure (Garcia-Mata and Lamattina 2002). Heat treatment of alfalfa cells, for example, has led to an increase in NO synthesis, while the application of exogenous NO improved cold tolerance in tomato, wheat, and maize (Neill et al. 2003). The observed effects were most likely associated with NO's antioxidative action, which amplifies the negative effects caused by the intensification of peroxidative metabolism in high temperatures (Neill et al. 2002).

Mackerness et al. (2001) demonstrated the involvement of NO in plant response to UV-B radiation, demonstrating poststress induction of chalcone synthase expression, a rise in NOS-type protein activity, and an increase in NO levels. According to the findings of Shi et al. (2005), NO positively shields plants against UV-B radiation, most likely through increased activity of the antioxidative system. NO-donor treatment of potato tubers prior to UV-B irradiation resulted in the development of approximately 50% more healthy leaves than plants not subjected to NO treatment (Neill et al. 2003). Exogenous NO has been shown to reduce the deleterious effects of heavy metals, ethylene, and herbicides on plants in response to alternative abiotic stresses (Kopyra and Gwóźdź 2003). The authors explained the protective effect as a result of NO-donor treatment of plant materials by the effect of NO on the elevation of activity of antioxidative enzymes, particularly SOD (Kopyra and Gwóźdź 2003).

According to the cited authors, such a chain of events could effectively reduce the amount of ROS produced during stress, thereby limiting oxidative stress in plant cells. Similarly to salinity stress, NO-donor treatment of rice seedlings resulted in loss minimization (Uhida et al. 2002). When NO was used, plant growth was increased, along with the maintenance of acceptable photosystem II activity, an increase in antioxidative protein activity, and thus the expression of specific salinity stress resistance genes. On the other hand, prolonged stress situations may result in the production of NO and NO-derived products, resulting in specific responses, referred to as nitrosative stress. Valderrama et al. (2007) found that salinity stress elicited the assembly of RNS, i.e. NO, GSNO, and RSNO, as well as an increase in tyrosine-nitrated proteins, which are sensitive markers of nitrosative stress, in olive leaves. Furthermore, they demonstrated that vascular tissues may play an important role in the distribution of NO-derived forms during nitrosative stress and in signaling processes. Tissue damage, which is usually due to a microorganism invading a cell, frequently results in NO production and H_2O_2 accumulation (Delledonne et al. 1998). According to Orozco-Cardenas and Ryan (2002), injury does not induce NO synthesis; however, the use of exogenous NO inhibits the method of NO generation and expression of wound-inducing genes.

It is important to note that many types of abiotic stress (cold and heat stress, salt and drought stress) increase polyamine (PA) synthesis (Bouchereau et al. 1999). Tun et al. (2006) discovered that PAs significantly increase NO generation. Work on *Arabidopsis* seedlings confirms that NO acts as a channel between PA-mediated stress responses and an alternative stress mediator, with NO as a stepping stone.

By scavenging ROS, NO plays a significant role in inhibitor defense against many abiotic stresses. Salinity stress has a negative impact on plant morphological traits and diffusion balance. Additionally, it promotes membrane disintegration, DNA damage, particle discharge, and death. NO has been shown to have a potential effect on diffusion stress tolerance as well as increased spermatophyte growth in rice, lupin, and cucumber when subjected to salt stress (Uchida et al. 2002; Kopyra and Gwóźdź 2003; Fan et al. 2007, 2013; Barakat et al. 2012). Similar evidence has been reported demonstrating the potential effect of NO in mitigating salinity stress in alfalfa, barley, jatropha, chickpea, and sunflower (Nabi et al. 2019). In *Oryza sativa*, a NO donor SNP inhibited accumulation, reduced ROS generation, and improved root growth (Kushwaha et al. 2019). In various plant species, the potential role of gas in mitigating serious metal toxicity has been investigated (Ahmad et al. 2018; Yuanjie et al. 2019; Bhuyan et al. 2020; Wei et al. 2020; Khator et al. 2021). As a result, it is clear that NO gas could be a potential mitigating agent for abiotic stresses.

1.4.1 Crosstalk of Nitric Oxide with Other Phytohormones in Plants to Confer Abiotic Stress Tolerance

NO is involved in advanced signal mechanisms, as well as synergistic collaboration with phytohormones and alternative secondary signal molecules, to confer stress tolerance in plants. NO has been linked to a variety of phytohormones, including gibberellins, brassinosteroids, and ABA. The interaction of NO with phytohormones has been studied in a variety of ways. NO and plant hormones work together to regulate a wide range of physiological responses in plants. By activating Ca^{2+} and calcium-dependent protein kinase via downstream signals, NO and auxin promote root development (Pagnussat et al. 2002). Similarly, the interaction of NO and auxin promotes Cd tolerance in the rosid dicot genus *Truncatula* by reducing auxin degradation (Xu et al. 2010). Furthermore, there is a growing of evidence pointing to the effect of NO in reducing serious metal toxicity (He et al. 2012; Yuan and Huang 2016; Wei et al. 2020). Iron deficiency, on the other hand, stimulates the assembly of auxin and increases NO levels, thereby upregulating ferric-chelate enzyme activity in *Arabidopsis*.

Various studies have emphasized the role of cytokinins in plant organic process processes, cellular division, and leaf senescence. Cytokinins and NO interact in a variety of ways, including synergistic and antagonistic responses. NO and cytokinins interact synergistically in response to drought stress causing leaf senescence, cellular division, and photosynthetic activity (Mishina et al. 2007; Shao et al. 2010; Shen et al. 2013). NO signaling via gibberellins induces multiple physiological responses in plants, including seed germination, root growth (Lozano-Juste and León 2011; Sanz et al. 2015), photosynthetic activity, and nutrient use potency. NO and gibberellins also have an antagonistic relationship because NO suppresses gibberellic acid signal events and signal transduction by promoting the accumulation of DELLA proteins (Asgher et al. 2017). Wu et al. (2014) discovered that gibberellins work antagonistically with NO to manage root growth in *Arabidopsis* at low and high P concentrations. Likewise, NO production is required for ABA-induced stomatal closure in guard cells (Neill et al. 2002). The role of NO–ABA interactions in drought stress and UV-B radiation stress has been well established in controlling stomatal closure and inhibitor defense machinery (Neill et al. 2008; Tossi et al. 2009).

Several studies have shown that ABA and NO interact in plant physiological responses and signaling mechanisms (Castillo et al. 2015; Asgher et al. 2017; Wang et al. 2020). Furthermore, NO and ethylene have an antagonistic interaction because NO inhibits ethylene synthesis and action by inhibiting leaf senescence and ripening (Leshem et al. 1998; Manjunatha et al. 2010). However, serious physiological responses to NO interactions with ethylene are being studied in *Arabidopsis*, *Cucumis*, and *Nicotiana* (Ederli et al. 2006). As a result, it is clear that the interaction of NO with phytohormones in abiotic stress tolerance is supported by a variety of evidence. Consider the interaction of NO with various hormones such as brassinosteroids, jasmonates, and polyamines (Liu et al. 2014; Lau et al. 2021; Nahar et al. 2016). The interaction of NO with hormones activates the advanced signal cascade, inducing a variety of responses to environmental stresses.

Despite the fact that there is conflicting evidence regarding the interference mechanism of NO with hormones, future research on the mechanisms of interaction of multiple hormones with NO and their potential pathways is required to be addressed in terms of stress responses and plant growth traits (Table 1.1).

Table 1.1 The physiological role of NO in plants under abiotic stress.

Plant species	Stressors	Physiological role	Reference
Wheat	Drought	Enhanced drought tolerance	Garcia-Mata and Lamattina 2001
Alianthus altissima	Drought	Enhanced antioxidant defense mechanism, proline and osmolyte metabolism	Filippou et al. 2014
Wheat	Drought	Enhanced seedling growth, high relative water content, mitigation of oxidative stress	Tian and Lei 2006
Crambe abyssinica	Drought	Enhanced NR activity and suppressed ROS and malondialdehyde content	Batista et al. 2018
Arabidopsis	Drought	Early drought responsive processes along with translational and transcriptional reprogramming	Ederli et al. 2019
Bean	UV-B radiation	Decreased H_2O_2 content, enhanced leaf growth, elevated antioxidant enzyme activity	Shi et al. 2005
Alfalfa	Salinity	Enhanced plant growth and seed germination	Wang and Han 2007
Chickpea	Salinity	Stimulates plant development and antioxidant enzyme activity	Ahmad et al. 2016
Avicennia marina	Salinity	Enhanced photosynthetic activity	Shen et al. 2018

(*Continued*)

Table 1.1 (Continued)

Plant species	Stressors	Physiological role	Reference
Zea mays	Salinity	Enhanced activity of tonoplast H^+- ATPase and gene for Na+/H+ antiporter	Zhang et al. 2006
Cucumis sativus	Salinity	Spermidine accumulation has increased	Fan et al. 2013a
Sunflower	Salinity	Seedling growth is improved, and antioxidant activity is increased and reduced ROS formation	Kaur and Bhatia 2016; Arora and Bhatia 2017
Jatropha	Salinity	Improved seedling growth with less oxidative stress and lower toxic ion deposition	Gadelha et al. 2017
Crocus sativus	Salinity	Increased growth due to osmolyte accumulation and antioxidant enzyme activity, as well as increased secondary metabolite synthesis	Babaei et al. 2020
Pea	Salinity	Enhanced chlorophyll content, nutrient uptake, and antioxidant enzyme activity	Dadasoghi et al. 2020
Mustard	Salinity	Increased synthesis of antioxidant enzymes, enzymes for N metabolism, photosynthesis and respiration, decreased H_2O_2, MDA content, and PCD	Sami et al. 2021
Vigna radiata	Salinity	Increased activity of proline, total amino acids, reducing sugars, modulates antioxidant enzyme activities, physiological traits	Roychoudhary et al. 2021
Tomato	Salinity	Enhanced activities of NO and ROS	Liu et al. 2015a
Spinach	Salinity	Enhanced secondary metabolites and activity of antioxidant enzymes	Du et al. 2015
Oryza sativa	As toxicity	Enhanced root growth and formation, reduced ROS generation, and As accumulation	Kushwaha et al. 2019
Arachis hypogea	Cd toxicity	Increased antioxidant enzyme activities, reduced ROS and Cd accumulation	Yuanjie et al. 2019
Arabidopsis	Cu toxicity	Improved cell viability	Peto et al. 2013
Tomato	Cd toxicity	Enhanced water uptake, reduced Cd uptake and oxidative damage	Ahmad et al. 2018
Cowpea, rye grass	Al toxicity	Enhanced chlorophyll content, antioxidant enzyme activities	Khairy et al. 2016; Sadeghipour 2016

(*Continued*)

Plant species	Stressors	Physiological role	Reference
Lablab purpureus L.	High temperature	Reduced lipid peroxidation and increased antioxidant activity, both enzymatic and nonenzymatic	Rai et al. 2018
Soybean	High temperature	Modulates activity of ascorbate peroxidase and peroxidase, reduced lipid peroxidation	Vital et al. 2019
Tomato	Low temperature	Enhanced germination, seedling root and shoot length, sugar content	Amooaghaie and Nikzad 2013
Juglans regia	Chilling stress	Enhanced chlorophyll, glutathione, and sugar content, reduced lipid peroxidation	Dong et al. 2018
Brassica	Herbicidal toxicity	Increase in NO content, low ROS formation, improved antioxidant activities	Hasanuzzman et al. 2018

1.5 NO and Gene Regulation in Plants

The involvement of NO in plant signaling pathways is well recognized, but it is necessary to decipher the NO signaling routes, its targets, and its inductive or repressive effects on gene expression levels. Polverari et al. (2003) investigated the NO-induced changes in expression profiles of 2500 *A. thaliana* transcripts and discovered NO-induced alterations in 120 transcripts. The comparison of 71 differentially expressed complementary DNAs with microarray data revealed that the majority of NO-regulated genes are also impacted by other biotic and abiotic stress situations. In addition, Polverari et al. (2003) discovered that NO generated several plant defense response modifying transcription factors such as WRKYs, EREBPs (ethylene responsive element binding proteins), numerous zinc finger proteins, and dehydration responsive element binding proteins (DREB1 and DREB 2). NO donors were administered to tobacco plants or cell suspension cultures, which promoted the expression of defense-related genes PAL and PR1. These are indicators for phenyl propanoid synthesis and SA-mediated signaling, respectively, and are encoded by cryptography. Each of these genes plays a significant role in disease resistance (Delledonne et al. 1998; Durner et al. 1998).

1.6 Conclusions and Future Prospects

Nitric oxide is a dynamic molecule that performs a variety of physiological functions in both plants and animals. NO signaling plays a part in an array of plant biological functions, growth and developmental patterns, and abiotic and biotic stress tolerance. NO biosynthesis involves a number of complex mechanisms, including both enzymatic and nonenzymatic processes. It is necessary to conduct research on the substrates and inhibitors of the NO synthase enzyme, as well as its possible action on plants, in order to identify NO-generating enzymes. The association of NO with secondary signaling molecules as well as various phytohormones, and their possible mechanisms and pathways, need to be investigated further.

Future research should focus on the spatial and temporal distribution of NO in plants, as well as its intercellular and intracellular effects. In the future, the role of *S*-nitrosylation in the plant system and in ABA during the NO signaling mechanism must be recognized. Although the journey of NO research has been pushed back over the past few decades, a path forward must still be explored. Agriculture's future requires effective research on the role of NO in increasing crop productivity and balancing ecological sustainability. The role of the NO and RNS-derived family in plant biology is being studied extensively. Elucidation of the role of NO in plant growth, seed germination, antioxidant defense mechanisms, post-harvest harvest management, and the ripening phenomenon will open the door for future NO signaling research. As a result, much more research activity in NO studies is required to explore nitric oxide-mediated responses in the plant kingdom, as NO research opportunities are limitless and endless.

References

Able, A.J. (2003). Role of reactive oxygen species in the response of barley to necrotrophic pathogens. *Protoplasma* 221: 137–143.

Ahmad, P., Abdel Latef, A.A., Hashem, A. et al. (2016). Nitric oxide mitigates salt stress by regulating levels of osmolytes and antioxidant enzymes in chickpea. *Frontiers in Plant Science* 7: 1–11. doi:10.3389/fpls.2016.00347.

Ahmad, P., Ahanger, M.A., Alyemeni, M.N. et al. (2018). Exogenous application of nitric oxide modulates osmolyte metabolism, antioxidants, enzymes of ascorbate-glutathione cycle and promotes growth under cadmium stress in tomato. *Protoplasma* 255 (1): 79–93.

Alamillo, J.M. and Garcia-Olmedo, F. (2001). Effects of urate, a natural inhibitor of peroxynitrite mediated toxicity, in the response of *Arabidopsis thaliana* to the bacterial pathogen *Pseudomonas syringae*. *The Plant Journal* 25: 529–540.

Albertos, P., Romero-Puertas, M.C., Tatematsu, K. et al. (2015). S-nitrosylation triggers ABI5 degradation to promote seed germination and seedling growth. *Nature Communications* 6: 8669.

Amooaghaie, R. and Nikzad, K. (2013). The role of nitric oxide in priming-induced low-temperature tolerance in two genotypes of tomato. *Seed Science Research* 23: b1–b4. doi:10.1017/S0960258513000068.

Aroca, Á., Serna, A., Gotor, C. et al. (2015). *S*-sulfhydration: a new post-translational modification in plant systems. *Plant Physiology* 168: 334–342.

Arora, D. and Bhatla, S.C. (2017). Melatonin and nitric oxide regulate sunflower seedling growth under salt stress accompanying differential expression of Cu/Zn SOD and Mn SOD. *Free Radical Biology and Medicine* 106: 315–328. doi:10.1016/j.freeradbiomed.2017.02.042.

Arteel, G.E., Briviba, K., and Sies, H. (1999). Protection against peroxynitrite. *FEBS Letters* 445: 226–230.

Asgher, M., Per, T.S., Masood, A. et al. (2017). Nitric oxide signaling and its crosstalk with other plant growth regulators in plant responses to abiotic stress. *Environmental Science and Pollution Research* 24: 2273–2285. doi:10.1007/s11356-016-7947-8.

Babaei, S., Niknam, V., and Behmanesh, M. (2020). Comparative effects of nitric oxide and salicylic acid on salinity tolerance in saffron (*Crocus sativus*). *Plant Biosystems* 155: 73–82. doi:10.1080/11263504.2020.1727975.

Barakat, A., Staton, M., Cheng, C.-H. et al. (2012). Chestnut resistance to the blight disease: insights from transcriptome analysis. *BMC Plant Biology* 12: 38. doi:10.1186/1471-2229-12-38.

Batista, P.F., Costa, A.C., Müller, C. et al. (2018). Nitric oxide mitigates the effect of water deficit in *Crambe abyssinica*. *Plant Physiology and Biochemistry*. doi:10.1016/j.plaphy.2018.06.012.

Belenghi, B., Romero-Puertas, M.C., Vercammen, D. et al. (2007). Metacaspase activity of *Arabidopsis thaliana* is regulated by S-nitrosylation of critical cysteine residue. *Journal of Biological Chemistry* 282: 1352–1358.

Bennett, M., Mehta, M., and Grant, M. (2005). Biophoton imaging: a non-destructive method for assaying R gene responses. *Molecular Plant–Microbe Interactions* 82: 95–102.

Bethke, P.C., Libourel, I.G., and Jones, R.L. (2007). Nitric oxide in seed dormancy and germination. In: *Annual Plant Reviews Vol. 27: Seed Development, Dormancy and Germination* (eds. K.J. Bradford and H. Nonogaki), 153–175. Oxford, UK: Blackwell Publishing Ltd.

Bhuyan, M.H.M.B., Hasanuzzaman, M., Parvin, K. et al. (2020). Nitric oxide and hydrogen sulfide: two intimate collaborators regulating plant defense against abiotic stress. *Plant Growth Regulation* 90: 409–424. doi:10.1007/s10725-020-00594-4.

Bouchereau, A., Aziz, A., Larher, F. et al. (1999). Polyamines and environmental challenges: recent developments. *Plant Science* 140: 103–125.

Bozhkov, P.V., Suarez, M.F., Filonova, L.H. et al. (2005). Cysteine protease mcll-Pa executes programmed cell death during plant embryogenesis. *Proceedings of the National Academy of Sciences of the United States of America* 102: 14463–14468.

Cao, N., Zhan, B., and Zhou, X. (2019). Nitric oxide as a downstream signaling molecule in brassinosteroid-mediated virus susceptibility to maize chlorotic mottle virus in maize. *Viruses* 11: 368. doi:10.3390/v11040368.

Castillo, M.C., Lozano-Juste, J., Gonzalez-Guzman, M. et al. (2015). Inactivation of PYR/PYL/RCAR ABA receptors by tyrosine nitration may enable rapid inhibition of ABA signaling by nitric oxide in plants. *Science Signaling* 8: ra89.

Chakraborty, N. and Acharya, K. (2017). "NO way"! Says the plant to abiotic stress. *Plant Gene* 11: 99–105.

Chen, Z.H., Wang, Y., Wang, J.W. et al. (2016). Nitrate reductase mutation alters potassium nutrition as well as nitric oxide-mediated control of guard cell ion channels in *Arabidopsis*. *New Phytologist* 209: 1456–1469. doi:10.1111/nph.13714.

Clarke, A., Desikan, R., Hurst, R.D., Hancock, J.T. et al. (2000). NO way back: nitric oxide and programmed cell death in *Arabidopsis thaliana* suspension cultures. *Plant Journal* 24: 667–677.

Corpas, F. and Palma, J. (2018). Assessing nitric oxide (NO) in higher plants: an outline. *Nitrogen* 1: 3. doi:10.3390/nitrogen1010003.

Corpas, F.J., Barroso, J.B., Carreras, A. et al. (2006). Constitutive arginine-dependent nitric oxide synthase activity in different organs of pea seedlings during plant development. *Planta* 224: 246–254.

Corpas, F.J., González-Gordo, S., Cañas, A. et al. (2019). Nitric oxide and hydrogen sulfide in plants: which comes first? *Journal of Experimental Botany* 70: 4391–4404.

Dadasoglu, E., Ekinci, M., Kul, R. et al. (2020). Nitric oxide enhances salt tolerance through regulating antioxidant enzyme activity and nutrient uptake in pea. *Legume Research* 44: 41–45. doi:10.18805/LR-540.

Delledonne, M. (2005). NO news is good news for plants. *Current Opinion in Plant Biology* 8: 390–396.

Delledonne, M., Xia, Y.J., Dixon, R.A. et al. (1998). Nitric oxide functions as a signal in plant disease resistance. *Nature* 394: 585–588.

Delledonne, M., Zeier, J., Marocco, A. et al. (2001). Signal interactions between nitric oxide and reactive oxygen intermediates in the plant hypersensitive disease resistance response. *Proceedings of the National Academy of Sciences of the United States of America* 98: 13454–13459.

Del Pozo, O. and Lam, E. (2003). Expression of the baculovirus p35 protein in tobacco inhibits hypersensitive response cell death and compromises N gene-mediated disease resistance in response to tobacco mosaic virus. *Molecular Plant–Microbe Interactions* 16: 485–494.

del Río, L.A., Corpas, F.J., and Barroso, J.B. (2004). Nitric oxide and nitric oxide synthase activity in plants. *Phytochemistry* 65: 783–792.

Deng, X.G., Zhu, T., Zou, L.J. et al. (2016). Orchestration of hydrogen peroxide and nitric oxide in brassinosteroid-mediated systemic virus resistance in *Nicotiana benthamiana*. *The Plant Journal* 85: 478–493.

Desikan, R., Griffiths, R., Hancock, J. et al. (2002). A new role for an old enzyme: nitrate reductase-mediated nitric oxide generation is required for abscisic acid-induced stomatal closure in *Arabidopsis thaliana*. *Proceedings of the National Academy of Sciences of the United States of America* 99: 16314–16318.

de Pinto, M.C., Tomassi, F., and de Gara, L. (2002). Changes in the antioxidant systems as a part of the signaling pathway responsible for the programmed cell death activated by nitric oxide and reactive oxygen species in tobacco bright-yellow 2 cells. *Plant Physiology* 130: 689–708.

Diaz, M., Achkor, H., Titarenko, E. et al. (2003). The gene encoding glutathione-dependent formaldehyde dehydrogenase/GSNO reductase is responsive to wounding, jasmonic acid and salicylic acid. *FEBS Letters* 543: 136–139.

Dickmann, M.B., Park, Y.K., Oltersdorf, T. et al. (2001). Abrogation of disease development in plants expressing animal antiapoptotic genes. *Proceedings of the National Academy of Sciences of the United States of America* 98: 6957–6962.

Dong, N., Li, Y., Qi, J. et al. (2018). Nitric oxide synthase dependent nitric oxide production enhances chilling tolerance of walnut shoots in vitro via involvement chlorophyll fluorescence and other physiological parameter levels. *Scientia Horticulturae* 230: 68–77.

D'Silva, I., Poirier, G.G., and Heath, M.C. (1998). Activation of cysteine proteases in cowpea plants during the hypersensitive response: a form of programmed cell death. *Experimental Cell Research* 245: 389–399.

Du, S., Liu, Y., Zhang, P. et al. (2015). Atmospheric application of trace amounts of nitric oxide enhances tolerance to salt stress and improves nutritional quality in spinach (*Spinacia oleracea* L.). *Food Chemistry* 173: 905–911. doi:10.1016/j.foodchem.2014.10.115.

Durner, J. and Klessig, D.F. (1999). Nitric oxide as a signal in plants. *Current Opinion in Plant Biology* 2: 369–374.

Durner, J., Wendehenne, D., and Klessig, D.F. (1998). Defense gene induction in tobacco by nitric oxide, cyclic GMP, and cyclic ADP ribose. *Proceedings of the National Academy of Sciences of the United States of America* 95: 10328–10333.

Dwivedi, P., Kumari, A., and Singh, B.N. (2016). Enhanced antioxidant system re-programmes oxidative stress through hypersensitive response in leaves of rice plant challenged with *Aspergillus* species. *International Journal of Agriculture, Environment and Biotechnology* 9 (2): 201–208.

Ederli, L., Bianchet, C., Paolocci, F. et al. (2019). Drought stress induces a biphasic NO accumulation in *Arabidopsis thaliana*. *Plant Signaling and Behavior* 14 (3): e1573098. doi:10.1080/15592324.2019.1573098.

Ederli, L., Morettini, R., Borgogni, A. et al. (2006). Interaction between nitric oxide and ethylene in the induction of alternative oxidase in ozone-treated tobacco plants. *Plant Physiology* 142: 595e608. doi:10.1104/pp.106.085472.

Fan, H., Guo, S., Jiao, Y. et al. (2007). Effects of exogenous nitric oxide on growth, active oxygen species metabolism, and photosynthetic characteristics in cucumber seedlings under NaCl stress. *Frontiers of Agriculture in China* 1: 308–314.doi:10.1007/s11703-007-0052-5.

Fan, H.F., Du, C.X., and Guo, S.R. (2013). Nitric oxide enhances salt tolerance in cucumber seedlings by regulating free polyamine content. *Environmental and Experimental Botany* 86: 52–59. doi:10.1016/j.envexpbot.2010.09.007.

Feechan, A., Kwon, E., Yun, B.W. et al. (2005). A central role for S-nitrosothiols in plant disease resistance. *PNAS* 102: 8054–8059.

Ferrer, M.A. and Ros-Barcelo, A. (1999). Differential effects of nitric oxide on peroxidase and H_2O_2 production by the xylem of *Zinnia elegans*. *Plant, Cell and Environment* 22: 891–897.

Filippou, P., Bouchagier, P., Skotti, E., and Fotopoulos, V. (2014). Proline and reactive oxygen/nitrogen species metabolism is involved in the tolerant response of the invasive plant species *Ailanthus altissima* to drought and salinity. *Environmental and Experimental Botany* 97: 1–10. doi:10.1016/j.envexpbot.2013.09.010.

Foissner, I., Wendehenne, D., Langebartels, C. et al. (2000). In vivo imaging of an elicitor-induced nitric oxide burst in tobacco. *Plant Journal* 23: 817–824.

Fröhlich, A. and Durner, J. (2011). The hunt for plant nitric oxide synthase (NOS): is one really needed? *Plant Science* 181: 401–404.

Gadelha, C.G., de Souza Miranda, R., Alencar, N.L.M. et al. (2017). Exogenous nitric oxide improves salt tolerance during establishment of *Jatropha curcas* seedlings by ameliorating oxidative damage and toxic ion accumulation. *Journal of Plant Physiology* 212: 69–79. doi:10.1016/j.jplph.2017.02.005.

Garcia, M.C., Gay, R., Sokolovski, S. et al. (2003). Nitric oxide regulates K^+ and Cl^- channels in guard cells through a subset of abscisic acid-evoked signaling pathways. *Proceedings of the National Academy of Sciences of the United States of America* 100: 11116–111121.

Garcia-Mata, C. and Lamattina, L. (2001). Nitric oxide induces stomatal closure and enhances the adaptive plant responses against drought stress. *Plant Physiology* 126: 1196–1204.

Garcia-Mata, C. and Lamattina, L. (2002). Nitric oxide and abscisic acid cross talk in guard cells. *Plant Physiology* 128: 790–792.

Garcia-Mata, C. and Lamattina, L. (2013). Gasotransmitters are emerging as new guard cell signaling molecules and regulators of leaf gas exchange. *Plant Science* 201–202: 66–73. doi:10.1016/j.plantsci.2012.11.007.

Garcia-Olmedo, F., Rodrigguez-Palenzulea, P., Molina, A. et al. (2001). Antibiotic activities of peptides, hydrogen peroxide and peroxynitrite in plant defence. *FEBS Letters* 489: 219–222.

Giba, Z., Grubisic, D., and Konjevic, R. (2007). Seeking the role of NO in breaking seed dormancy. In: *Nitric Oxide in Plant Growth, Development and Stress Physiology* (eds. L. Lamattina and J. Polacco), 91–111. Berlin, Heidelberg, Germany: Springer.

Gouvea, C.M.C.P., Souza, J.F., and Magalhaes, M.I.S. (1997). NO-releasing substances that induce growth elongation in maize root segments. *Plant Growth Regulation* 21: 183–187.

Greco, M., Chiappetta, A., Bruno, L. et al. (2012). In *Posidonia oceanica* cadmium induces changes in DNA methylation and chromatin patterning. *Journal of Experimental Botany* 63: 695–709. doi:10.1093/jxb/err313.

Grubisic, D., Giba, Z., and Konjevic, R. (1992). The effect of organic nitrates in phytochrome controlled germination of *Paulownia tomentosa* seeds. *Photochemistry and Photobiology* 56: 629–632.

Gupta, K.J., Igamberdiev, A.U., Manjunatha, G.et al. (2011). The emerging roles of nitric oxide (NO) in plant mitochondria. *Plant Science* 181: 520–526.

Gupta, K.J. and Kaiser, W.M. (2010). Production and scavenging of nitric oxide by barley root mitochondria. *Plant and Cell Physiology* 51: 576–584.

Hasanuzzaman, M., Nahar, K., Alam, M.M. et al. (2018). Exogenous nitric oxide pretreatment protects *Brassica napus* L. seedlings from paraquat toxicity through the modulation of antioxidant defense and glyoxalase systems. *Plant Physiology and Biochemistry* 126: 173–186.

Hatsugai, N., Kuroyanagi, M., Yamada, K. et al. (2004). A plant vacuolar protease, VPE, mediates virus-induced hypersensitive cell death. *Science* 305: 855–858.

Hayat, S., Yadav, S., Ali, B. et al. (2010). Interactive effect of nitric oxide and brassinosteroids on photosynthesis and the antioxidant system of *Lycopersicon esculentum*. *Russian Journal of Plant Physiology* 57: 212–221.

He, H.Y., He, L.F., Gu, M.H. et al. (2012). Nitric oxide improves aluminum tolerance by regulating hormonal equilibrium in the root apices of rye and wheat. *Plant Science* 183: 123–130.

Hebelstrup, K., Shah, J., and Igamberdiev, A. (2013). The role of nitric oxide and hemoglobin in plant development and morphogenesis. *Physiologia Plantarum* 148: 457–469. doi:10.1111/ppl.12062.

Hichri, I., Boscari, A., Castella, C. et al. (2015). Nitric oxide: a multifaceted regulator of the nitrogen-fixing symbiosis. *Journal of Experimental Botany* 66: 2877–2887. doi:10.1093/jxb/erv051.

Huang, X., Stettmaier, K., Michel, C. et al. (2004). Nitric oxide is induced by wounding and influences jasmonic acid signaling in *Arabidopsis thaliana*. *Planta* 218: 938–946.

Hung, K.T. and Kao, C.H. (2003). Nitric oxide counteracts the senescence of rice leaves induced by abscisic acid. *Journal of Plant Physiology* 160: 871–879.

Jovanovic, V., Giba, Z., Djokovic, D. et al. (2005). Gibberellic acid nitrite stimulates germination of two species of light-requiring seeds via the nitric oxide pathway. *Annals of the New York Academy of Sciences* 1048: 476–481.

Kaur, H. and Bhatla, S.C. (2016). Melatonin and nitric oxide modulate glutathione content and glutathione reductase activity in sunflower seedling cotyledons accompanying salt stress. *Nitric Oxide* 59: 42–53. doi:10.1016/j.niox.2016.07.001.

Khairy, A.I.H., Oh, M.J., Lee, S.M. et al. (2016). Nitric oxide overcomes Cd and Cu toxicity in vitro-grown tobacco plants through increasing contents and activities of rubisco and rubisco activase. *Biochimie Open* 2: 41e51.

Khator, K., Saxena, I., and Shekhawat, G.S. (2021). Nitric oxide induced Cd tolerance and phytoremediation potential of *B. juncea* by the modulation of antioxidant defense system and ROS detoxification. *BioMetals* 34: 15–32. doi:10.1007/s10534-020-00259-9.

Klepper, L.A. (1979). Nitric oxide (NO) and nitrogen dioxide (NO_2) emissions from herbicide-treated soybean plants. *Atmospheric Environment* 13: 537–542.

Kohli, S.K. et al. (2019). A current scenario on role of brassinosteroids in plant defense triggered in response to biotic challenges. In: *Brassinosteroids: Plant Growth and Development* (eds. S. Hayat, M. Yusuf, R. Bhardwaj, and A. Bajguz). Singapore: Springer. doi:10.1007/978-981-13-6058-9_13.

Kopyra, M. and Gwóźdź, E.A. (2003). Nitric oxide stimulates seed germination and counteracts the inhibitory effect of heavy metals and salinity on root growth of *Lupinus luteus*. *Plant Physiology and Biochemistry* 41: 1011–1017.

Kovacs, I., Durner, J., and Lindermayr, C. (2015). Crosstalk between nitric oxide and glutathione is required for NONEXPRESSOR OF PATHOGENESIS-RELATED GENES 1 (NPR1)-dependent defense signaling in *Arabidopsis thaliana*. *The New Phytologist* 208: 860–872. doi:10.1111/nph.13502.

Kushwaha, B.K., Singh, S., Tripathi, D.K. et al. (2019). New adventitious root formation and primary root biomass accumulation are regulated by nitric oxide and reactive oxygen species in rice seedlings under arsenate stress. *Journal of Hazardous Materials* 361: 134–140.

Lamattina, L., García-Mata, C., Graziano, M. et al. (2003). Nitric oxide: the versatility of an extensive signal molecule. *Annual Review of Plant Biology* 54: 109–136.

Lau, S.-E., Hamdan, M.F., Pua, T.-L. et al. (2021). Plant nitric oxide signaling under drought stress. *Plants* 10 (2): 360. doi:10.3390/plants10020360.

Lee, H.Y., Byeon, Y., Tan, D.X. et al. (2015). *Arabidopsis* serotonin N-acetyltransferase knockout mutant plants exhibit decreased melatonin and salicylic acid levels resulting in susceptibility to an avirulent pathogen. *Journal of Pineal Research* 58: 291–299.

Lee, K., Choi, G.H., and Back, K. (2017). Cadmium-induced melatonin synthesis in rice requires light, hydrogen peroxide, and nitric oxide: key regulatory roles for tryptophan decarboxylase and caffeic acid O-methyltransferase. *Journal of Pineal Research* 63: e12441.

Leshem, Y.Y., Wills, R.B.H., and Ku, V.-V.-V. (1998). Evidence for the function of the free radical gas-nitric oxide (NO) as an endogenous maturation and senescence regulating factor in higher plants. *Plant Physiology and Biochemistry* 36: 825–833.

Levine, A., Tenhaken, R., Dixon, R. et al. (1994). H_2O_2 from the oxidative burst orchestrates the plant hypersensitive disease resistance response. *Cell* 79: 583–593.

Li, X., Zhang, L., Ahammed, G.J. et al. (2017). Nitric oxide mediates brassinosteroid-induced flavonoid biosynthesis in *Camellia sinensis* L. *Journal of Plant Physiology* 214. doi:10.1016/j.jplph.2017.04.005.

Liu, S., Dong, Y., Xu, L. et al. (2014). Effects of foliar applications of nitric oxide and salicylic acid on salt-induced changes in photosynthesis and antioxidative metabolism of cotton seedlings. *Plant Growth Regulation* 73: 67–78.

Liu, W., Li, R.-J., Han, T.-T. et al. (2015a). Salt stress reduces root meristem size by nitric oxide-mediated modulation of auxin accumulation and signaling in *Arabidopsis*. *Plant Physiology* 168: 343–356. doi:10.1104/pp.15.00030.

Lozano-Juste, J. and León, J. (2011). Nitric oxide regulates DELLA content and PIF expression to promote photomorphogenesis in Arabidopsis. *Plant Physiology* 156: 1410–1423.

Mackerness, S.A.H., John, C.F., Jordan, B. et al. (2001). Early signaling components in ultraviolet-B responses: distinct roles for different reactive oxygen species and nitric oxide. *FEBS Letters* 489: 237–242.

Manjunatha, G., Lokesh, V., and Neelwarne, B. (2010). Nitric oxide in fruit ripening: trends and opportunities. *Biotechnology Advances* 2: 489–499.

Mishina, T.E., Lamb, C., and Zeier, J. (2007). Expression of a nitric oxide degrading enzyme induces a senescence programme in *Arabidopsis*. *Plant Cell and Environment* 30: 39–52.

Modolo, L.V., Augusto, O., Almeida, I.M.G. et al. (2006). Decreased arginine and nitrite levels in nitrate reductase-deficient *Arabidopsis thaliana* plants impair nitric oxide synthesis and the hypersensitive response to *Pseudomonas syringae*. *Plant Science* 171: 34–40.

Modolo, L.V., Cunha, F.Q., Braga, M.R. et al. (2002). Nitric oxide synthase mediated phytoalexin accumulation in soybean cotyledons in response to the *Diaporthe phaseolorum* f. sp. *eridionalis* elicitor. *Plant Physiology* 130: 1288–1297.

Moreau, M., Lindermayr, C., Durner, J. et al. (2010). NO synthesis and signaling in plants – where do we stand? *Plant Physiology* 138: 372–383.

Mur, L.A.J., Carver, T.L.W., and Prats, E. (2005). NO way to live; the various roles of nitric oxide in plant–pathogen interaction. *Journal of Experimental Botany* 57: 489–505.

Murgia, I., Delledonne, M., and Soave, C. (2002). Nitric oxide mediates iron-induced ferritin accumulation in *Arabidopsis*. *Plant Journal* 30: 521–528.

Nabi, R.B.S., Tayadea, R., Hussain, A. et al. (2019). Nitric oxide regulates plant responses to drought, salinity, and heavy metal stress. *Environmental and Experimental Botany* 161: 120–133.

Nahar, K., Hasanuzzaman, M., Alam, M.M. et al. (2016). Polyamine and nitric oxide crosstalk: antagonistic effects on cadmium toxicity in mung bean plants through upregulating the metal detoxification, antioxidant defense and methylglyoxal detoxification systems. *Ecotoxicology and Environmental Safety* 126: 245–255.

Navarre, D.A., Wendehenne, D., Durner, J. et al. (2000). Nitric oxide modulates the activity of tobacco aconitese. *Plant Physiology* 122: 573–582.

Neill, S., Barros, R., Bright, J. et al. (2008). Nitric oxide, stomatal closure, and abiotic stress. *Journal of Experimental Botany* 59: 165–176.

Neill, S.J., Desikan, R., Clarke, A. et al. (2002). Nitric oxide is a novel component of abscisic acid signalling in stomatal guard cells. *Plant Physiology* 128: 13–16.

Neill, S.J., Desikan, R., and Hancock, J.T. (2003). Nitric oxide signalling in plants. *New Phytologist* 159: 11–35.

Noritake, T., Kawakita, K., and Doke, N. (1996). Nitric oxide induces phytoalexin accumulation in potato tuber tissues. *Plant Cell Physiology* 37: 113–116.

Noronha-Dutra, A.A., Epperlein, M.M., and Woolf, N. (1993). Reaction of nitric oxide with hydrogen peroxide to produce potentially cytotoxic singlet oxygen as a model for nitric oxide-mediated killing. *FEBS Letters* 321: 59–62.

Orozco-Cardenas, M.L. and Ryan, C.A. (2002). Nitric oxide negatively modulates wound signaling in tomato plants. *Plant Physiology* 130: 487–493.

Pagnussat, G.C., Simontacchi, M., Puntarulo, S. et al. (2002). Nitric oxide is required for root organogenesis. *Plant Physiology* 129: 954–956.

Palmer, R.M.J., Ferrige, A.J., and Moncada, S. (1987). Nitric oxide accounts for the biological activity of endothelium-derived relaxing factor. *Nature* 327: 524–526.

Parani, M., Rudrabhatla, S., Myers, R. et al. (2004). Microarray analysis of nitric oxide responsive transcripts in *Arabidopsis*. *Plant Biotechnology Journal* 2: 359–366.

Pedroso, M.C., Magalhaes, J.R., and Durzan, D. (2000). A nitric oxide burst precedes apoptosis in angiosperm and gymnosperm callus cells and foliar tissues. *Journal of Experimental Botany* 51: 1027–1036.

Peto, A., Lehotai, N., Feigl, G. et al. (2013). Nitric oxide contributes to copper tolerance by influencing ROS metabolism in *Arabidopsis*. *Plant Cell Reports* 32: 1913e1923.

Polverari, A., Molesini, B., Pezzotti, M. et al. (2003). Nitric oxide-mediated transcriptional changes in *Arabidopsis thaliana*. *Molecular Plant–Microbe Interactions* 16: 1094–1111.

Prado, A.M., Colaço, R., Moreno, N. et al. (2008). Targeting of pollen tubes to ovules is dependent on nitric oxide (NO) signaling. *Molecular Plant* 1: 703–714.

Prats, E., Mur, L.A.J., Sanderson, R. et al. (2005). Nitric oxide contributes both to papilla-based resistance and the hypersensitive response in barley attacked by *Blumeria graminis* f. sp. *Hordei*. *Molecular Plant Pathology* 6: 65–78.

Procházková, D. and Wilhelmová, N. (2011). Nitric oxide, reactive nitrogen species and associated enzymes during plant senescence. *Nitric Oxide* 24: 61–65.

Rai, K.K., Rai, N., and Rai, S.P. (2018). Salicylic acid and nitric oxide alleviate high temperature induced oxidative damage in *Lablab purpureus* L. plants by regulating bio-physical processes and DNA methylation. *Plant Physiology and Biochemistry* 128: 72–88.

Rockel, P., Strube, F., Rockel, A. et al. (2002). Regulation of nitric oxide (NO) production by plant nitrate reductase in vivo and in vitro. *Journal of Experimental Botany* 53: 103–110.

Rojo, E., Martin, R., Carter, C. et al. (2004). VPEg exhibits a caspase-like activity that contributes to defense against pathogens. *Current Biology* 14: 1897–1906.

Romero-Puertas, M.C., Perazzolli, M., Zago, E.D. et al. (2004). Nitric oxide signaling functions in plant–pathogen interactions. *Cellular Microbiology* 6: 795–803.

Rőszer, T. (2012a). Nitric oxide synthesis in leaf peroxisomes and in plant-type mitochondria. In: *The Biology of Subcellular Nitric Oxide* (ed. T. Rőszer), 67–80. New York: Springer.

Rőszer, T. (2012b). Nitric oxide synthesis in the chloroplast. In: *The Biology of Subcellular Nitric Oxide* (ed. T. Rőszer), 49–66. New York: Springer.

Rőszer, T. (2014). Biosynthesis of nitric oxide in plants. In: *Nitric Oxide in Plants: Metabolism and Role in Stress Physiology* (eds. M. Khan, M. Mobin, F. Mohammad, and F. Corpas). Cham, Switzerland: Springer. doi:10.1007/978-3-319-06710-0_2.

Roychoudhury, A., Singh, A., Aftab, T. et al. (2021). Seedling priming with sodium nitroprusside rescues *Vigna radiata* from salinity stress-induced oxidative damages. *Journal of Plant Growth Regulation* 1: 1–12: doi:10.1007/s00344-021-10328-z.

Sadeghipour, O. (2016). Pretreatment with nitric oxide reduces lead toxicity in cowpea (*Vigna unguiculata* L. *Walp.*). *Archives of Biological Sciences* 68: 165e175.

Sami, F., Siddiqui, H., Alam, P. et al. (2021). Nitric oxide mitigates the salt-induced oxidative damage in mustard by upregulating the activity of various enzymes. *Journal of Plant Growth Regulation*. 213: 1–13. doi:10.1007/s00344-021-10331-4.

Sanz, L., Albertos, P., Mateos, I. et al. (2015). Nitric oxide (NO) and phytohormones crosstalk during early plant development. *Journal of Experimental Botany* 66: 2857–2868.

Seligman, K., Saviani, E.E., Oliveira, H.C. et al. (2008). Floral transition and nitric oxide emission during flower development in *Arabidopsis thaliana* is affected in nitrate reductase-deficient plants. *Plant and Cell Physiology* 49 (7): 1112–1121.

Shao, R., Wang, K., and Shangguan, Z. (2010). Cytokinin-induced photosynthetic adaptability of *Zea mays* L. to drought stress associated with nitric oxide signal: probed by ESR spectroscopy and fast OJIP fluorescence rise. *Journal of Plant Physiology* 167: 472–479.

Shen, Q., Wang, Y.T., Tian, H. et al. (2013). Nitric oxide mediates cytokinin functions in cell proliferation and meristem maintenance in *Arabidopsis*. *Molecular Plant* 6: 1214–1225.

Shen, Z.J., Chen, J., Ghoto, K. et al. (2018). Proteomic analysis on mangrove plant *Avicennia marina* leaves reveals nitric oxide enhances the salt tolerance by up-regulating photosynthetic and energy metabolic protein expression. *Tree Physiology* 38: 1605–1622.

Shi, C., Qi, C., Ren, H. et al. (2015). Ethylene mediates brassinosteroid induced stomatal closure via Gα protein-activated hydrogen peroxide and nitric oxide production in *Arabidopsis*. *The Plant Journal* 82: 280–301.

Shi, H., Ye, T., Zhu, J.K. et al. (2014). Constitutive production of nitric oxide leads to enhanced drought stress resistance and extensive transcriptional reprogramming in *Arabidopsis*. *Journal of Experimental Botany* 65: 4119–4131. doi:10.1093/jxb/eru184.

Shi, S., Wang, G., Wang, Y. et al. (2005). Protective effect of nitric oxide against oxidative stress under ultraviolet-B irradiation. *Nitric Oxide* 13: 1–9.

Smirnoff, N. (2000). Ascorbic acid: metabolism and functions of a multi-facetted molecule. *Current Opinion in Plant Biology* 3: 229–235.

Song, F. and Goodman, R.M. (2001). Activity of nitric oxide is dependent on, but is partially required for function of salicylic acid in the signaling pathway in tobacco systemic acquired resistance. *Molecular Plant–Microbe Interactions* 12: 1458–1462.

Tada, Y., Mori, T., Shinogi, T. et al. (2004). Nitric oxide and reactive oxygen species do not elicit hypersensitive cell death but induce apoptosis in the adjacent cells during the defense response of oat. *Molecular Plant–Microbe Interactions* 17: 245–253.

Tian, X. and Lei, Y. (2006). Nitric oxide treatment alleviates drought stress in wheat seedlings. *Biologia Plantarum* 50: 775–778. doi:10.1007/s10535-006-0129-7.

Tischner, R., Planchet, E., and Kaiser, W.M. (2004). Mitochondrial electron transport as a source for nitric oxide in the unicellular green alga *Chlorella sorokiniana*. *FEBS Letters* 576: 151–155.

Tossi, V., Lamattina, L., and Cassia, R. (2009). An increase in the concentration of abscisic acid is critical for nitric oxide-mediated plant adaptive responses to UV-B irradiation. *New Phytologist* 18: 871–879.

Tun, N.N., Santa-Catarina, C., Begum, T. et al. (2006). Polyamines induce rapid biosynthesis of nitric oxide (NO) in *Arabidopsis thaliana* seedlings. *Plant and Cell Physiology* 47: 346–354.

Uchida, A., Jagendorf, A.T., Hibino, T. et al. (2002). Effects of hydrogen peroxide and nitric oxide on both salt and heat stress tolerance in rice. *Plant Science* 163 (3): 515–523.

Uhida, A., Jagendorf, T., Hibino, T. et al. (2002). Effects of hydrogen peroxide and nitric oxide on both salt and heat stress tolerance in rice. *Plant Science* 163: 515–523.

Valderrama, R., Corpas, F.J., Carreras, A. et al. (2007). Nitrosative stress in plants. *FEBS Letters* 581: 453–461.

Vital, R.G., Müller, C., Fabia, B. et al. (2019). Nitric oxide increases the physiological and biochemical stability of soybean plants under high temperature. *Agronomy* 9: 412. doi:10.3390/agronomy9080412.

Vleehouwers, V.G.A.A., van Dooijeweert, W., Govers, F. et al. (2000). The hypersensitive response is associated with host and nonhost resistance to *Phytophtora infestans*. *Planta* 210: 853–864.

Wang, X., Li, Q., Yang, M. et al. (2020). Crosstalk between hydrogen peroxide and nitric oxide mediates priming-induced drought tolerance in wheat. *Journal of Agronomy and Crop Science* 207 (2): 224–235. doi:10.1111/jac.12458.

Wang, X.S. and Han, J.G. (2007). Effects of NaCl and silicon on ion distribution in the roots, shoots and leaves of two alfalfa cultivars with different salt tolerance. *Journal of Soil Science and Plant Nutrition* 53 (3): 278–285. doi:10.1111/j.1747-0765.2007.00135.x.

Wei, L., Zhang, J., Wang, C. et al. (2020). Recent progress in the knowledge on the alleviating effect of nitric oxide on heavy metal stress in plants. *Plant Physiology and Biochemistry* 147: 161–171. doi:10.1016/j.plaphy.2019.12.021.

Wendehenne, D., Pugin, A., Klessig, D. et al. (2001). Nitric oxide: comparative synthesis and signaling in animal and plant cells. *Trends in Plant Science* 6: 177–183.

Wu, A.P., Gong, L., Chen, X. et al. (2014). Interactions between nitric oxide, gibberellic acid, and phosphorus regulate primary root growth in *Arabidopsis*. *Plant Biology* 58: 335–340.

Xu, J., Wang, W., Yin, H. et al. (2010). Exogenous nitric oxide improves antioxidative capacity and reduces auxin degradation in roots of *Medicago truncatula* seedlings under cadmium stress. *Plant Soil* 32: 321–330.

Yamasaki, H., Shimoji, H., Ohshiro, Y. et al. (2001). Inhibitory effects of nitric oxide on oxidative phosphorylation in plant mitochondria. *Nitric Oxide – Biology and Chemistry* 5: 261–270.

Yuan, H.M. and Huang, X. (2016). Inhibition of root meristem growth by cadmium involves nitric oxide-mediated repression of auxin accumulation and signalling in *Arabidopsis*. *Plant Cell and Environment* 39: 120–135.

Yuanjie, D., Wei-Feng, C., Xiaoying, B. et al. (2019). Effects of exogenous nitric oxide and 24-epibrassinolide on physiological characteristics of peanut under cadmium stress. *Pedosphere* 29: 45–59. doi:10.1016/S1002-0160(17)60376.

Zafra, A., Rodríguez-García, M.I., and Alché, J.D. (2010). Cellular localization of ROS and NO in olive reproductive tissues during flower development. *BMC Plant Biology* 10: 36–53.

Zago, E., Morsa, S., Dat, J.F. et al. (2006). Nitric oxide and hydrogen peroxide responsive gene regulation during cell death induction in tobacco. *Plant Physiology* 141: 40411.

Zeier, J., Delledonne, M., Mishina, T. et al. (2004). Genetic elucidation of nitric oxide signaling in incompatible plant–pathogen interactions. *Plant Physiology* 136: 2875–2886.

Zhang, Y., Wang, L., Liu, Y. et al. (2006). Nitric oxide enhances salt tolerance in maize seedlings through increasing activities of proton-pump and Na+/H+ antiport in the tonoplast. *Planta* 224: 545–555.

Zhao, J., Fujita, K., and Sakai, K. (2005). Oxidative stress in plant cell culture: a role in production of beta-thujaplicin by *Cupressus lusitanica* cultures. *Biotechnology and Bioengineering* 90: 621.

Zhao, L., Chen, L., Gu, P. et al. (2019). Exogenous application of melatonin improves plant resistance to virus infection. *Plant Pathology*. 68: 1287–1296. doi:10.1111/ppa.13057.

Zhao, Z., Chen, G., and Zhang, C. (2001). Interaction between reactive oxygen species and nitric oxide in drought-induced abscisic acid synthesis in root tips of wheat seedlings. *Australian Journal of Plant Physiology* 28: 1055–1061.

Zhao, Z., Zhang, F., Guo, J. et al. (2004). Nitric oxide functions as a signal in salt resistance in the calluses from two ecotypes of reed. *Plant Physiology* 134: 849–857.

Zou, L.J., Deng, X.G., Zhang, L.E. et al. (2018). Nitric oxide as a signaling molecule in brassinosteroid mediated virus resistance to Cucumber mosaic virus in *Arabidopsis thaliana*. *Physiologia Plantarum* 163: 196–210.

2

Regulation of NO Biosynthesis Under Abiotic Stresses and Modulation Due to Osmolytes

Kaneez Fatima[1], *Fozia Sardar*[2], *and Asma Imran*[2]

[1]*Department of Life Sciences, University of Management and Technology (UMT), Lahore, Pakistan*
[2]*National Institute for Biotechnology and Genetic Engineering–Campus–Pakistan Institute of Engineering and Applied Sciences (NIBGE-C-PIEAS), Faisalabad, Pakistan*

2.1 Introduction

Nitrogen monoxide/nitric oxide (NO) is a small, fat-soluble gaseous radical that has been well recognized for a long period to be involved in the nitrogen cycle (Jedelská et al. 2021). Earlier it was considered a gaseous free radical and toxic air pollutant (Gupta et al. 2011; Astier et al. 2018). However, over the past 30 years, NO has turned out to be the center of many biological processes in animals and plants as it has been reported as a signaling molecule. This has shifted research interest and understanding of NO as a noxious by-product to it being seen as an important molecule. NO has multiple functions at diverse levels including chemical, cellular, organ, and tissue (Santolini et al. 2017; Corpas et al. 2019; Vishwakarma et al. 2019). In animals, NO acts as a vital regulator and plays an important role in reproduction; while, in plants, it is involved in various functional developments, for instance, germination, seed dormancy (Probert 2000; Bethke et al. 2006; Arc et al. 2013; Nagel et al. 2019), defense responses, flowering/leaf senescence, and protection against environmental stresses (Lora et al. 2019; Abedi et al. 2021).

Nitric oxide is available as three exchangeable species: the cation of nitrosonium (NO^+), NO, and the anion of nitroxyl (NO^-) (Butler et al. 1995). It is water soluble and by the addition of ferrous salts, its solubility is enhanced. Therefore, it can easily diffuse in the aqueous part of a cell including the cytoplasm, and also can freely move through the lipid bilayer. It is believed that once it is produced, it can move to the adjacent cells (Del Río et al. 2004; Delledonne 2005).

2.2 Biosynthesis of NO

There are numerous prospective sources of nitric oxide in plants (Figure 2.1) and its production is highly dependent on the type of plant, tissues/cells, plant growth environment, and stimulation of the signaling pathway under specific conditions (González-Moscoso et al. 2021). It is believed that the synthesis of NO can be achieved via two major routes: nitrate reductase and the oxidative pathway (Bethke et al. 2004; Besson-Bard et al. 2008). These are discussed in Sections 2.2.1 and 2.2.2.

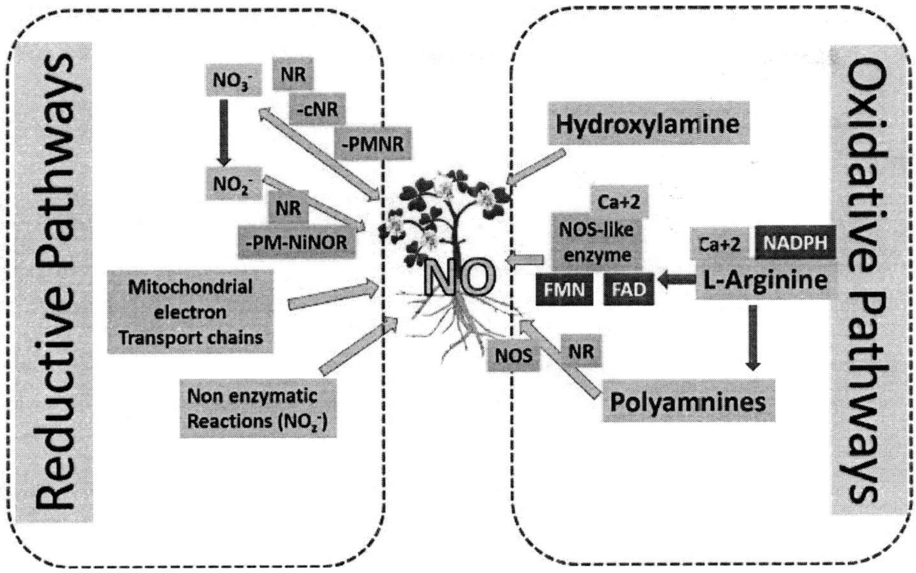

Figure 2.1 Biosynthesis pathways of nitric oxide in plants.

2.2.1 Nitrate Reductase

In higher plants, nitrate reductase (NR, EC 1.7.1.1) is a cytosolic enzyme that facilitates the assimilation and metabolism of nitrogen (Ahmad et al. 2021). This enzyme utilizes nicotinamide adenine dinucleotide (NADH) (electron donor), molybdopterin, and heme/flavin adenine dinucleotide (FAD) (cofactors) to activate the reduction of nitrate into nitrite (Kaya et al. 2020b).Nitrate reductase is generally present as a homodimer in tetrameric form, depending upon the plant species. In addition to its primary role, nitrate reductase is involved in the production of nitrite: its activity level is quite low and makes up 1% of total nitrate-reducing ability (Mohn et al. 2019).

As well as cytoplasmic-based nitrate reductase, membrane-associated nitrite – NO reductase (Ni-NOR) – is also involved in NO production in plants. It is exclusively present in rhizome, which utilizes NAD(P)H (electron donor) to yield the NO from nitrite. It works at low oxygen pressure and in conjunction with membrane-linked nitrate reductase, which produces nitrite from nitrate. In their structural features, both cytoplasmic and membrane-bound reductase enzymes exhibit the presence of a cofactor named molybdenum (Moco) (Chamizo-Ampudia et al. 2017; Gao et al. 2019). In plants, some other Moco-containing enzymes exist, including (i) aldehyde oxidases (AOs), (ii) sulfite oxidases (SOs), and (iii) xanthine oxidases (XOs), all of which produce NO from nitrite (Bethke et al. 2004).

2.2.2 Mechanisms of Oxidative NO Synthesis

As plants can survive and grow without nitrite and nitrate, they should have nitrite-independent, oxidative mechanisms for NO synthesis. Indeed, analogous to animal nitric oxide synthase (NOS) (EC 1.14.13.39), the plant enzyme is not dependent on nitrite and is involved in the deamination of L-arginine into L-citrulline and NO via NADPH and O_2 and necessitating Ca^{2+}/calmodulin (Rőszer 2012). Enzymatic oxidation has been witnessed in the peroxisomes (leaf), chloroplasts of the tracheophytes, and in green algae. In chloroplasts, the oxidation of L-arginine to NO requires NADPH and is independent of Ca^{2+}availability. While in the leaf peroxisomes, the conversion of L-arginine/L-citrulline requires Ca^{2+}/calmodulin, flavin mononucleotide (FMN), FAD, and NADPH (Mur et al. 2013).

2.2.3 Nonenzymatic Synthesis of NO

In addition to enzymatic methods, a nonenzymatic pathway for the production of NO from NO_2 has been reported. This sort of NO production is favored by low pH, e.g., in the apoplast of the growing seeds, where NO can be released from a protonated form of NO_2, i.e., nitrous acid (HNO_2). Another possible mechanism is the release of NO from *S*-nitrosoglutathione (GSNO). In the peroxisomes, both NO and the NO-derived peroxynitrite can interact with glutathione to synthesize GSNO (Neill et al. 2003; Courtois et al. 2008; Wilson et al. 2008; Palavan-Unsal and Arisan 2009).

2.3 NO Signaling and Gene Regulation Under Abiotic Stress

In animals, NO is involved in the regulation and the expression of different genes related to several different pathways, either directly by interacting with receptors involved in the signal transduction pathway, or indirectly by moderating the activity of transcription factors, or influencing the stability and translation of messenger RNA (mRNA) (Kolbert et al. 2021). Recently, a protein named "AtNOS1" exhibiting NOS activity has been recognized. This protein belongs to the NOS family (a novel family) and displays some similarity to previously reported mammalian NOS. It consists of a guanosine-5′-triphosphate (GTP)-binding domain and may have GTPase activity (Li et al. 2020). AtNOS1 activity is highly dependent upon Ca^{2+}-calmodulin/$NADPH^-$; however, it is independent of FAD, FMN, and BH4. Studies have provided a genetic basis for a role of AtNOS1 as a source of NO in abscisic acid (ABA)-induced closure of stomata, in the suppression of flowering, and the defense responses induced by lipopolysaccharides (Wang et al. 2020). Under abiotic stress, ABA production is the key stress response in plants. The signaling between ABA and NO governs the major molecular mechanisms that incorporate external signals to modify internal systems leading to plant adaptations against stress (Falak et al. 2021). NO interacts with ABA to modulate the gene expression and the protein function. For example, ABA prevents water loss through stomatal closure and contributes to the synthesis of osmolytes. It has been observed that in the guard cells of *Arabidopsis thaliana*, ABA increases H_2O_2 production, which in turn regulates the production of NO, which subsequently regulates the stomatal closure (Li et al. 2020; Wang et al. 2020; Falak et al. 2021).

Salicylic acid (SA) is another organic plant compound that contributes to the modulation of seed germination, vegetative defense, and response to changing environment. The association between NO and the non-expressors of pathogenesis-related proteins, i.e. *NPR1*, *NPR3*, and *NPR4* (SA receptor genes), is well recognized for plant defense (Prakash et al.

2021). These NPR proteins regulate various transcription factors (TFs) and endure post-translational changes (modulated by SA). *S*-nitrosylation of NPR proteins modifies NO signaling through oligomerization of NPR (Falak et al. 2021; Prakash et al. 2021). NO also controls the expression of *NPR1* during plant defense. In *Arabidopsis*, the SA-dependent defense response has been shown to be triggered by NO. For instance, the application of the NO donor "GSNO" (*S*-nitrosoglutathione) enhanced the expression of *NPR1* and its analogous protein, which subsequently improved the SA content and induced *PR* genes, conferring resistance against *Pseudomonas*. Moreover, a rapid change in the levels of glutathione, which is important for the activation of *NPR1* and accumulation of SA, has been induced by NO (Fatima et al. 2021; Hediji et al. 2021; Kaur et al. 2021).

2.4 Role of NO in Metabolic and Developmental Pathways

NO has been shown to play an important role in the modulation of various physiological processes of plants mentioned in Table 2.1. Some of those processes include germination of seed and pollen, plant growth and development, closing and opening of stomata, senescence, flower development, and fruit ripening (Kopyra 2004; Delledonne 2005; Wilson et al. 2008). NO was documented to induce germination in light-dependent *Arabidopsis* and lettuce seeds, which indicates that NO takes part in root development (Li et al. 2016). Various NO donors stimulated root tip enlargement of *Zea mays*, in a dosage-related mode, and its length was prohibited by a NO scavenger (methylene blue) (Kopyra 2004). It has been proposed that NO and indole-3-acetic acid share some common steps in cell signaling, as they both stimulate the same plant response. Various studies have been suggested that stress regulates the endogenous levels of indole-3-acetic acid. As well as root growth and development, NO was found to be involved in the development and realignment of pollen tubes, best studied in *A. thaliana* (Pagnussat et al. 2003, 2004). It has been reported that NO has also been involved in the synthesis of several biological constituents including saponin and phytoalexin (Pagnussat et al. 2003, 2004).

Table 2.1 Role of nitric oxide (NO) during abiotic stress.

Abiotic stress	NO-mediated effect	Plant species	Reference
Drought	Abscisic acid signaling and closure of stomata	*Pisum sativum*	Syed Nabi et al. 2019
Salt	Increased tolerance of root growth	*Oryza sativa* *Nicotiana tabacum*	van Zelm et al. 2020
Heavy metals	Heightened response in terms of root growth and oxidative stress	*Lupinus* *Chlorella vulgaris*	Solórzano et al. 2020
Temperature	Release of NO and increased tolerance of seedlings	*Oryza sativa* *Nicotiana tabacum*	Syed Nabi et al. 2019
Ultraviolet radiation	Nitric oxide synthase activity	*Arabidopsis thaliana*	Tossi et al. 2009

Plants have established multifaceted innate/induced and hypersensitive mechanisms (HR) to protect themselves from pathogenic microorganisms, which subsequently terminate in systemic acquired resistance (SAR). An initial significant event is the elevation of $Ca^{2+}$by using the plasma membrane cyclic nucleotide-gated ion channels (CNGCs). This is consistent with a role for cyclic guanosine monophosphate (cGMP) in plants, and various putative guanylate cyclase activity proteins including phytosulfokine receptor (PSKR). In PSKR, the cytosolic guanylate cyclase domain is activated by its biologically active ligand, a sulfonated phytosulfokine. An increase in Ca^{2+}triggers the creation of SA, NO, and reactive oxygen species (ROS), which stimulates apoptosis in the proximity of the infection, thereby restraining pathogen progression (Heath 2000; Mur et al. 2019; Noman et al. 2020).

2.5 Role of NO in Overcoming Abiotic Stress

Plants are static creatures confined to the particular place where the seed sprouts. During its life cycle, a plant experiences a diverse and changing environment. With time, plants have established various mechanisms to handle the different kinds of stress environments. Although stresses are numerous, plant's reactions commonly comprise similar components and signaling pathways. NO is an important constituent in numerous plant acclimation responses to biotic and abiotic strains (Figure 2.2) (Verma et al. 2016; Nowicka et al. 2018; Gong et al. 2020). The response mechanisms for NO signaling are generally similar for all types of stresses. Figure 2.3 contains details of the NO signaling pathway to scavenge ROS during different abiotic stresses.

2.5.1 Drought and Low Mineral Nutrient Supply

During water deficit, plant functioning (plant growth and photosynthesis) is affected adversely. However, through stomatal closure, plants can overcome a temporary drought. But in long-term drought situations, suppressed leaf expansion, leaf excision, and changes in root morphology have been observed. The vulnerability of different plants to water deficiency is largely dependent upon their structural and physiological modifications to water stress (Santisree et al. 2015).

It has been speculated that slight water deficiencies lead to an increased level of NO production in the roots of cucumber plant. Pretreatment with extrinsic NO, e.g. 100 μM sodium nitroprusside (SNP) and GSNO was capable of neutralizing drought-induced membrane impairment and peroxidation of lipid in water-deficient plants. SNP (200 μM), an NO donor, has also been recognized to employ a shielding effect (improved growth, rise in water content, and less oxidative destruction) in wheat plantlets under polyethylene glycol-induced water stress (Zeppel et al. 2015). The capability of NO to manage drought stress might be linked to its direct effect as an antioxidant, the effects on root physiology, and a role in closure of stomata. In guard cells, NO is involved in the stimulation of intracellular Ca^{2+}release, and regulates Ca^{2+}-sensitive K^+ and Cl^- channels in the plasma membrane (Simontacchi et al. 2015).

In agriculture, one limitation is the low accessibility of some essential nutrients, which delivers varied surroundings to the plant. Plants have numerous mechanisms to extract minerals from the soil including major changes in the root growth pattern, the improved

Figure 2.2 General role of nitric oxide molecule in plant growth.

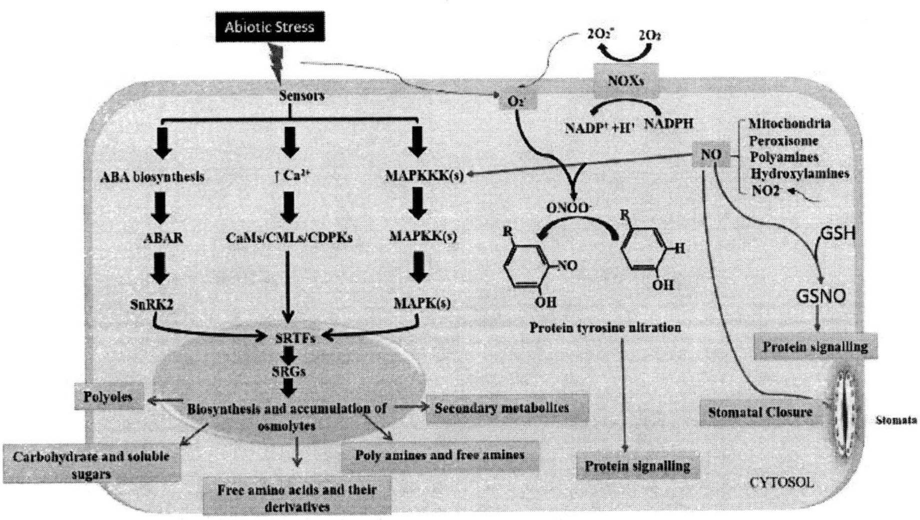

Figure 2.3 Activation of nitric oxide signaling and osmolyte pathways in plant cell under abiotic stress.

ability to obtain nutrients from depleted environments, and also the release of different substances that facilitate nutrient availability in the close proximity of roots, thus favoring their accumulation by roots. Recent study indicates that NO is involved in the modulation of these aforementioned mechanisms (Lambers et al. 2011). For example, in white lupin (*Lupinus albus*) at low phosphorus levels, the contribution of NO in plant responses has been observed and confirmed. Along with this plant, in other dicots and monocots, phosphorus deficiency mediates root branching followed by the release of exudates, comprising organic acids (Wang et al. 2014), which facilitate plants to increase P acquisition. From this perspective, it has been indicated that white lupin roots exposed to P deficiency exhibited heightened accumulation of NO, which was directly linked to the clustering of roots and increased exudation of citrate compounds. Sequentially, the addition of NO donors (e.g. SNP and GSNO) leads to the propagation of cluster roots, whereas NO scavengers (e.g. cPTIO [2-4-carboxyphenyl-4,4,5,5-tetramethylimidazoline-1-oxyl-3-oxide]) eradicate this progression (Palavan-Unsal and Arisan 2009).

2.5.2 Salinity

In addition to drought and low mineral content, salinity is one of the most prolonged and unfavorable conditions that has a negative effect on plant health. Significantly, not all plants react in a similar manner to those compounds (lethal osmotic constituents) due to the existence of an array of resistance mechanisms that assist the plant to overcome the stress. These adaptive mechanisms might consist of accumulation of organic components, and the regulation of antioxidant machinery (van Zelm et al. 2020). It has been detected that rice plants exposed to SNP (NO donor) had improved overall growth. This shielding influence exerted by exogenous application of NO was related to the preservation of high relative moisture content and chlorophyll, whereas ionic discharge was sustained at a low level. Likewise, a clear effect on the accumulation of Na^+ and K^+ was also witnessed. These results proposed that NO has a protective effect on plants including managing water levels, sustaining ionic balance, and minimizing the damage imposed during early stages of the salinity response (Suzuki et al. 2014; Bashir et al. 2019; Hernández 2019).

2.5.3 Ultraviolet Radiation

Sunlight is the principal source of energy for plants, and a portion of this energy exists in the form of ultraviolet (UV) radiation. It is typically partitioned into three categories: (i) UV-A (315–400 nm), (ii) UV-B (280–315 nm), and (iii) UV-C (200–280 nm). It is well known that the atmospheric gases completely absorb UV-C radiation, while UV-B rays are moderately absorbed by tropospheric ozone, and only a trivial fraction is transferred to the surface, and UV-A is scarcely absorbed by the ozone. Since the destruction of the ozone layer, the levels of UV-B radiation rise on the Earth's surface. UV-B has high energy and induces production of ROS that employ harmful morphological and molecular changes in plants (Simontacchi et al. 2015). Chloroplasts are highly affected by UV-B radiation and the light-induced impairment is directed primarily to photosystem II (PSII), with the inhibition of the respiratory chain and oxidative damage to the D1 protein. UV-B treatment increased leakage of ions, hydrogen peroxide levels, and oxidation of thylakoid membrane protein in bean (*Phaseolus vulgaris*) leaves. It was also observed that the proficiency of PSII photochemistry (F_v/F_m) and the considerable production of PSII was decreased under UV radiation. But after exposure to NO, carbonyl groups of thylakoid membrane and levels of H_2O_2 were decreased. Thus, NO could exert a protective role against protein oxidation under stress conditions as was previously reported regarding lipid oxidation (Vanhaelewyn et al. 2016; Mannucci et al. 2020).

UV-B adversely affects the chloroplast and mesophyll cells of a plant. It has been reported that UV-B increases the level of plant hormone ABA, activates the NADPH oxidase, and generates H_2O_2. Pretreatment with apocynin (an inhibitor of NADPH oxidase), minimizes the UV-B-induced oxidative damage by reducing the breakdown of chlorophyll (Tewari et al. 2013; Bajguz 2014).

Ozone is formed by a photochemical reaction in NOx and volatile organic compounds (VOCs) and enters the plants through stomata where it accumulates in parenchyma tissues. As a pollutant, it affects the plant by generating oxidative stress via releasing ROS, e.g. superoxide O^{2-}, singlet oxygen, hydroxyl radicals, and hydrogen peroxide in the plant cells. This oxidative stress can further damage the DNA, carbohydrate, lipids, and proteins (Vaultier and Jolivet 2015; Li et al. 2018). The contribution of NO in response to UV-B rays acts through ABA-mediated pathways (Xu et al. 2012). It has been proposed that in plant cells, UV-B radiation stress leads to an increase in ABA levels, which stimulates an increase in cytosolic Ca^{2+} concentration that ultimately starts NO production via NOS and/or NOS-like activities. This increase in NO production increase a plant's tolerance for higher doses of UV-B indirectly by protecting cell redox homeostasis from uncontrolled generation of ROS and associated deleterious effects provoked by UV-B radiation.

2.5.4 Heavy Metals

Present industrial development has been achieved at the cost of the contamination of natural resources. The increasing concentration of heavy metals (HMs) in the environment has led to stress and toxicity for the biotic component of the ecosystem (plant, animals, and microorganisms) (Delledonne 2005). In the rhizosphere, heavy metal concentration changes the plant physiology, morphology, and cellular and biochemical functions, thus creating negative impacts on plant growth, cell division, differentiation, reproduction, photosynthesis, and antioxidant activities (Sharma and Dietz 2009).

The mechanisms of toxicity are composed of functional groups that block essential molecules, e.g. enzymes, nutrient/ion transport, polynucleotides, displacement/replacement of ions from cells, disruption of cell and organelle membrane integrity, denaturation, and inactivation of enzymes (Hall 2002). How the heavy metals affect plants and how plants tolerate these stresses are still under research (Brouquisse 2019; Syed Nabi et al. 2019). However, many research studies have validated the role of NO in relieving heavy metal stress (Solórzano et al. 2020) either produced endogenously or applied exogenously. NO can ameliorate the toxicity of heavy metal in two possible ways: (i) scavenging the ROS overproduction directly in the cell by increasing enzymatic and nonenzymatic antioxidants; or (ii) acting as a signaling molecule to modify the antioxidative expression of the gene (Verma et al. 2013; Zhao et al. 2016). Furthermore, it might change root cell wall components, which leads to increased heavy metal accumulation in plant roots, with heavy metal levels consequently decreased in aerial parts.

ROS consist of superoxide anions, H_2O_2, hydroxyl free radicals, and oxygen-free species (Singh et al. 2016), which are formed in various organelles (mitochondria and chloroplasts) of cells (Singh et al. 2016). Hydrogen peroxide can produce hydroxyl free radicals and oxygen-free species by Harber Weiss and Fenton reactions under heavy metals like Cu^+ and Fe^{2+} (redox-active metals) thus causing oxidative strain in the plant.

This oxidative stress can be ameliorated by enzymatic antioxidants, i.e. catalase (CAT), superoxide dismutase (SOD), glutathione peroxidase (GPX), and ascorbate peroxidase (APX), and nonenzymatic antioxidants like ascorbic acid (AsA), glutathione (GSH), phenolic compounds, tocopherols, and other nonprotein amino acids that are capable of scavenging ROS (Zhao et al. 2016). Studies have revealed that endogenous and exogenous sources of NO can activate the production of antioxidant species both by enzymatic and nonenzymatic pathways. Different types of strategies are related to the presence of NO in the plants, e.g. protection of phospholipid bilayers, cell wall relaxation, and whole growth of plants. The main role in the maintenance of osmotic pressure, the viscosity of cytoplasm, and the protection of chloroplast membranes as well as chlorophyll pigment under heavy metal stress is played by NO (Ahmad et al. 2018; Syed Nabi et al. 2019). Induction of metallochaperones (genes in the heavy metal-associated domain) by NO has been shown to result in the site-specific and safe transport of metallic ions within the cell. Fourteen new heavy metal-associated domains have been discovered. Genes in these domains have shown distinct expression of NO donor *S*-nitroso-L-cysteine (CySNO) in RNA sequence-based transcriptomic studies (Imran et al. 2016; Syed Nabi et al. 2019).

Endogenous concentration of NO in a plant varies under stress depending upon factors like type of metal, plant species, specific type of tissue/organ exposed to heavy metals, exposure duration, and method of quantification of NO (Syed Nabi et al. 2019). In *Arabidopsis*, lead (Pb) stress leads to the overproduction of endogenous NO, which starts a chain of different catalytic reactions in peroxisomes that cause lateral root enlargement (Chen et al. 2016). Heavy metals can increase or decrease endogenous NO, e.g.in *Arabidopsis* exogenous melatonin reduced the Al-induced NO production, hence hindering root elongation and arresting the cell cycle (Zhang et al. 2019). However, in the case of soya bean exposure to Cd for 72 h, increased production of endogenous NO has been detected (Kopyra et al. 2006).

Both endogenous and exogenous NO application have lessened heavy metal stress in different plants including rice, wheat, pea, and many others (Bartha-Dima et al. 2005; Singh et al. 2008; Xiong et al. 2009; Zhao et al. 2016). Exogenous NO increases the pectin and hemicellulose in the root cell wall, which results in increased accumulation of Cd in the roots and a decrease of Cd in aerial parts of the plant (Bartha-Dima et al. 2005; Rodríguez-Serrano et al. 2009; Xiong et al. 2009). Ultimately, NO mediates the transportation of Cd in root cell walls and simultaneously reduces its distribution in the soluble fractions of root and leaves. This shielding effect of NO against heavy metal strain is also supported by differential production of AsA in different plant tissues (Zhao et al. 2016). Heavy metals such as Cd cause oxidative stress in plants by activating ROS in plants, which ultimately cause lipid peroxidation. In this condition NO provides a protective function by reacting with lipids, hence inhibiting the oxidation of lipids (Verma et al. 2013). The defense mechanisms of NO against heavy metal stress depend upon many factors like plant species, organ, or tissue of the affected plant, concentration of heavy metals and exposure duration, and the source of NO (in the case of exogenous NO) (Chen et al. 2016).

2.6 NO Modulation and Regulation by Osmolytes Under Abiotic Stress

Plants are sessile and confined to particular site where seed germinates and plant grow. From germination till new seed development, plants withstand a fluctuating and heterogeneous environment. Among several abiotic factors, water stress subsists in various forms, along with salt, drought and temperature stress. These stresses have negative impact on cellular and developmental processes and also impose high osmotic stress (Bajguz 2014; Brouquisse 2019).

To tolerate hazardous osmotic and oxidative stress, one universal method acquired by plants is the synthesis of "compatible solutes" or "osmolytes," which act by driving a gradient for the uptake of water and play an important part in maintaining the turgor pressure. Osmolytes are low molecular weight, nontoxic, highly soluble organic compounds that are utilized by plant cells to sustain turgor pressure by osmotic adjustments. They also help in redox metabolism by eliminating excess levels of ROS and reinstating the balance of cellular redox. Osmotic adjustment has been characterized as the central dogma in stress physiology. Osmoprotective compounds are diverse in nature and include amino acids (glycine betaine, proectoine, pipecolic acid), simple sugars (glucose and fructose), complex sugars (raffinose, fructans), and sulfonium compounds.

Salinity as a condition has some of the most diverse consequences for plant development due to its deleterious effect on the regulation of ions and osmotic balance. During salt stress, NO increases the action of antioxidant enzymes and also protect plants by elevating the levels of ascorbate and glutathione (antioxidant metabolites) and proline/soluble sugar (osmolytes) in the cytoplasm and other organelles. Improved accumulation of soluble sugars and soluble proteins in reaction to saline stress was documented by Amini and Ehsanpour (2005) in *Nitraria tangutorum*. Proline is one of the most important osmolytes that facilitates cell osmoregulation under salinity. It also has a shielding effect on photosynthetic machinery, acts as energy storage, and protects the cell from oxidative damage. Under salt stress, soluble proteins play an important part in osmotic adjustment and can provide a storage form (nitrate) of N_2. The elevation in soluble protein content may be due to the enhanced production of particular stress-related proteins. Soluble sugars act as essential osmolytes to retain the cell homeostasis (Ahmad et al. 2016; Khator and Shekhawat 2019; Shams et al. 2019). Exogenous application of NO has also been known to induce the *P5CS1*gene encoding δ1-pyrroline-5-carboxylate synthetase, a key enzyme involved in proline synthesis. Exogenous application of NO leads to accumulation of proline, total soluble proteins, and sugars, which may improve salt tolerance of cells through osmotic regulation (Amini and Ehsanpour 2005; Khator and Shekhawat 2019). As a result, the augmented osmotic pressure in the cells improves water uptake, and consequently plant growth and biomass yield. Furthermore, various studies have revealed that in conditions of high salinity, plant cells may experience an imbalance of the inner cellular ions, thus causing osmotic stress and ion toxicity. Increasing Na^+ elimination and K^+ absorption, thus sustaining the optimal ratio of K^+/Na^+, is essential to endure salt stress (Amini and Ehsanpour 2005; Ahmad et al. 2016; Shams et al. 2019).

During periods of biological strain in plants, the production of signaling molecules (NO, H_2S, and CO) has been observed. There are several reports indicating that NO prompts the development of an essential osmolyte like proline in wheat (Wang et al. 2019), cabbage, and rye grass (Lou et al. 2015) by enhancing the expression of *P5C5*and down regulating the *proDH* genes. It has been suggested that NO enhances the accumulation of osmolytes, e.g. proline and glycine betaine, in tomato during cadmium stress when applied externally. Additionally, NO application confines the uptake of Cd and boosts the buildup of nutrients in various portions of tomato plants (Ahmad et al. 2018). The activities of SOD, catalase, ascorbate peroxidase, glutathione reductase, monodehydroascorbate reductase, dehydroascorbate reductase (antioxidant enzymes), and other enzymatic activities of the ascorbate–glutathione cycle are also improved by the application of nitric oxide (Kaya et al. 2020a).

Water deficiency is one of the leading reasons for low agri-productivity and addition of NO has been shown to have an ameliorating effect during water stress in wheat. It has been proposed that SNP applied to wheat seedlings under polyethylene glycol-induced drought stress led to enhanced plantlet growth and high relative water contents and lessened the oxidative destruction to the seedlings under stress (Tian and Lei 2006). In this case, whether NO is protective or toxic to plants was found to be concentration dependent. A low concentration (0.2 mM) inhibits the loss of water, lowers the content of H_2O_2, and leads to higher antioxidative enzyme activity, which supports the wheat seedling growth, while high concentrations (2 mM) of SNP have the opposite effect.

2.7 Conclusions

Many studies have stated that NO has a presence and role in plant growth, development, and defense responses. It is widely recognized that NO is a second messenger and key regulator in plants, and at low concentrations, it acts as a signal to induce or stabilize the expression of many antioxidative enzymes such as SOD, CAT, etc. It also reacts with ROS, acting as a chain breaker, and exhibits antioxidant properties. It also reacts with lipid alkoxyl and peroxyl radicals and stops the proliferation of radical-mediated oxidation of lipids. Thus, NO helps plants to survive under stressful conditions, but its role is mainly concentration dependent.

References

Abedi, S., Iranbakhsh, A., Ardebili, Z.O. et al. (2021). Nitric oxide and selenium nanoparticles confer changes in growth, metabolism, antioxidant machinery, gene expression, and flowering in chicory (*Cichorium intybus* L.): potential benefits and risk assessment. *Environmental Science and Pollution Research* 28: 3136–3148.

Ahmad, A., Khan, W.U., Shah, A.A. et al. (2021). Synergistic effects of nitric oxide and silicon on promoting plant growth, oxidative stress tolerance and reduction of arsenic uptake in *Brassica juncea*. *Chemosphere* 262: 128384.

Ahmad, P., Abdel Latef, A.A., Hashem, A. et al. (2016). Nitric oxide mitigates salt stress by regulating levels of osmolytes and antioxidant enzymes in chickpea. *Frontiers in Plant Science* 7: 347.

Ahmad, P., Ahanger, M.A., Alyemeni, M.N. et al. (2018). Exogenous application of nitric oxide modulates osmolyte metabolism, antioxidants, enzymes of ascorbate-glutathione cycle and promotes growth under cadmium stress in tomato. *Protoplasma* 255: 79–93.

Amini, F. and Ehsanpour, A.A. (2005). Soluble proteins, proline, carbohydrates and Na^+/K^+ changes in two tomato (*Lycopersicon esculentum* Mill.) cultivars under in vitro salt stress. *American Journal of Biochemistry and Biotechnology* 1: 204–208.

Arc, E., Galland, M., Godin, B. et al. (2013). Nitric oxide implication in the control of seed dormancy and germination. *Frontiers in Plant Science* 4: 346.

Astier, J., Gross, I., and Durner, J. (2018). Nitric oxide production in plants: an update. *Journal of Experimental Botany* 69: 3401–3411.

Bajguz, A. (2014). Nitric oxide: role in plants under abiotic stress. In: *Physiological Mechanisms and Adaptation Strategies in Plants under Changing Environment* (eds. P. Ahmad and M.R. Wani), 137–159. London: Springer.

Bartha-Dima, B., Kolbert, Z., and Erdei, L. (2005). Nitric oxide production induced by heavy metals in *Brassica juncea* L. Czern. and *Pisum sativum* L. *Acta Biologica Szegediensis* 49: 9–12.

Bashir, K., Matsui, A., Rasheed, S. et al. (2019). Recent advances in the characterization of plant transcriptomes in response to drought, salinity, heat, and cold stress. *F1000 Research* 8: 658.

Besson-Bard, A., Pugin, A., and Wendehenne, D. (2008). New insights into nitric oxide signaling in plants. *Annual Review of Plant Biology* 59: 21–39.

Bethke, P.C., Badger, M.R., and Jones, R.L. (2004). Apoplastic synthesis of nitric oxide by plant tissues. *The Plant Cell* 16: 332–341.

Bethke, P.C., Libourel, I.G., and Jones, R.L. (2006). Nitric oxide reduces seed dormancy in *Arabidopsis*. *Journal of Experimental Botany* 57: 517–526.

Brouquisse, R. (2019). Multifaceted roles of nitric oxide in plants. *Journal of Experimental Botany* 70: 4319–4322.

Butler, A.R., Flitney, F.W., and Williams, D.L.H. (1995). NO, nitrosonium ions, nitroxide ions, nitrosothiols and iron-nitrosyls in biology: a chemist's perspective. *Trends in Pharmacological Sciences* 16: 18–22.

Chamizo-Ampudia, A., Sanz-Luque, E., Llamas, A. et al. (2017). Nitrate reductase regulates plant nitric oxide homeostasis. *Trends in Plant Science* 22: 163–174.

Chen, Z.H., Wang, Y., Wang, J.W. et al. (2016). Nitrate reductase mutation alters potassium nutrition as well as nitric oxide-mediated control of guard cell ion channels in *Arabidopsis*. *The New Phytologist* 209: 1456–1469.

Corpas, F.J., González-Gordo, S., Cañas, A. et al. (2019). Nitric oxide and hydrogen sulfide in plants: which comes first? *Journal of Experimental Botany* 70: 4391–4404.

Courtois, C., Besson, A., Dahan, J. et al. (2008). Nitric oxide signalling in plants: interplays with Ca^{2+} and protein kinases. *Journal of Experimental Botany* 59: 155–163.

Delledonne, M. (2005). NO news is good news for plants. *Current Opinion in Plant Biology* 8: 390–396.

Del Río, L.A., Corpas, F.J., and Barroso, J.B. (2004).Nitric oxide and nitric oxide synthase activity in plants. *Phytochemistry* 65: 783–792.

Falak, N., Imran, Q.M., Hussain, A. et al. (2021). Transcription factors as the "blitzkrieg" of plant defense: a pragmatic view of nitric oxide's role in gene regulation. *International Journal of Molecular Sciences* 22: 522.

Fatima, A., Husain, T., Suhel, M. et al. (2021). Implication of nitric oxide under salinity stress: the possible interaction with other signaling molecules. *Journal of Plant Growth Regulation*. https://doi.org/10.1007/s00344-020-10255-5.

Gao, Z., Wang, Y., Chen, G. et al. (2019). The indica nitrate reductase gene OsNR2 allele enhances rice yield potential and nitrogen use efficiency. *Nature Communications* 10: 1–10.

Gong, Z., Xiong, L., Shi, H. et al. (2020). Plant abiotic stress response and nutrient use efficiency. *Science China Life Sciences* 63: 635–674.

González-Moscoso, M., González-García, Y., Martínez-Villegas, N.V. et al. (2021). Nitric oxide modified growth, nutrient uptake and the antioxidant defense system in tomato seedlings stressed with arsenic. *Theoretical and Experimental Plant Physiology* 33: 205–233.

Gupta, K.J., Fernie, A.R., Kaiser, W.M. et al. (2011). On the origins of nitric oxide. *Trends in Plant Science* 16: 160–168.

Hall, J.L. (2002). Cellular mechanisms for heavy metal detoxification and tolerance. *Journal of Experimental Botany* 53: 1–11.

Heath, M.C. (2000). Hypersensitive response-related death. *Plant Molecular Biology* 44: 321–334.

Hediji, H., Kharbech, O., Massoud, M.B. et al. (2021). Salicylic acid mitigates cadmium toxicity in bean (*Phaseolus vulgaris* L.) seedlings by modulating cellular redox status. *Environmental and Experimental Botany* 186: 104432.

Hernández, J.A. (2019). Salinity tolerance in plants: trends and perspectives. *International Journal of Molecular Sciences*20.

Imran, Q.M., Falak, N., Hussain, A. et al. (2016). Nitric oxide responsive heavy metal-associated gene AtHMAD1 contributes to development and disease resistance in *Arabidopsis thaliana*. *Frontiers in Plant Science* 7: 1712.

Jedelská, T., Luhová, L., and Petřivalský, M. (2021). Nitric oxide signalling in plant interactions with pathogenic fungi and oomycetes. *Journal of Experimental Botany* 72: 848–863.

Kaur, G., Sharma, P., Rathee, S. et al. (2021). Salicylic acid pre-treatment modulates Pb 2+-induced DNA damage vis-à-vis oxidative stress in *Allium cepa* roots. *Environmental Science and Pollution Research* 28 (37): 51989–52000.

Kaya, C., Ashraf, M., Alyemeni, M.N. et al. (2020a). Responses of nitric oxide and hydrogen sulfide in regulating oxidative defence system in wheat plants grown under cadmium stress. *Physiologia Plantarum* 168: 345–360.

Kaya, C., Ashraf, M., Alyemeni, M.N. et al. (2020b). The role of nitrate reductase in brassinosteroid-induced endogenous nitric oxide generation to improve cadmium stress tolerance of pepper plants by upregulating the ascorbate-glutathione cycle. *Ecotoxicology and Environmental Safety* 196: 110483.

Khator, K. and Shekhawat, G.S. (2019). Nitric oxide improved salt stress tolerance by osmolyte accumulation and activation of antioxidant defense system in seedling of *B. juncea* (L.) Czern. *Vegetos* 32: 583–592.

Kolbert, Z., Szőllősi, R., Feigl, G. et al. (2021). Nitric oxide signalling in plant nanobiology: current status and perspectives. *Journal of Experimental Botany* 72: 928–940.

Kopyra, M. (2004). The role of nitric oxide in plant growth regulation and responses to abiotic stresses. *Acta Physiologiae Plantarum* 26: 459–473.

Kopyra, M., Stachoń-Wilk, M., and Gwóźdź, E.A. (2006). Effects of exogenous nitric oxide on the antioxidant capacity of cadmium-treated soybean cell suspension. *Acta Physiologiae Plantarum* 28: 525–536.

Lambers, H., Finnegan, P.M., Laliberté, E. et al. (2011). Update on phosphorus nutrition in Proteaceae. Phosphorus nutrition of proteaceae in severely phosphorus-impoverished soils: are there lessons to be learned for future crops? *Plant Physiology* 156: 1058–1066.

Li, C., Song, Y., Guo, L. et al. (2018). Nitric oxide alleviates wheat yield reduction by protecting photosynthetic system from oxidation of ozone pollution. *Environmental Pollution* 236: 296–303.

Li, X., Pan, Y., Chang, B. et al. (2016). NO promotes seed germination and seedling growth under high salt may depend on EIN3 protein in *Arabidopsis*. *Frontiers in Plant Science* 6: 1203.

Li, X., Wu, Z., Xiao, S. et al. (2020). Characterization of abscisic acid (ABA) receptors and analysis of genes that regulate rutin biosynthesis in response to ABA in *Fagopyrum tataricum*. *Plant Physiology and Biochemistry* 157: 432–440.

Lora, J., Laux, T., and Hormaza, J.I. (2019). The role of the integuments in pollen tube guidance in flowering plants. *New Phytologist* 221: 1074–1089.

Lou, Y., Yang, Y., Hu, L. et al. (2015). Exogenous glycinebetaine alleviates the detrimental effect of Cd stress on perennial ryegrass. *Ecotoxicology* 24: 1330–1340.

Mannucci, A., Mariotti, L., Castagna, A. et al. (2020). Hormone profile changes occur in roots and leaves of Micro-Tom tomato plants when exposing the aerial part to low doses of UV-B radiation. *Plant Physiology and Biochemistry* 148: 291–301.

Mohn, M.A., Thaqi, B., and Fischer-Schrader, K. (2019). Isoform-specific NO synthesis by *Arabidopsis thaliana* nitrate reductase. *Plants* 8: 67.

Mur, L.A., Kumari, A., Brotman, Y. et al. (2019). Nitrite and nitric oxide are important in the adjustment of primary metabolism during the hypersensitive response in tobacco. *Journal of Experimental Botany* 70: 4571–4582.

Mur, L.A., Mandon, J., Persijn, S. et al. (2013). Nitric oxide in plants: an assessment of the current state of knowledge. *AoB Plants* 5.

Nagel, M., Alqudah, A.M., Bailly, M. et al. (2019). Novel loci and a role for nitric oxide for seed dormancy and preharvest sprouting in barley. *Plant, Cell & Environment* 42: 1318–1327.

Neill, S.J., Desikan, R., and Hancock, J.T. (2003). Nitric oxide signalling in plants. *New Phytologist* 159: 11–35.

Noman, A., Aqeel, M., Qari, S.H. et al. (2020). Plant hypersensitive response vs pathogen ingression: death of few gives life to others. *Microbial Pathogenesis* 104224.

Nowicka, B., Ciura, J., Szymańska, R. et al. (2018). Improving photosynthesis, plant productivity and abiotic stress tolerance – current trends and future perspectives. *Journal of Plant Physiology* 231: 415–433.

Pagnussat, G.C., Lanteri, M.L., and Lamattina, L. (2003). Nitric oxide and cyclic GMP are messengers in the indole acetic acid-induced adventitious rooting process. *Plant Physiology* 132: 1241–1248.

Pagnussat, G.C., Lanteri, M.L., Lombardo, M.C. et al. (2004). Nitric oxide mediates the indole acetic acid induction activation of a mitogen-activated protein kinase cascade involved in adventitious root development. *Plant Physiology* 135: 279–286.

Palavan-Unsal, N. and Arisan, D. (2009). Nitric oxide signalling in plants. *The Botanical Review* 75: 203–229.

Prakash, V., Singh, V.P., Tripathi, D.K. et al. (2021). Nitric oxide (NO) and salicylic acid (SA): a framework for their relationship in plant development under abiotic stress. *Plant Biology* 23: 39–49.

Probert, R.J. (2000). The role of temperature in the regulation of seed dormancy and germination. *Seeds: The Ecology of Regeneration in Plant Communities* 2: 261–292.

Rodríguez-Serrano, M., Romero-Puertas, M.C., Pazmiño, D.M. et al. (2009). Cellular response of pea plants to cadmium toxicity: cross talk between reactive oxygen species, nitric oxide, and calcium. *Plant Physiology* 150: 229–243.

Rőszer, T. (2012). Nitric oxide synthesis in leaf peroxisomes and in plant-type mitochondria. In: *The Biology of Subcellular Nitric Oxide*, 67–80. Dordrecht, the Netherlands: Springer.

Santisree, P., Bhatnagar-Mathur, P., and Sharma, K.K. (2015). NO to drought-multifunctional role of nitric oxide in plant drought: do we have all the answers? *Plant Science: An International Journal of Experimental Plant Biology* 239: 44–55.

Santolini, J., André, F., Jeandroz, S. et al. (2017). Nitric oxide synthase in plants: where do we stand? *Nitric Oxide* 63: 30–38.

Shams, M., Ekinci, M., Ors, S. et al. (2019). Nitric oxide mitigates salt stress effects of pepper seedlings by altering nutrient uptake, enzyme activity and osmolyte accumulation. *Physiology and Molecular Biology of Plants* 25: 1149–1161.

Sharma, S.S. and Dietz, K.-J. (2009). The relationship between metal toxicity and cellular redox imbalance. *Trends in Plant Science* 14: 43–50.

Simontacchi, M., Galatro, A., Ramos-Artuso, F. et al. (2015). Plant survival in a changing environment: the role of nitric oxide in plant responses to abiotic stress. *Frontiers in Plant Science* 6: 977.

Singh, H.P., Batish, D.R., Kaur, G. et al. (2008). Nitric oxide (as sodium nitroprusside) supplementation ameliorates Cd toxicity in hydroponically grown wheat roots. *Environmental and Experimental Botany* 63: 158–167.

Singh, R., Singh, S., Parihar, P. et al. (2016). Reactive oxygen species (ROS): beneficial companions of plants' developmental processes. *Frontiers in Plant Science* 7.

Solórzano, E., Corpas, F.J., González-Gordo, S. et al. (2020). Reactive oxygen species (ROS) metabolism and nitric oxide (NO) content in roots and shoots of rice (*Oryza sativa* L.) plants under arsenic-induced stress. *Agronomy* 10: 1014.

Suzuki, N., Rivero, R.M., Shulaev, V. et al. (2014). Abiotic and biotic stress combinations. *F1000 Research* 203: 32–43.

Syed Nabi, R., Tayade, R., Hussain, A. et al. (2019). Nitric oxide regulates plant responses to drought, salinity, and heavy metal stress. *Environmental and Experimental Botany* 161: 120–133.

Tewari, R.K., Prommer, J., and Watanabe, M. (2013). Endogenous nitric oxide generation in protoplast chloroplasts. *Plant Cell Reports* 32: 31–44.

Tian, X. and Lei, Y. (2006). Nitric oxide treatment alleviates drought stress in wheat seedlings. *Biologia Plantarum* 50: 775–778.

Tossi, V., Lamattina, L., and Cassia, R. (2009). An increase in the concentration of abscisic acid is critical for nitric oxide-mediated plant adaptive responses to UV-B irradiation. *New Phytologist* 181: 871–879.

van Zelm, E., Zhang, Y., and Testerink, C. (2020). Salt tolerance mechanisms of plants. *Annual Review of Plant Biology* 71: 403–433.

Vanhaelewyn, L., Prinsen, E., Van Der Straeten, D. et al. (2016). Hormone-controlled UV-B responses in plants. *Journal of Experimental Botany* 67: 4469–4482.

Vaultier, M.-N. and Jolivet, Y. (2015). Ozone sensing and early signaling in plants: an outline from the cloud. *Environmental and Experimental Botany* 114: 144–152.

Verma, K., Mehta, S.K., and Shekhawat, G.S. (2013). Nitric oxide (NO) counteracts cadmium induced cytotoxic processes mediated by reactive oxygen species (ROS) in *Brassica juncea*: cross-talk between ROS, NO and antioxidant responses. *Biometals* 26: 255–269.

Verma, V., Ravindran, P., and Kumar, P.P. (2016). Plant hormone-mediated regulation of stress responses. *BMC Plant Biology* 16: 86.

Vishwakarma, A., Wany, A., Pandey, S. et al. (2019). Current approaches to measure nitric oxide in plants. *Journal of Experimental Botany* 70: 4333–4343.

Wang, X., Mao, Z., Zhang, J. et al. (2019). Osmolyte accumulation plays important roles in the drought priming induced tolerance to post-anthesis drought stress in winter wheat (*Triticum aestivum* L.). *Environmental and Experimental Botany* 166: 103804.

Wang, Z., Ma, R., Zhao, M. et al. (2020). NO and ABA interaction regulates tuber dormancy and sprouting in potato. *Frontiers in Plant Science* 11: 103804.

Wang, Z., Straub, D., Yang, H. et al. (2014). The regulatory network of cluster-root function and development in phosphate-deficient white lupin (*Lupinus albus*) identified by transcriptome sequencing. *Physiologia Plantarum* 151: 323–338.

Wilson, I.D., Neill, S.J., and Hancock, J.T. (2008). Nitric oxide synthesis and signaling in plants. *Plant, Cell & Environment* 31: 622–631.

Xiong, J., An, L., Lu, H. et al. (2009). Exogenous nitric oxide enhances cadmium tolerance of rice by increasing pectin and hemicellulose contents in root cell wall. *Planta* 230: 755–765.

Xu, M., Zhu, Y., Dong, J. et al. (2012). Ozone induces flavonol production of *Ginkgo biloba* cells dependently on nitrate reductase-mediated nitric oxide signaling. *Environmental and Experimental Botany* 75: 114–119.

Zeppel, M.J., Harrison, S.P., Adams, H.D. et al. (2015). Drought and resprouting plants. *The New Phytologist* 206: 583–589.

Zhang, J., Li, D., Wei, J. et al. (2019). Melatonin alleviates aluminum-induced root growth inhibition by interfering with nitric oxide production in *Arabidopsis*. *Environmental and Experimental Botany* 161: 157–165.

Zhao, H., Jin, Q., Wang, Y. et al. (2016). Effects of nitric oxide on alleviating cadmium stress in *Typha angustifolia*. *Plant Growth Regulation* 78: 243–251.

3

Role of Nitric Oxide in Abiotic Stress

Arajmand Frukh[1], *Saima Liaqat*[1], *and Altaf Ahmad*[2]

[1]*Department of Botany, School of Chemical and Life Sciences, Jamia Hamdard, New Delhi, India*

[2]*Department of Botany, Aligarh Muslim University, Uttar Pradesh, India*

3.1 Introduction

Abiotic stress is one of the main factors affecting growth and productivity of plants due to the process of overproduction of reactive oxygen species (ROS) and reactive nitrogen species (RNS) (Del Río 2015). When produced in low concentrations these molecules trigger the defense mechanisms of plants to sustain the redox homeostasis. However, prolonged abiotic stress conditions elevate the contents of ROS and RNS, which leads to the disruption of cellular macromolecules such as lipids, nucleic acids, and proteins, resulting in nitro-oxidative stress and finally causing cell death (Di Meo et al. 2016). In this context, it is necessary to understand the processes involved in the defense mechanisms of plants that enable them to withstand adverse environmental conditions. A number of studies have suggested nitric oxide (NO) and RNS play a role in plant defenses against various adverse environmental constraints.

NO is released during oxidative metabolism as a free radical from toxic by-products and acts as a major signaling biomolecule (Zhou et al. 2021). NO is both beneficial and harmful to plants in a dose-dependent manner. The endogenous synthesis of this signaling molecule (NO) at low concentrations plays a vital role in enhancing the antioxidant defense mechanism of plants (Nabi et al. 2019). Besides this, the synergistic interactions of NO with other phytohormones, such as auxins, cytokinins, gibberellins, abscisic acid, ethylene, jasmonic acid, and salicylic acid are considered as the most effective and promising weapons against abiotic stress (Fatima et al. 2021). The exogenous application of NO at low concentrations promotes plant growth, seed germination, photosynthesis, growth of pollen, and delays leaf senescence. Contrary to this, higher levels of NO cause nitrosative stress to plants, which ultimately leads to cell death (Sami et al. 2018). Therefore, NO has dual effects acting both as a positive and a negative regulator of stress responses.

A lot of research over the past 40 years on NO has led to a better understanding of the behavior of the multifunctionality of this signaling molecule. It has been observed that the bioactivity of NO is achieved primarily by posttranslational modifications (PTMs) that include *S*-nitrosylation and tyrosine nitration (Gupta et al. 2020). Throughout the years, the *S*-nitrosylation function of NO has evolved as the main signaling pathway to combat abiotic stress (Santisree et al. 2020). Therefore, in this chapter, we focus on current knowledge on the effects of NO and *S*-nitrosylation in abiotic stress responses in plants.

3.2 Biosynthesis of NO in Plants

NO is a redox gaseous signaling molecule present in all the eukaryotes; however, its synthesis, signaling, and effects are very specific within the plant kingdom in various aspects (Domingos et al. 2015). Although, advanced research has led to an understanding of the synthesis and signaling of NO in animals, in the case of plants this fundamental question still needs to be answered. NO synthesis in plants is achieved by both reductive and oxidative routes (Parankusam et al. 2017) as described in Figure 3.1. The reductive pathway involves nitrite reduction to NO, while the oxidative pathway involves oxidation of aminated molecules.

3.2.1 Reductive Route of NO Production

The nitrite reduction pathway for the synthesis of NO is considered as the most reliable and evidenced pathway in plants (Astier et al. 2018). This occurs both nonenzymatically

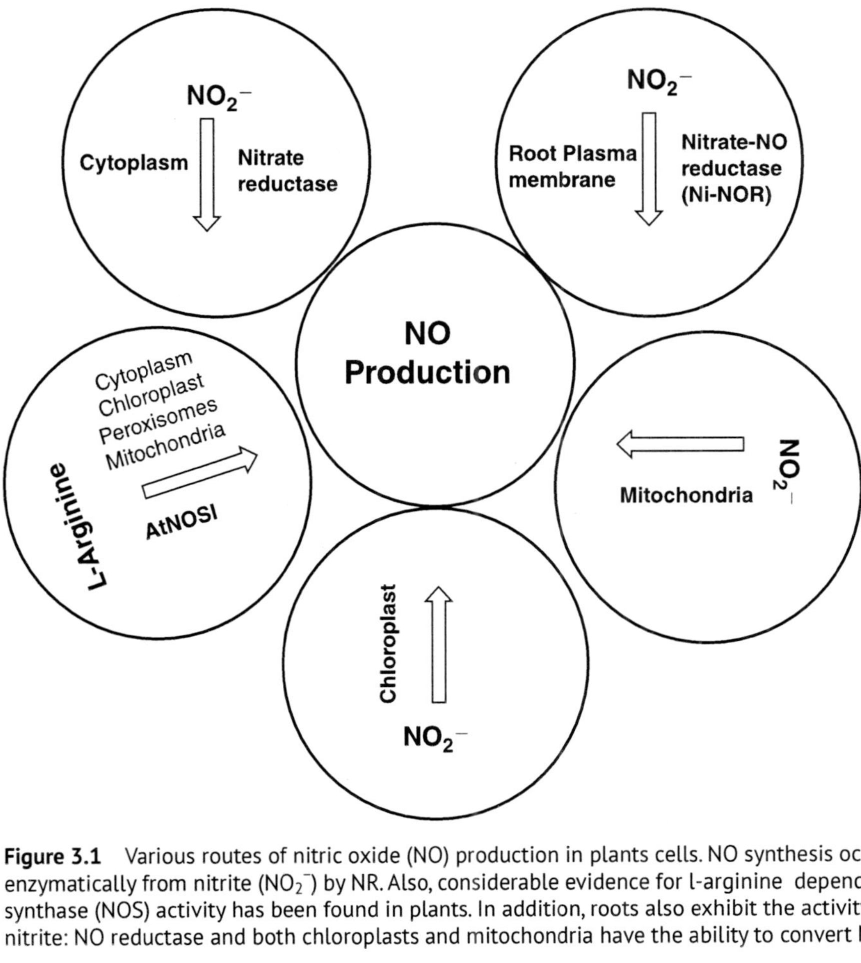

Figure 3.1 Various routes of nitric oxide (NO) production in plants cells. NO synthesis occurs enzymatically from nitrite (NO_2^-) by NR. Also, considerable evidence for l-arginine dependent NO synthase (NOS) activity has been found in plants. In addition, roots also exhibit the activity of a nitrite: NO reductase and both chloroplasts and mitochondria have the ability to convert NO_2^- to NO.

and enzymatically. The nonenzymatic pathway involves NO synthesis in adverse conditions such as a highly reducing environment or low pH. Although, these conditions are seldom encountered they lead to efficient and expeditious synthesis of NO from nitrite (Mur et al. 2013).

Apart from the nonenzymatic synthesis of NO, several proteins are involved in the enzymatic synthesis of NO from nitrites. The enzymatic synthesis of NO occurs with the involvement of a multifunctional cytoplasmic enzyme known as nitrate reductase (Lopes-Oliveira et al. 2021) that is also involved in assimilation and metabolism of nitrogen. It catalyzes the first rate-limiting step in nitrogen assimilation, which involves the reduction of nitrate to nitrite using nicotinamide adenine dinucleotide (NADH) electron donor. Additionally, nitrate reductase possesses nitrite:NO reductase activity (Ni-NR activity) resulting in the formation of NO (Mohn et al. 2019). Under normal conditions, the nitrate-reducing capacity of NR is only 1% and hence the second reaction level is relatively low. It has also been observed that the NR activity is relatively higher for nitrites than that of nitrates and that for efficient Ni-NR activity, accumulation of nitrite is required (Astier et al. 2018). This can be achieved by anoxic or acidic environments that increase the nitrite levels, and in these conditions NO release can increase up to 100-fold (Crawford 2006). Despite all these specific requirements, NR-dependent NO production plays an important role in various processes, such as salt or osmotic stress, cold stress, stomatal movements, and hormone responses (Zhao et al. 2009).

In addition to cytoplasmic NR, production of NO in plants by plasma membrane-bound nitrite reductase (Ni:NOR) has been reported. As well as this, NO is produced from nitrite through the mitochondrial electron transport chain in mitochondria (Gupta et al. 2011). This reaction occurs mainly in mitochondrial membrane, involving complex III and IV, and it has been observed in plant species such as pea, tobacco, and barley (Gupta et al. 2018).

3.2.2 Oxidative Route of NO Production

In addition to the reductive pathway of NO production, an arginine-dependent mechanism has emerged in plants similar to that of animal nitric oxide synthase (NOS) (Goyal et al. 2021). In the case of animals, a short redox pathway involves the conversion of L-arginine conversion to L-citrulline through hydroxyl-arginine, a trapped intermediate, simultaneously releasing NO. In the case of plants, a homolog of L-arginine, N^G-nitro-L-arginine methyl ester (L-NAME) protein was found, suggesting the presence of NOS-like enzymes in plants (Wilson et al. 2008). However, many plant researchers have been unsuccessful in analyzing citrulline in their assays, and thus doubts arise regarding the specificity of L-NAME in plants (Wilson et al. 2008).

However, the first NOS was characterized in the plant kingdom at the beginning of 2010 in the green alga, *Ostreococcus tauri* (Foresi et al. 2010). This enzyme was found to have ~43% similarity in its sequence to that of the human eNOS sequence and was found to produce NO similar to that of animal NOS, i.e. from L-arginine. More recently, NOS activity has been reported in the leaves, roots, and stems of *Arabidopsis* seedlings (Domingos et al. 2015).

3.3 NO Signaling in Plants

Signaling refers to the series of events involved in transformation of stimuli, which is mainly accomplished by reception (a signal molecule binds to a specific receptor), transduction (secondary messengers are released into the cytosol), and induction (cellular process alteration) (Newton et al. 2016). Signaling molecules should be structurally simple, highly diffusible, and also should be able to influence the secondary messengers directly. NO possesses all of these properties and due to the presence of an unpaired electron it is highly reactive leading to the formation of a variety of intermediates (Kükürt et al. 2021).

The molecular mechanism of NO's function as a signaling molecule and its activity being translated into various biological functions is under extensive research. Unlike animals, a NO–cGMP pathway does not exist in plants; therefore, the NO signal is perceived by PTMs of the target proteins. NO signaling has been proposed to occur mainly through two processes: one that involves activation of GC by NO, producing the secondary messenger cyclic guanosine monophosphate (cGMP); and the other by means of PTM of proteins, which occurs either by tyrosine nitration or via *S*-nitrosylation of cysteine residues (Martínez-Ruiz et al. 2011). The most significant of the two NO signaling processes is *S*-nitrosylation (Singh et al. 2021).

3.3.1 *S*-nitrosylation of Cysteine Residues

S-nitrosylation, also known as *S*-nitrosation, is one of the most explored and interpreted NO-dependent PTMs in plants. *S*-nitrosylation is a process in which an NO moiety binds to cysteinyl residue (Cys) by a thiol group of a target protein, resulting in the formation of *S*-nitrosothiol (SNO) (Evangelista et al. 2013). This PTM leads to alteration of enzymatic activity or function of the concerned target protein. NO is present in three reactive states showing different reactivities with thiol groups. These reactive states include nitric oxide radical ($NO^{\bullet}$), nitrosonium cation (NO^+), and nitroxyl anion (NO^-) (Astier and Lindermayr 2012). Among the three reactive groups, NO^+ confers strong reactivity toward R-SH (sulfhydryl group) species while $NO^{\bullet}$ does not directly interact with thiols. This molecule acts as a precursor of many higher nitrogen oxides as well as being directly involved in *S*-nitrosylation activity and thus effectively intervening in the process of *S*-nitrosylation of proteins (Kovacs and Lindermayr 2013). Additionally, *S*-nitrosylation has also been found to occur by a process known as a transnitrosylation reaction that involves the exchange of an NO moiety from an *S*-nitrosylated protein. In this context, the major protein with transnitrosylase activity in plants appears to be S-nitrosoglutathione (GSNO), modulating the total SNO content (Fernando et al. 2019).

Although the majority of the plant proteins contain Cys *S*-nitrosylation is confined to only specific Cys residues. The main cause of this specificity appears to be the presence of targeted Cys near the acidic and basic amino acids, and also the presence of Cys in the hydrophobic pocket, which is favorable for the concentration of nitrosylating agents (Marino and Gladyshev 2010). Moreover, the increase in the nucleophilicity of Cys residue by the conformational and allosteric mechanisms favors the process of *S*-nitrosylation (Astier and Lindermayr 2012). Analyses of salt-stressed *Arabidopsis thaliana* (Fares et al.

2011) and NO gas (Lindermayr et al. 2005), protein extracts after GSNO treatment of *Kalanchoe pinnata* (Abat et al. 2008), and also in low temperature exposed *Brassica juncea* (Abat and Deswal 2009) provide evidence that *S*-nitrosylation is a major PTM in plants.

3.3.2 Tyrosine Nitration

Protein tyrosine nitration (PTN) promotes conformational changes leading to target protein activation or inhibition through covalent and specific modification of tyrosine (Tyr) residues (Kolbert et al. 2017). The tyrosine nitration occurs when a reaction takes place between a tyrosine residue of a target protein and a nitrating agent. It involves addition of a nitro group (NO_2) to the phenolic hydroxyl group at the *ortho* position resulting in the formation of 3-nitrotyrosine. The precursor of NO_2 (peroxynitrite, $ONOO^-$) is a powerful oxidative agent formed from the reaction of NO and a superoxide anion (Radi 2013). The biological significance of PTN is not well established in plants. Only a few reports have revealed tyrosine nitrated proteins in plants. These include studies on *Pseudomonas syringae*-infected *A. thaliana* (Romero-Puertas et al. 2007), hypocotyls of sunflower exposed to high temperature (Chaki et al. 2009), and the leaves and roots of citrus plants exposed to salt stress or NO or H_2O_2 treatments (Tanou et al. 2012).

3.4 NO Cross-talk with Phytohormones During Abiotic Stress Responses in Plants

NO, along with phytohormones, plays a very important role in various metabolic and physiological processes occurring in plants. Although, interactions between NO and virtually all major phytohormones are both synergistic and antagonistic, their interactions are substantially more limited when they are exposed to severe environmental conditions (Adimulam et al. 2017), which are described in the next few sections.

3.4.1 NO–Auxin Interaction

Auxins (AUXs) are the key controllers of several metabolic processes and much evidence highlights their role in modulating root architecture (Sukumar et al. 2013). NO has also been found to be a central signaling molecule affecting the regulation of root architecture and a strong interconnection between signaling of NO and AUX pathways has been observed in regulation of growth, development, and morphology of plant roots (Sun et al. 2018). NO has also been found to be involved in the downstream signaling pathway of AUXs in cucumber to increase adventitious root development (Pagnussat et al. 2004). In the case of *Medicago truncatula*, Cd tolerance was induced by supplementation of NO, which inhibited IAA oxidase activity and hence decreased degradation of AUX (Xu et al. 2010). Moreover, NO supplementation led to an increase in IAA content to induce aluminum (Al) stress tolerance in wheat (Faria-Lopes et al. 2019). NO directly affects the auxin receptor protein, TRANSPORT INHIBITOR RESPONSE 1 (TIR1), by *S*-nitrosation at distinct (Cys 140 and Cys 480) residues, thus influencing AUX perception and signaling. This PTM promotes the interaction of TIR1 with AUXIN/INDOLE-3-ACETIC ACID (AUX/IAA) proteins resulting in the facilitation of degradation of AUX/IAA and AUX signal transduction as a result of repression of AUX-responsive reporter genes (Sun et al. 2021).

3.4.2 NO–Gibberellic Acid Interaction

Gibberellins (GAs) are involved in the enhancement of various physiological processes such as leaf expansion, seed germination, stem elongation, and flower initiation (Davière and Achard 2013). Their biosynthesis is very important for growth of plants and researchers have also found a correlation between GA biosynthesis and NO. NO transcribes *GA3ox1* and *GA3ox2*, key biosynthetic enzymes for active GA synthesis (Sanz et al. 2015). As well as this, a strong interplay between GAs and NO influences germination of seeds, primary root growth, and hypocotyl inhibition during de-etiolation (Beligni and Lamattina 2000). In addition, the coordinated activity of these two signaling molecules confers Al (He et al. 2019) and salt stress (Fatma et al. 2016) tolerance in plants. The molecular interaction of NO with GA involves the promotion of accumulation of DELLA proteins by NO, thus repressing the GA signaling (Asgher et al. 2017).

3.4.3 NO–Cytokinin Interaction

Cytokinins (CKs) are the key players involved in various processes such as cell division, leaf senescence, and chloroplast biogenesis promoting the growth and development of plants (Hönig et al. 2018). NO and CKs also act both synergistically and antagonistically based on the plant cell status (Freschi 2013). The synergistic effect has been observed in NO-deficient mutants or transgenic plants, which exhibited reduced sensitivity to inhibition of root growth when treated with CKs (Sanz et al. 2015). NO deficiency also leads to the CYCLIN-D3;1 (CYCD3;1) cell cycle involving gene inhibition. These genes are transcriptionally activated by CK and hence NO and CK interaction has also been found to regulate plant cell division and subsequent initiation of callus formation from somatic plant tissues with NO deficiency (Shen et al. 2013). In various processes, the synergistic interaction of CKs and NO has displayed control of leaf senescence (Mishina et al. 2007), cell division, and differentiation (Shen et al. 2013), and led to the adaptability of plants to drought stress (Shao et al. 2010).

On the contrary, reports have revealed an antagonistic interaction of NO and CK such as in *Vicia faba* seedlings, where stomatal opening in darkness was triggered by CK-induced reduction in NO levels (Xiao-Ping and Xi-Gui 2006). NO inhibits phosphorylation activity of HISTIDINE PHOSPHOTRANSFER PROTEIN 1 (AHP1) by *S*-nitrosation at Cys115 thereby negatively regulating cytokinin signaling (Sun et al. 2021).

3.4.4 NO–Abscisic Acid Interaction

Abscisic acid (ABA) has a key role in inhibition of seed germination and the induction of seed dormancy in plants (Tuan et al. 2018). Contrarily, NO relieves dormancy and promotes seed germination. Extensive research has been carried out on the interactions between these two molecules at the synthesis level (Bethke et al. 2004). In the case of *Arabidopsis*, exogenous supplementation of ABA led to an increase in endogenous NO content (Bright et al. 2006). This accumulation suggested the underlying mechanism of NO-associated amelioration of the ABA repressor effect. Thus, dormant seeds treated with ABA and NO at the same time exhibited reduced ABA sensitivity and increased germination rates in *Arabidopsis* (Bethke et al. 2006). In addition, with exogenous application of NO, the accumulation of CYP707A2 protein, which is involved in ABA catabolism, was correlated with decreased ABA content (Liu et al. 2009). Open stomata 1/sucrose nonfermenting 1-related protein kinase 2–6 and ABA-insensitive 5 result in negative regulation of NO in ABA signaling (Sun et al. 2021).

3.4.5 NO–Ethylene Interaction

Ethylene (ET) is a gaseous phytohormone involved in the regulation of various processes of growth and development in plants (Iqbal et al. 2017). It also enables plants to develop adaptations against many biotic and abiotic stresses. The prevailing interaction between NO and ET is considered to be antagonistic (Freschi 2013). Reports suggest that NO inhibits biosynthesis and in this way hinders the process of senescence in leaves and flowers, and also fruit ripening. In the case of *Persea americana* and *Fragaria anannasa* during the ripening initiation process reduced NO and increased ET emission rates were observed (Kolbert et al. 2019). NO negatively regulates ethylene production by binding to 1-aminocyclopropane-1-carboxylic acid oxidase (ACO) resulting in the formation of a binary complex (ACO-NO) that in turn forms a stable ternary complex with ACC. This ternary complex negatively regulates ethylene production (Asgher et al. 2017).

3.4.6 NO Interaction with Other Plant Hormones

Apart from interacting with the major five phytohormones, NO cross-talk with other plant hormones, including jasmonic acid, salicylic acid, and brassinosteroids has also been reported. Most of these interactions have been found to occur during plant defense responses against biotic stress.

NO-induced PTM of NON-EXPRESSER OF PATHOGENESIS-RELATED GENE1 (NPR_1) at cys156 NPR1 via *S*-nitrosylation (Tada et al. 2008) plays a key integrative role in coordinated plant immunity responses (Yu et al. 2012). However, the *S*-nitrosylation of cytosolic NPR_1 leads to the suppression of responses that are activated by jasmonic acid (Spoel et al. 2003). In parallel, during the later events of bacterial infection, *S*-nitrosylation at cys280 of SA-BINDING PROTEIN 3 (SABP3) antagonizes the elucidation of plant immunity responses. It represses the capacity to bind SA, thereby representing a negative feedback loop that seems to be essential for the appropriate SA-modulated regulation of plant defense against biotic challenges (Wang et al. 2009). NO and brassinosteroid interaction has been found in the leaf mesophyll cells of maize. Brassinosteroids induced an increase in the NO levels in mesophyll cells of maize leaves acting as an intermediate in the brassinosteroid-induced ABA biosynthesis (Zhang et al. 2011). Additionally, in *Arabidopsis*, brassinosteroid-induced NO production mediates the plant root architecture modification process (Tossi et al. 2013).

3.5 NO Function in Abiotic Stress

Due to its multifunctional role in plant growth and regulation of plant cellular mechanisms NO has gained a significant place in plant science (Siddiqui et al. 2011). Moreover, the role of NO in plant tolerance mechanisms against various abiotic stress challenges has been paid much attention during recent years. It plays a dominant role in defending against some of the major abiotic stress conditions including salt, drought, temperature (high and low), and heavy metal stress (Siddiqui et al. 2011), which are discussed in the next few sections.

3.5.1 NO in Drought and Salinity Stress

Drought stress is accompanied by the loss of water due to increased transpiration rate and is controlled by shrunken stomata. This closure of the stomata and also the decreased CO_2 conductance results in hampering of photosynthesis by Rubisco deactivation. NO is involved in the regulation of the secondary messenger, cGMP, which is involved in stomatal regulation. The plant defense genes are activated by cGMP and also by cyclic adenine dinucleotide phosphate ribose (cADPR), which also acts as a messenger of NO (Sidana et al. 2015). NO upregulates transcription factors DREB1A/CBF3 involved in stress (ABA-dependent or independent) responses and thus decreases the osmotic stress in *Arabidopsis*. It is known to act upstream of the SnRK2-induced ABA-dependent pathway. NO also mediates the effects of oxidative and nitrosative stress by induction of heat shock protein 70 (HSP 70), hence protecting the cells from apoptosis.

A study on wheat (*Triticum aestivum*) seedlings exposed to drought stress induced by polyethylene glycol (PEG), sodium nitroprusside (SNP; nitric oxide donor) examined the effects of nitric oxide on drought-stressed wheat plants. The net results of the study were that 0.2 mM SNP increased the growth, maintained the relative water content, and also reduced the oxidative damage in the wheat seedlings, but a higher concentration of 2 mM SNP increased the ROS, aggravating the stress. The increase in ROS is a result of the ineffectiveness of the antioxidant system due to higher NO concentrations. NO acts as a signaling molecule for the activation of antioxidant enzymes and acts directly on active oxygen and lipids. Also, the effect of NO is concentration dependent, wherein a high concentration of SNP leads to the inhibition of wheat seedlings (Tian and Lei 2006). Another study on wheat by García-Mata and Lamattina (2001) indicated that 150 μm SNP retained more water (15%) than in wheat pretreated with water/NO_2^-/NO_3^- after being exposed to seven days of drought. NO was also able to increase stomatal closure by 35% in *Tradescantia* sp. (monocotyledonous) and by 30% and 65% in *Salpichroa organifolia* and fava bean (*Vicia faba*), dicotyledonous species, respectively (García-Mata and Lamattina 2001). The NO-specific scavenger carboxy-PTIO (2-(4-carboxyphenyl)-4,4,5,5-tetramethylimidazoline-1-oxyl-3-oxide) restored stomatal aperture, transpiration rate, and ion leakage to their respective control values.

The effect of SNP and carboxy-PTIO on proline accumulation and NO content in rice leaves under drought stress was studied. Drought stress increased the proline and NO content whereas the foliar spraying of SNP inhibited the proline accumulation and carboxy-PTIO acted antagonistically to SNP in accumulation of proline. Therefore, under drought stress endogenous NO is expendable for proline accumulation in rice leaves (Xiong et al. 2012). Transgenic barley plants (UHb) were used to investigate the role of NO in drought stress. UHb transgenic plants overexpressed the barley nonsymbiotic hemoglobin gene *HvHb1* responsible for oxidation of NO to NO_3^-. Drought tolerance in UHb barley plants was correlated with reduced NO production. Also, there was an interplay between NO and polyamine production. UHb plants increased polyamines (particularly spermidine) and this was correlated with the accumulation of amino acid precursors and also with the increased expression of the biosynthesis gene of a specific polyamine. It has been suggested that ethylene is associated with NO signaling and that is also related to polyamine metabolism. This was proved in UHb plants by decreased amounts of in vivo ethylene and also in the first step of ethylene biosynthesis by the decreased expression of the aminocyclopropane-1-carboxylic acid synthase gene. This suggests NO–ethylene regulates the polyamine synthesis that is associated with the drought tolerance of barley (Montilla-Bascón et al. 2017). NO also increased growth, accumulation of compatible solutes, and antioxidant (enzymatic and nonenzymatic) mechanisms in soybean against drought stress. NO-induced growth is associated with phenylalanine ammonia-lyase and tyrosine ammonia-lyase (Rezayian et al. 2020). Transgenic *Arabidopsis* (constitutively expressing nNOS, neuronal NO synthase) had increased antioxidant enzymes and osmolyte concentration. An increase in expression of genes (ABA receptors *AtPYL4* and *AtPYL5*) by increased endogenous NO content stemming from transformation of nNOS and stress-mediated ROS accumulation was reported. Hence, NO is also involved in the drought stress response in *Arabidopsis* by modulating various physiological and molecular mechanisms (Shi et al. 2014). Another study in Dunkeld and Cyclone, two canola (*Brassica napus* L.) cultivars, revealed that the foliar application of NO improved growth, oxidative defense mechanisms (CAT), osmoprotectant accumulation (gibberellin, phenolics, total soluble proteins), and decreased lipid peroxidation (decreased malondialdehyde [MDA] and peroxidase [POD] activity). However, other physio-biochemical attributes such as hydrogen peroxide (H_2O_2), proline, and superoxide dismutase (SOD) showed no significant response to the NO application (Akram et al. 2018). NO also had a significant effect on the growth of adventitious roots of marigold (*Tagetes erecta* L.) under drought stress. Plants responded significantly in a dose-dependent manner to NO (donor SNP) application and the most promising effects were shown at 10 μM NO by Liao et al. (2012). They suggested that endogenous NO promotes rooting and H_2O_2 further leads to rooting mechanisms in plants under drought stress. Also, NO and H_2O_2 reduced mesophyll cell destruction and thus improved the photosynthetic performance and alleviated the negative effects of drought stress. A study on Indian mustard (*Brassica juncea*) showed that cv. Pusa Jagannath (PJN) activated nonenzymatic antioxidants and Varuna (VAR) activated both enzymatic and nonenzymatic antioxidants to combat oxidative stress in drought stress. Also, the individual nitric oxide treatment was more effective in mitigating the drought stress than ABA treatments (Sahay et al. 2019).

NO has also been reported to be beneficial in salinity-stressed plants. The effects of NO on salinity-stressed plants was evaluated in chickpea (*Cicer arietinum* L.) using *S*-nitroso-*N*-acetylpenicillamine (SNAP) as an NO donor. The deleterious effects of salinity stress were alleviated by NO application by enhancing leaf relative water content (RWC), photosynthetic pigments, antioxidant enzymes, and osmolyte accumulation (Ahmad et al. 2016). NO-induced accumulation of proteins – more importantly defense proteins, increased antioxidant enzyme activity, and enhanced cell division and energy metabolism (Bai et al. 2011). Proteomic and western blotting techniques revealed the tolerance of maize seedlings under salinity stress is associated with activation of G-protein-associated antioxidant enzyme activity, protein accumulation, and also the controlling of H_2O_2 levels by NO. The G-protein signaling works upstream of NO biogenesis and is an early event in the diverse downstream defense responses (Bai et al. 2011).

Salt stress-affected mustard leaves were used to study the role of nitric oxide (SNP donor) on leaf chlorophyll (Chl) content, carbonic anhydrase (CA), nitrate reductase (NR), and leaf RWC. After 24 h of salt stress NO led to improvement in the levels of NR, CA, Chl, and leaf RWC, along with enhancement of the activities of antioxidative enzymes (superoxide dismutase [SOD], catalase [CAT], peroxidase [POX], ascorbate peroxidase [APX], and glutathione reductase [GR]). Osmolyte (proline and glycine betaine) accumulation increased concomitantly with the NO treatment accompanied by decreased H_2O_2 content, thiobarbituric acid reactive substances (TBARS), and electrolyte leakage (Khan et al. 2012). Another interesting study on application of SNP in cucumber roots revealed that at 50 μM it stimulated ROS-scavenging enzymes accompanied by decreased H_2O_2 accumulation in mitochondria resulting in the reduced lipid peroxidation of mitochondria induced by NaCl. It also reduced the inhibition in H^+-ATPase and H^+-PPase in plasma membrane/tonoplast caused by NaCl (Shi et al. 2007). Another important finding of the study was that an analog of SNP (sodium ferrocyanide) does not release NO and therefore acted in an antagonistic manner to SNP. Also, the NO scavenger (hemoglobin) led to the retrogression of the effects of SNP (Shi et al. 2007). Another study by Fatma et al. (2016) reported an increase in the activities of CAT, APX, GR, and ATP-sulfurylase (ATPS) by 100 μM NO (as sodium nitroprusside), which therefore helped in minimizing oxidative stress caused by salinity stress in mustard. They also suggested an increase in cysteine (Cys) and decreased glutathione (GSH) was a result of the plants receiving NO in addition to the enhanced *S*-assimilation (Fatma et al. 2016). An important study in pak choi (*Brassica chinensis* L.) indicated that exogenous SNP (a NO donor) acted in a dose-dependent manner in alleviating the negative effects of NaCl stress. It enhanced the activities of SOD, CAT, and APX, along with a reduction in MDA, H_2O_2,and also the rate of the superoxide anion radical (O_2^-), thereby preventing oxidative damage. Furthermore, it increased proline, soluble sugar in radicles and plumules, and in addition pretreatment with SNP (10 μM) also led to increased K^+ and decreased Na^+ contents, thereby improving K^+/Na^+ ratio (Ren et al. 2020).

3.5.2 NO in Temperature Stress

Temperature (high and low) stress imparts negative effects on plant growth and productivity. High temperature stress can cause oxidative damage, lipid peroxidation, loss of pigments, protein degradation, enzyme inactivation, and damage to macromolecules (Awasthi et al. 2015). Similarly, low temperature induces many physiological and biochemical changes and disturbs ROS homeostasis in plants (Ahanger et al. 2017). The role of NO in alleviating heat and cold stress has been reported in many studies. Although, an increase in NO content has been observed with the increased temperature, which enhanced the plant adaptation to stress, the emission of NO depends on the duration and severity of heat treatment (Yu et al. 2014). For instance, heat shock at 35°C exhibited less NO burst in a *Nicotiana tabacum* cell suspension compared with 55°C heat shock (Locato et al. 2008). Also, heat treatment at 45°C in *Medicago sativa* and *N. tabacum* leaves for a short duration increased NO levels (Parankusam et al. 2017). Besides the endogenous emission of NO, exogenously applied NO has also been found to minimize the cellular damage induced by excess heat, further highlighting the association of NO and plant heat response (Hasanuzzaman et al. 2013). In the case of low temperature, the exogenous application of NO has been shown to induce tolerance in maize, wheat, and tomato (Neill et al. 2003). Cold acclimation was found to be associated with the increased levels of NO promoted by gene upregulation of NR-dependent NO synthesis in *A. thaliana* (Zhao et al. 2009).

3.5.3 NO in Heavy Metal Stress

Heavy metals are a major source of biosphere contamination (Masindi and Muedi 2018). Several reports to evaluate the effects of heavy metal and remedies to these effects in plants have been conducted, and, in these studies, NO has been found to play a vital role in amplification of antioxidant enzyme activities, which in turn alleviates heavy metal toxicity (Terrón-Camero et al. 2019). It was reported that exogenous application of NO donor SNP reduced copper (Cu) toxicity in rice (Mostofa et al. 2014). In cell cultures of soybean, the mechanism of detoxification and antioxidative properties of NO were found under both Cd (cadmium) and Cu (copper) stress (Singh et al. 2008). Furthermore, in *Hibiscus moschetuos*, NO also led to a reduction in aluminum (Al^{3+}) toxicity (Tian et al. 2007). Application of SNP in tomato plants increased ROS-scavenging enzyme activity along with reduced accumulation of H_2O_2, and also significantly mitigated the effect of $CuCl_2$-induced growth inhibition (Cui et al. 2010). Wheat under Al^{3+} stress exposed to SNP showed enhancement in the activities of SOD, CAT, APX, and increased protein content, with a decline in the contents of H_2O_2 and MDA (Zhang et al. 2008). Also, exogenous application of NO enhanced accumulation of Cd in root cell walls, with increased hemicellulose and pectin content in the root cell walls and decreased accumulation of Cd in leaves of rice leading to Cd tolerance (Xiong et al. 2009).

3.6 Conclusions

We know that NO plays an important role in multiple physiological processes in plants. However, the work related to NO signaling, NO-mediated gene regulation, and biosynthesis of NO pathways in plants and their correlation with various environmental disturbances is still scarce and needs special attention with respect to increasing environmental threats to crop productivity.

References

Abat, J.K. and Deswal, R. (2009). Differential modulation of S-nitrosoproteome of *Brassica juncea* by low temperature: change in S-nitrosylation of Rubisco is responsible for the inactivation of its carboxylase activity. *Proteomics* 9 (18): 4368–4380.

Abat, J.K., Mattoo, A.K., and Deswal, R. (2008). S-nitrosylated proteins of a medicinal CAM plant *Kalanchoe pinnata*-ribulose-1,5-bisphosphate carboxylase/oxygenase activity targeted for inhibition. *FEBS Journal* 275 (11): 2862–2872.

Adimulam, S.S., Bhatnagar-Mathur, P., and Santisree, P. (2017). Interaction of nitric oxide with phytohormones under drought stress. *Journal of Plant Studies* 6 (1): 58–61.

Ahanger, M.A., Tomar, N.S., Tittal, M. et al. (2017). Plant growth under water/salt stress: ROS production; antioxidants and significance of added potassium under such conditions. *Physiology and Molecular Biology of Plants* 23 (4): 731–744.

Ahmad, P., Abdel Latef, A.A., Hashem, A. et al. (2016). Nitric oxide mitigates salt stress by regulating levels of osmolytes and antioxidant enzymes in chickpea. *Frontiers in Plant Science* 7: 347.

Akram, N.A., Iqbal, M., Muhammad, A. et al. (2018). Aminolevulinic acid and nitric oxide regulate oxidative defense and secondary metabolisms in canola (*Brassica napus* L.) under drought stress. *Protoplasma* 255 (1): 163–174.

Asgher, M., Per, T.S., Masood, A. et al. (2017). Nitric oxide signaling and its crosstalk with other plant growth regulators in plant responses to abiotic stress. *Environmental Science and Pollution Research* 24 (3): 2273–2285.

Astier, J., Gross, I., and Durner, J. (2018). Nitric oxide production in plants: an update. *Journal of Experimental Botany* 69 (14): 3401–3411.

Astier, J. and Lindermayr, C. (2012). Nitric oxide-dependent posttranslational modification in plants: an update. *International Journal of Molecular Sciences* 13 (11): 15193–15208.

Awasthi, R., Bhandari, K., and Nayyar, H. (2015). Temperature stress and redox homeostasis in agricultural crops. *Frontiers in Environmental Science* 3: 11.

Bai, X., Yang, L., Yang, Y. et al. (2011). Deciphering the protective role of nitric oxide against salt stress at the physiological and proteomic levels in maize. *Journal of Proteome Research* 10 (10): 4349–4364.

Beligni, M.V. and Lamattina, L. (2000). Nitric oxide stimulates seed germination and de-etiolation, and inhibits hypocotyl elongation, three light-inducible responses in plants. *Planta* 210 (2): 215–221.

Bethke, P.C., Gubler, F., Jacobsen, J.V. et al. (2004). Dormancy of *Arabidopsis* seeds and barley grains can be broken by nitric oxide. *Planta* 219 (5): 847–855.

Bethke, P.C., Libourel, I.G., Reinöhl, V. et al. (2006). Sodium nitroprusside, cyanide, nitrite, and nitrate break *Arabidopsis* seed dormancy in a nitric oxide-dependent manner. *Planta* 223 (4): 805–812.

Bright, J., Desikan, R., Hancock, J.T. et al. (2006). ABA-induced NO generation and stomatal closure in *Arabidopsis* are dependent on H_2O_2 synthesis. *The Plant Journal* 45 (1): 113–122.

Chaki, M., Valderrama, R., Fernández-Ocaña, A.M. et al. (2009). Protein targets of tyrosine nitration in sunflower (*Helianthus annuus* L.) hypocotyls. *Journal of Experimental Botany* 60 (15): 4221–4234.

Crawford, N.M. (2006). Mechanisms for nitric oxide synthesis in plants. *Journal of Experimental Botany* 57 (3): 471–478.

Cui, X.M., Zhang, Y.K., Wu, X.B. et al. (2010). The investigation of the alleviated effect of copper toxicity by exogenous nitric oxide in tomato plants. *Plant Soil and Environment* 56 (6): 274–281.

Davière, J.M. and Achard, P. (2013). Gibberellin signaling in plants. *Development* 140 (6): 1147–1151.

Del Río, L.A. (2015). ROS and RNS in plant physiology: an overview. *Journal of Experimental Botany* 66 (10): 2827–2837.

Di Meo, S., Reed, T.T., Venditti, P. et al. (2016). Role of ROS and RNS sources in physiological and pathological conditions. *Oxidative Medicine and Cellular Longevity* 2016: 1245049.

Domingos, P., Prado, A.M., Wong, A. et al. (2015). Nitric oxide: a multitasked signaling gas in plants. *Molecular Plant* 8 (4): 506–520.

Evangelista, A.M., Kohr, M.J., and Murphy, E. (2013). S-nitrosylation: specificity, occupancy, and interaction with other post-translational modifications. *Antioxidants & Redox Signaling* 19 (11): 1209–1219.

Fares, A., Rossignol, M., and Peltier, J.B. (2011). Proteomics investigation of endogenous S-nitrosylation in *Arabidopsis*. *Biochemical and Biophysical Research Communications* 416 (3–4): 331–336.

Faria-Lopes, R.O., Muniz, D.R., Chaves, I.S. et al. (2019). Nitric oxide precursors prevent Al-triggered auxin flow inhibition in *Triticum aestivum* roots. *Journal of Advanced Research* 15: 27–36.

Fatima, A., Husain, T., Suhel, M. et al. (2021). Implication of nitric oxide under salinity stress: the possible interaction with other signaling molecules. *Journal of Plant Growth Regulation* 1–15.

Fatma, M., Masood, A., Per, T.S. et al. (2016). Nitric oxide alleviates salt stress inhibited photosynthetic performance by interacting with sulfur assimilation in mustard. *Frontiers in Plant Science* 7: 521.

Fernando, V., Zheng, X., Walia, Y., Sharma, V. et al. (2019). S-Nitrosylation: an emerging paradigm of redox signaling. *Antioxidants* 8 (9): 404.

Foresi, N., Correa-Aragunde, N., Parisi, G. et al. (2010). Characterization of a nitric oxide synthase from the plant kingdom: NO generation from the green alga *Ostreococcus tauri* is light irradiance and growth phase dependent. *The Plant Cell* 22 (11): 3816–3830.

García-Mata, C. and Lamattina, L. (2001). Nitric oxide induces stomatal closure and enhances the adaptive plant responses against drought stress. *Plant Physiology* 126 (3): 1196–1204.

Goyal, V., Jhanghel, D., and Mehrotra, S. (2021). Emerging warriors against salinity in plants: nitric oxide and hydrogen sulphide. *Physiologia Plantarum* 171 (4): 896–908.

Gupta, K.J., Fernie, A.R., Kaiser, W.M. et al. (2011). On the origins of nitric oxide. *Trends in Plant Science* 16 (3): 160–168.

Gupta, K.J., Kolbert, Z., Durner, J. et al. (2020). Regulating the regulator: nitric oxide control of post-translational modifications. *New Phytologist* 227 (5): 1319–1325.

Gupta, K.J., Kumari, A., Florez-Sarasa, I. et al. (2018). Interaction of nitric oxide with the components of the plant mitochondrial electron transport chain. *Journal of Experimental Botany* 69 (14): 3413–3424.

Hasanuzzaman, M., Nahar, K., Alam, M. et al. (2013). Physiological, biochemical, and molecular mechanisms of heat stress tolerance in plants. *International Journal of Molecular Sciences* 14 (5): 9643–9684.

He, H., Li, Y., and He, L.F. (2019). Role of nitric oxide and hydrogen sulfide in plant aluminum tolerance. *Biometals* 32 (1): 1–9.

Hönig, M., Plíhalová, L., Husičková, A. et al. (2018). Role of cytokinins in senescence, antioxidant defence and photosynthesis. *International Journal of Molecular Sciences* 19 (12): 4045.

Iqbal, N., Khan, N.A., Ferrante, A. et al. (2017). Ethylene role in plant growth, development and senescence: interaction with other phytohormones. *Frontiers in Plant Science* 8: 475.

Khan, M.N., Siddiqui, M.H., Mohammad, F. et al. (2012). Interactive role of nitric oxide and calcium chloride in enhancing tolerance to salt stress. *Nitric Oxide* 27 (4): 210–218.

Kolbert, Z., Feigl, G., Bordé, Á. et al. (2017). Protein tyrosine nitration in plants: present knowledge, computational prediction and future perspectives. *Plant Physiology and Biochemistry* 113: 56–63.

Kolbert, Z., Feigl, G., Freschi, L. et al. (2019). Gasotransmitters in action: nitric oxide-ethylene crosstalk during plant growth and abiotic stress responses. *Antioxidants* 8 (6): 167.

Kovacs, I. and Lindermayr, C. (2013). Nitric oxide-based protein modification: formation and site-specificity of protein S-nitrosylation. *Frontiers in Plant Science* 4: 137.

Kükürt, A., Gelen, V., Başer, Ö.F. et al. (2021). Thiols: role in oxidative stress-related disorders. In: *Lipid Peroxidation*. IntechOpen.

Liao, W.B., Huang, G.B., Yu, J.H. et al. (2012). Nitric oxide and hydrogen peroxide alleviate drought stress in marigold explants and promote its adventitious root development. *Plant Physiology and Biochemistry* 58: 6–15.

Lindermayr, C., Saalbach, G., and Durner, J. (2005). Proteomic identification of S-nitrosylated proteins in *Arabidopsis*. *Plant Physiology* 137 (3): 921–930.

Liu, Y., Shi, L., Ye, N. et al. (2009). Nitric oxide-induced rapid decrease of abscisic acid concentration is required in breaking seed dormancy in *Arabidopsis*. *New Phytologist* 183 (4): 1030–1042.

Locato, V., Gadaleta, C., De Gara, L. et al. (2008). Production of reactive species and modulation of antioxidant network in response to heat shock: a critical balance for cell fate. *Plant, Cell & Environment* 31 (11): 1606–1619.

Lopes-Oliveira, P.J., Oliveira, H.C., Kolbert, Z. et al. (2021). The light and dark sides of nitric oxide: multifaceted roles of nitric oxide in plant responses to light. *Journal of Experimental Botany* 72 (3): 885–903.

Marino, S.M. and Gladyshev, V.N. (2010). Structural analysis of cysteine S-nitrosylation: a modified acid-based motif and the emerging role of trans-nitrosylation. *Journal of Molecular Biology* 395 (4): 844–859.

Martínez-Ruiz, A., Cadenas, S., and Lamas, S. (2011). Nitric oxide signaling: classical, less classical, and nonclassical mechanisms. *Free Radical Biology and Medicine* 51 (1): 17–29.

Masindi, V. and Muedi, K.L. (2018). Environmental contamination by heavy metals. *Heavy Metals* 10: 115–132.

Mishina, T.E., Lamb, C., and Zeier, J. (2007). Expression of a nitric oxide degrading enzyme induces a senescence programme in *Arabidopsis*. *Plant Cell & Environment* 30 (1): 39–52.

Mohn, M.A., Thaqi, B., and Fischer-Schrader, K. (2019). Isoform-specific NO synthesis by *Arabidopsis thaliana* nitrate reductase. *Plants* 8 (3): 67.

Montilla-Bascón, G., Rubiales, D., Hebelstrup, K.H. et al. (2017). Reduced nitric oxide levels during drought stress promote drought tolerance in barley and is associated with elevated polyamine biosynthesis. *Scientific Reports* 7 (1): 1–15.

Mostofa, M.G., Seraj, Z.I., and Fujita, M. (2014). Exogenous sodium nitroprusside and glutathione alleviate copper toxicity by reducing copper uptake and oxidative damage in rice (*Oryza sativa* L.) seedlings. *Protoplasma* 251 (6): 1373–1386.

Mur, L.A., Mandon, J., Persijn, S. et al. (2013). Nitric oxide in plants: an assessment of the current state of knowledge. *AoB Plants* 5:: 3131313231gcg.

Nabi, R.B.S., Tayade, R., Hussain, A. et al. (2019). Nitric oxide regulates plant responses to drought, salinity, and heavy metal stress. *Environmental and Experimental Botany* 161: 120–133.

Neill, S.J., Desikan, R., and Hancock, J.T. (2003). Nitric oxide signalling in plants. *New Phytologist* 159 (1): 11–35.

Newton, A.C., Bootman, M.D., and Scott, J.D. (2016). Second messengers. *Cold Spring Harbor Perspectives in Biology* 8 (8): a005926.

Pagnussat, G.C., Lanteri, M.L., Lombardo, M.C. et al. (2004). Nitric oxide mediates the indole acetic acid induction activation of a mitogen-activated protein kinase cascade involved in adventitious root development. *Plant Physiology* 135 (1): 279–286.

Parankusam, S., Adimulam, S.S., Bhatnagar-Mathur, P. et al. (2017). Nitric oxide (NO) in plant heat stress tolerance: current knowledge and perspectives. *Frontiers in Plant Science* 8: 1582.

Radi, R. (2013). Protein tyrosine nitration: biochemical mechanisms and structural basis of functional effects. *Accounts of Chemical Research* 46 (2): 550–559.

Ren, Y., Wang, W., He, J. et al. (2020). Nitric oxide alleviates salt stress in seed germination and early seedling growth of pak choi (*Brassica chinensis* L.) by enhancing physiological and biochemical parameters. *Ecotoxicology and Environmental Safety* 187: 109785.

Rezayian, M., Ebrahimzadeh, H., and Niknam, V. (2020). Nitric oxide stimulates antioxidant system and osmotic adjustment in soybean under drought stress. *Journal of Soil Science and Plant Nutrition* 20: 1122–1132.

Romero-Puertas, M.C., Laxa, M., Matte, A. et al. (2007). S-nitrosylation of peroxiredoxin II E promotes peroxynitrite-mediated tyrosine nitration. *The Plant Cell* 19 (12): 4120–4130.

Sahay, S., Khan, E., and Gupta, M. (2019). Nitric oxide and abscisic acid protects against PEG-induced drought stress differentially in *Brassica* genotypes by combining the role of stress modulators, markers and antioxidants. *Nitric Oxide* 89: 81–92.

Sami, F., Faizan, M., Faraz, A. et al. (2018). Nitric oxide-mediated integrative alterations in plant metabolism to confer abiotic stress tolerance, NO crosstalk with phytohormones and NO-mediated post translational modifications in modulating diverse plant stress. *Nitric Oxide* 73: 22–38.

Santisree, P., Sanivarapu, H., Gundavarapu, S. et al. (2020). Nitric oxide as a signal in inducing secondary metabolites during plant stress. In: *Co-Evolution of Secondary Metabolites* (eds. J.M. Mérillon and K.G. Ramawat), 593–621, Cham, Switzerland: Springer.

Sanz, L., Albertos, P., Mateos, I. et al. (2015). Nitric oxide (NO) and phytohormones crosstalk during early plant development. *Journal of Experimental Botany* 66 (10): 2857–2868.

Shao, R., Wang, K., and Shangguan, Z. (2010). Cytokinin-induced photosynthetic adaptability of *Zea mays* L. to drought stress associated with nitric oxide signal: probed by ESR spectroscopy and fast OJIP fluorescence rise. *Journal of Plant Physiology* 167 (6): 472–479.

Shen, Q., Wang, Y.T., Tian, H. et al. (2013). Nitric oxide mediates cytokinin functions in cell proliferation and meristem maintenance in *Arabidopsis*. *Molecular Plant* 6 (4): 1214–1225.

Shi, H., Ye, T., Zhu, J.K. et al. (2014). Constitutive production of nitric oxide leads to enhanced drought stress resistance and extensive transcriptional reprogramming in *Arabidopsis*. *Journal of Experimental Botany* 65 (15): 4119–4131.

Shi, Q., Ding, F., Wang, X. et al. (2007). Exogenous nitric oxide protect cucumber roots against oxidative stress induced by salt stress. *Plant Physiology and Biochemistry* 45 (8): 542–550.

Sidana, S., Bose, J., Shabala, L. et al. (2015). Nitric oxide in drought stress signalling and tolerance in plants. In: *Nitric Oxide Action in Abiotic Stress Responses in Plants* (eds. M. Khan, M. Mobin, F. Mohammad, and F.J. Corpas), 95–114. Cham, Switzerland: Springer.

Siddiqui, M.H., Al-Whaibi, M.H., and Basalah, M.O. (2011). Role of nitric oxide in tolerance of plants to abiotic stress. *Protoplasma* 248 (3): 447–455.

Singh, H.P., Batish, D.R., Kaur, G. et al. (2008). Nitric oxide (as sodium nitroprusside) supplementation ameliorates Cd toxicity in hydroponically grown wheat roots. *Environmental and Experimental Botany* 63 (1–3): 158–167.

Singh, K., Shukla, I., Tiwari, A.K. et al. (2021). Physiological, biochemical, and molecular mechanism of nitric oxide-mediated abiotic stress tolerance. In: *Plant Growth Regulators: Signalling Under Stress Conditions* (eds. T. Aftab and K.R. Hakeem), 217. Cham, Switzerland: Springer.

Spoel, S.H., Koornneef, A., Claessens, S.M. et al. (2003). NPR1 modulates cross-talk between salicylate-and jasmonate-dependent defense pathways through a novel function in the cytosol. *The Plant Cell* 15 (3): 760–770.

Sukumar, P., Legue, V., Vayssieres, A. et al. (2013). Involvement of auxin pathways in modulating root architecture during beneficial plant–microorganism interactions. *Plant, Cell & Environment* 36 (5): 909–919.

Sun, C., Zhang, Y., Liu, L. et al. (2021). Molecular functions of nitric oxide and its potential applications in horticultural crops. *Horticulture Research* 8 (1): 1–17.

Sun, H., Feng, F., Liu, J. et al. (2018). Nitric oxide affects rice root growth by regulating auxin transport under nitrate supply. *Frontiers in Plant Science* 9: 659.

Tada, Y., Spoel, S.H., Pajerowska-Mukhtar, K. et al. (2008). Plant immunity requires conformational charges of NPR1 via S-nitrosylation and thioredoxins. *Science* 321 (5891): 952–956.

Tanou, G., Filippou, P., Belghazi, M. et al. (2012). Oxidative and nitrosative-based signaling and associated post-translational modifications orchestrate the acclimation of citrus plants to salinity stress. *The Plant Journal* 72 (4): 585–599.

Terrón-Camero, L.C., Peláez-Vico, M.Á., Del-Val, C. et al. (2019). Role of nitric oxide in plant responses to heavy metal stress: exogenous application versus endogenous production. *Journal of Experimental Botany* 70 (17): 4477–4488.

Tian, Q.Y., Sun, D.H., Zhao, M.G. et al. (2007). Inhibition of nitric oxide synthase (NOS) underlies aluminum-induced inhibition of root elongation in *Hibiscus moscheutos*. *New Phytologist* 174 (2): 322–331.

Tian, X. and Lei, Y. (2006). Nitric oxide treatment alleviates drought stress in wheat seedlings. *Biologia Plantarum* 50 (4): 775–778.

Tossi, V., Lamattina, L., and Cassia, R. (2013). Pharmacological and genetical evidence supporting nitric oxide requirement for 2,4-epibrassinolide regulation of root architecture in *Arabidopsis thaliana*. *Plant Signaling & Behavior* 8 (7): e24712.

Tuan, P.A., Kumar, R., Rehal, P.K. et al. (2018). Molecular mechanisms underlying abscisic acid/gibberellin balance in the control of seed dormancy and germination in cereals. *Frontiers in Plant Science* 9: 668.

Wang, Y.Q., Feechan, A., Yun, B.W. et al. (2009). S-nitrosylation of AtSABP3 antagonizes the expression of plant immunity. *Journal of Biological Chemistry* 284 (4): 2131–2137.

Wilson, I.D., Neill, S.J., and Hancock, J.T. (2008). Nitric oxide synthesis and signalling in plants. *Plant, Cell & Environment* 31 (5): 622–631.

Xiao-Ping, S. and Xi-Gui, S. (2006). Cytokinin-and auxin-induced stomatal opening is related to the change of nitric oxide levels in guard cells in broad bean. *Physiologia Plantarum* 128 (3): 569–579.

Xiong, J., An, L., Lu, H. et al. (2009). Exogenous nitric oxide enhances cadmium tolerance of rice by increasing pectin and hemicellulose contents in root cell wall. *Planta* 230 (4): 755–765.

Xiong, J., Zhang, L., Fu, G. et al. (2012). Drought-induced proline accumulation is uninvolved with increased nitric oxide, which alleviates drought stress by decreasing transpiration in rice. *Journal of Plant Research* 125 (1): 155–164.

Xu, J., Wang, W., Yin, H. et al. (2010). Exogenous nitric oxide improves antioxidative capacity and reduces auxin degradation in roots of *Medicago truncatula* seedlings under cadmium stress. *Plant and Soil* 326 (1): 321–330.

Yu, M., Lamattina, L., Spoel, S.H. et al. (2014). Nitric oxide function in plant biology: a redox cue in deconvolution. *New Phytologist* 202 (4): 1142–1156.

Yu, M., Yun, B.W., Spoel, S.H. et al. (2012). A sleigh ride through the SNO: regulation of plant immune function by protein S-nitrosylation. *Current Opinion in Plant Biology* 15 (4): 424–430.

Zhang, A., Zhang, J., Zhang, J. et al. (2011). Nitric oxide mediates brassinosteroid-induced ABA biosynthesis involved in oxidative stress tolerance in maize leaves. *Plant and Cell Physiology* 52 (1): 181–192.

Zhang, H., Li, Y.H., Hu, L.Y. et al. (2008). Effects of exogenous nitric oxide donor on antioxidant metabolism in wheat leaves under aluminum stress. *Russian Journal of Plant Physiology* 55 (4): 469–474.

Zhao, M.G., Chen, L., Zhang, L.L. et al. (2009). Nitric reductase-dependent nitric oxide production is involved in cold acclimation and freezing tolerance in Arabidopsis. *Plant Physiology* 151 (2): 755–767.

Zhou, X., Joshi, S., Khare, T. et al. (2021). Nitric oxide, crosstalk with stress regulators and plant abiotic stress tolerance. *Plant Cell Reports* 40: 1395–1414.

4

Molecular Approaches for Designing NO-mediated Stress Tolerance Pathways

Mobina Ulfat[1], Habib-ur-Rehman Athar[2], Zafar Ullah Zafar[2], and Muhammad Ashraf[3]

[1]*Department of Botany, Lahore College for Women University, Lahore, Pakistan*
[2]*Institute of Pure and Applied Biology, Bahauddin Zakariya University, Multan, Pakistan*
[3]*The University of Agriculture, Faisalabad, Pakistan*

4.1 Introduction

The odorless and colorless gas discovered by Joseph Priestley (1772), initially named "nitrous air," was eventually called nitric oxide (NO). Initially, NO gas appeared to be a toxic pollutant until it was observed to be emitted by herbicide-treated intact plant leaves of *Glycine max* along with NO_2 (Klepper 1979). Gradually, NO gained major interest in research related to plants as well as animals, and in 1992, NO was crowned as "Molecule of the Year" by *Science* magazine. Later on, a joint discovery by F. Murad, L.J. Ignarro, and R.F. Furchgott won the Noble Prize in Physiology/Medicine (1998), reporting that NO is produced as a signaling molecule by the mammalian species. Numerous other related studies unveiled the biological significance of NO in plants as well as animals. NO research has gradually gained significance in plants.

4.2 Biosynthesis of NO in Plants

In plants, biosynthesis of NO can take any of the routes among the nitric oxide synthase (NOS) pathway (Del Río et al. 2004), the nitrate reductase (NR) pathway (Rockel and Kaiser 2002; Tejada-Jimenez et al. 2019), or the pathway using xanthine dehydrogenase/oxidase (Li et al. 2001). Nonenzymatic synthesis of NO was also found while studying aleurone layers (apoplast) in *Hordeum vulgare* (Bethke et al. 2004)

4.3 NOS

NO biosynthesis occurs via nicotinamide adenine dinucleotide phosphate (NADPH)-dependent oxidation of L-arginine (L-Arg) through NOS. In the plant kingdom, the first NOS enzyme was figured out in *Ostreococcus tauri* (a green alga) (Foresi et al. 2010). A transfer DNA insertion mutant in *Arabidopsis thaliana* that was based on snail protein homology provided initial evidence regarding synthesis of NO in higher plants via a gene encoding NOS-like activity. However, in *Arabidopsis* it was concluded that *AtNOS1* has no involvement in synthesis of NO (Gupta Kapuganti et al. 2010) but appears to be a regulator of NO in the cells (Besson-Bard et al. 2008) via NADPH-dependent oxidation of L-Arg through NO synthase. Moreover, NO is also synthesized nonenzymatically by: (i) nitrous acid (HNO_2) derived from protonated nitrate; (ii) carotenoid-mediated reduction of nitrite (NO_2) to nitric oxide, when illuminated; and (iii) oxidation of salicyl-hydroxymate and hydroxylamine (Figure 4.1).

4.4 NR

Synthesis of NO by NR has already been established in *Triticum aestivum*, *Zea mays*, *Spinacea oleracea*, *Helianthus annuus*, *Cucumis sativus*, *Malaxis monophyllos*, *Chlamydomonas reinhardtii*, and *Aloe vera*. Two genes, i.e. *NIA1* and *NIA2*, have been reported in *Arabidopsis* as responsible for producing NR. In plants, NR is meant for NADPH-dependent synthesis of NO from nitrate. However, the nitrogen obtained from nitrate can only be available to NR after it has been transported into the cell/tissue/body of the plant from surrounding media through specialized proteins involved in the sensing of the nitrate followed by its uptake leading to distribution throughout the plant body and finally storage. In higher plants, several transporter protein families are involved in transport of nitrate, including NRT1/PTR/NPF (NITRATE TRANSPORTER 1/PEPTIDE TRANSPORTER/NITRATE PEPTIDE TRANSPORTER family), NRT2/NNP (NITRATE TRANSPORTER 2/NITRATE NITRITE PORTER), CLC (CHLORIDE CHANNELS), SLAC1/SLAH (SLOW ANION CHANNEL 1/SLOW ANION CHANNEL-ASSOCIATED 1 HOMOLOG 3) and ALMT (ALUMINUM-ACTIVATED MALATE TRANSPORTERS). NAR1, NRT2, and NRT1/NPF are found in *Chlamydomonas*.

The high-affinity transport (HAT) system ensures the transport of nitrite (whether absorbed from the medium or produced from nitrate) from the cytosol to the chloroplast.

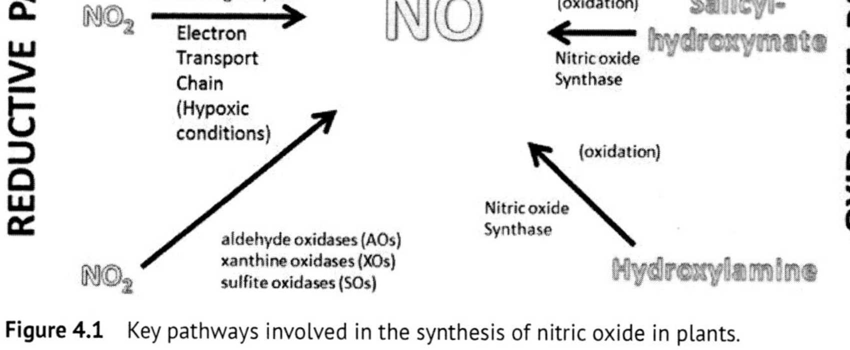

Figure 4.1 Key pathways involved in the synthesis of nitric oxide in plants.

NAR1 facilitates HAT of nitrite in *Chlamydomonas*. In higher plants, CLC family transporters facilitate the transport of nitrite to chloroplasts (Tejada-Jimenez et al. 2019), and NRT1 family transporters enable its export from the chloroplasts (Fan et al. 2009). Similarly, in *Arabidopsis*, HAT of nitrite and nitrate is mediated by NRT1 (dual-affinity nitrite/nitrate transporter) and NRT2 (in association with NAR2 as an accessory protein) (Mariscal et al. 2006). In the chloroplast, NiR (nitrite reductase) produces ammonium after reduction of nitrite. *NIT2* appears to be a master regulatory gene controlling (NAR1/NAR2/NRT2/NR/NiR) in *Chlamydomonas* (Camargo et al. 2007). In higher plants, *NLPs* appear to act as regulatory genes (Yanagisawa 2014). Ammonium is finally incorporated into the organic carbon skeleton as glutamate with the help of glutamate synthase/glutamine synthetase (GS/GOGAT) as reported in wheat spikes (Kumar et al. 2009). Cytosolic nitrite concentrations are lower than nitrate concentrations, which are relatively high (1–6 mM). Moreover, the level of nitrate in the cytosol is more stable compared with in the vacuole (5–75 mM). External nitrate concentrations can be 10 000-fold enhanced (Miller and Smith 1996). This ensures efficiency of nitrate assimilation, along with nitrate signaling (Fernandez and Galvan 2008).

NR reduces nitrate to nitrite, using NAD(P)H as an electron donor. NR is a 200-KDa protein with two subunits, each carrying heme b557, flavin adenine dinucleotide (FAD), and molybdenum as prosthetic groups. Molybdenum is chelated by dithiolene and bound to pyranopterin forming a molybdenum cofactor (Moco). Two hinge regions join the domains and are protease-sensitive. These domains act as redox centers, and within the active site flow of electrons occurs as NAD(P)H FAD heme Moco (Iobbi-Nivol and Leimkühler 2013).

In addition, several enzymes can produce NO by reducing nitrite, including aldehyde oxidases (AOs), xanthine oxidases (XOs), and sulfite oxidases (SOs). These molybdoenzymes in plants catalyze the synthesis of NO while reducing nitrite in vitro and under anaerobic conditions (Wang, Krizowski et al. 2015). In mammals, NO formation occurs either through arginine oxidation under aerobic conditions or through nitrite reduction during anaerobic conditions (Chamizo-Ampudia et al. 2016). Plant SOs possesses a mild nitrite reductase activity as compared with mammalian AOs and xanthine oxidoreductase (XOR) (Schrader et al. 2003). NO production from nitrite is also associated with the NR bound on the plasma membrane and the inner mitochondrial membrane, in hypoxic environments (Ostroukhova and Ermilova 2019). This reaction occurs in plant tissues such as roots that are exposed to hypoxic conditions and helps in protection of the ETC and overall metabolism in the mitochondria if there is a poor supply of oxygen (Tejada-Jimenez et al. 2019).

4.5 NO-controlled Functions in Plants

After a great deal of research NO was thought to be a plant defense modulator related to pathogen attack. Research reports also established the role of NO in various processes during plant development including seed germination (Bethke et al. 2004; Patel et al. 2017), root formation (Correa-Aragunde et al. 2004; Sun et al. 2017), development of seed (Patel et al. 2017), photosynthesis (Astier et al. 2012), stomatal movements (Neill et al. 2002), senescence (Corpas et al. 2004), etc. NO has also been reported to cause availability and accumulation of Fe and also can enhance levels of chlorophyll in plants grown with a limited supply of Fe (Graziano et al. 2002; Graziano and Lamattina 2005). Other research emphasized the crucial role of NO in moderating responses of plants against abiotic (Corpas et al. 2011) and biotic stresses (Gaupels et al. 2011). Moreover, both during and after the incidence of stress, NO has been observed to cause enhanced expression of various defense proteins (Uchida et al. 2002). Another interesting role of NO is as an antioxidant that helps to decrease levels of reactive oxygen species (ROS) and maintain cellular homeostasis, thereby reducing oxidative damage caused by the prevailing stress (Nabi et al. 2019). However, the functional status of NO is not yet well understood and NO research is extremely challenging due to the short life span and complex chemical nature of this diffusible free radical.

Plants experience many abiotic stresses like heat, drought, salinity, and toxic concentrations of metals at various stages of their life cycle. Plants produce various signaling molecules that are meant to trigger a cascade of responses that can lead to acclimation of the plant to the stress or will result in programmed cell death (Pedroso et al. 2000). Hence, in plants, NO appears to be a small signaling molecule modulating various physiological processes. However, the exact function NO is performing in various plants is still not clear but it appears to be species specific (Santisree et al. 2015). Apparently, exogenous application of NO in plants alleviates the negative effects of stress and leads to improved defense against oxidative stress in wheat plants (Noreen et al. 2009; Hasanuzzaman and Fujita 2013). Furthermore, important posttranslational modifications are mediated by NO including *S*-nitrosylation and tyrosine nitration (Astier and Lindermayr 2012).

4.6 NO-mediated Posttranslational Modifications

NO acts as a signaling effector molecule that tends to modify biological molecules, especially proteins, under normal and stressed environments (Astier and Lindermayr 2012). Chloroplastic and mitochondrial protein complexes are the usual molecular targets of NO (Mannick 2001; Palmieri et al. 2010). In addition other targets include lipids (Rubbo 2000), thiols, nucleic acids, phenolics (Joshi et al. 1999), hormone receptors (Terrile et al. 2012), and free radicals like superoxides (Patel et al. 1999). Reactive nitrogen species (RNS) that are derived from NO interact with biological molecules and tend to modify their structure and function. Modifications in structure lead to altered stability of proteins and activation or inhibition of the specific activities. In the case of transcription factors, NO-induced modifications can have an influence on DNA binding and lead to altered expression of genes. Generally, NO-mediated modifications occur as posttranslational changes either through the nitration of tyrosine residues or through the *S*-nitrosation of the cysteine residues.

In proteins, *S*-nitrosation of cysteine residues results in the formation of *S*-nitrosothiol (SNO). However, this *S*-nitrosation depends on the primary structure and conformation of the target protein, and proximity between the protein and NO molecules (Lamotte et al. 2015). This modification can be achieved via the action of free NO/N_2O_3, nonenzymatically, or through trans-nitrosation from *S*-nitrosoglutathione (GSNO) to other molecules as transfer reactions (Kneeshaw et al. 2014).

In the metalloproteins, NO causes the nitrosation of transition metals (zinc, copper, and iron) triggering conformational changes, altering protein activity (Astier and Lindermayr 2012), and thus creating another class of posttranslational modification. Superoxide dismutase (SOD), a metallo-enzyme, is thought to convert superoxide radicals to H_2O_2 (Gutteridge and Halliwell 2000). Based on metals like Zn, Mn, Cu, or Fe, there are three main types of SODs (Zelko et al. 2002). Upon peroxynitrite-mediated nitration, eukaryotic SODs are inactivated (Martinez et al. 2014). However, SOD activities are enhanced after treating with GSNO (Sehrawat et al. 2013). In order to analyze the incidence of posttranslational modifications in various isozymes of SOD as mediated by NO, in vitro studies on *A. thaliana* were conducted. Observations revealed that nitration caused inhibition of activities in CuZn-SOD3, Fe-SOD3, and Mn-SOD1 to varying degrees; however, activities of SODs were not affected by *S*-nitrosylation (Holzmeister et al. 2015).

Catalase (a peroxisomal enzyme that regulates levels of H_2O_2) was analyzed in vitro in order to estimate the effects mediated by various NO donors (Clark et al. 2000). It was established that catalases in plants tend to undergo *S*-nitrosylation and nitration, in vitro, causing inhibition of enzyme activity (Chaki et al. 2015). Inhibition results in reduced capacity for H_2O_2 removal and is therefore correlated with increased oxidative metabolism.

NO-mediated posttranslational modifications by nitration of tyrosine residues occur mainly due to peroxynitrite ($ONOO^-$) being formed after reacting with the superoxide radical (O_2^-) or nitrogen dioxide radical ($NO_2^\bullet$) being formed after reaction of NO with oxidants like H_2O_2 and O_2^-(Radi 2004). Not all tyrosine residues within a protein molecule are equally susceptible to nitration. Susceptibility of the residues depends on how exposed they are to the redox environment according to their conformation (Corpas et al. 2013) along with the nature of the nearby amino acid units (Lozano-Juste et al. 2011).

4.7 NO and Ion Homeostasis

NO regulates the negative influences of salinity stress and causes increased resistance in plants, playing a crucial role in the maintenance of ion homeostasis (Hasanuzzaman et al. 2018). NO causes increased activity of vacuolar H^+-PPase and H^+-ATPase when subjected to high salt stress (Zhang et al. 2006). Na^+ compartmentation is facilitated by Na^+/H^+ antiport and tonoplastic H^+-ATPase, whose activity is reported to be increased by exogenously applied NO (Parihar et al. 2015). However, application of NO under salinity caused a decrease in Na^+ contents as well as an increase in K^+ contents in seashore marshmallow shoots. It has also been reported to cause enhanced expression of H^+-ATPases on plasma membrane (Zhao et al. 2004).

In leaves, sheaths, and roots of *Zea mays*, a decreased ratio of Na^+ or increased ratio of K^+ is reported conferring salinity resistance (Zhang et al. 2006). Wang et al. (2009) have described that SNPs (NO donor) have a role in repairing salt stress-induced injury by retaining the higher ratio of K^+/Na^+ in wild-type *Arabidopsis* callus. The NO-specific scavengers (2-(4-carboxyphenyl)-4,4,5,5-tetramethylimidazoline1-oxyl-3-oxide [cPTIO], 2-phenyl-4,4,5,5-tetramethylimidazoline1-oxyl-3-oxide [PTIO], etc.) and inhibitors (N(G)-nitro-L-arginine methyl ester [L-NAME], etc.) have proved the role of vacuolar H^+-PPase and H^+-ATPase as well as plasma membrane H^+-ATPases in increasing the ratio of K^+/Na^+ in cytosol, thus showing that NO mediates resistance to salinity in plants (Zhang et al. 2006; Siddiqui et al. 2017).

In abiotic stress responses,Ca^{2+} signals and NO promote each other (Liu et al. 2018) indicating that in the Ca^{2+} and NO signaling pathway, NO controls nearly every kind of Ca^{2+} channel and acts as a compound that mobilizes Ca^{2+} (Jeandroz et al. 2013). NO and with its derivatives (SNOs, $ONOO^-$, etc.) are also interlinked with other pathways and act differently in different pathways (Begara-Morales et al. 2018).

Under salinity stress, NO has been observed to cause H_2S accumulation in barley in a dose-dependent manner. NO is also reported to induce H_2S synthesis in wheat seedlings subjected to osmotic stress. Under stress, H_2S not only acts as an antioxidant but also causes cellular homeostasis via accumulation of osmolytes (da-Silva and Modolo 2018) and facilitates plants to tolerate elevated levels of lipid peroxidation (Hossain et al. 2016).

4.8 NO and Phytohormones

NO mediates altered plant metabolism through posttranslational modifications and influence of various phytohormones (Sami et al. 2018). SA and NO facilitate plant defense against biotic stress (Durner and Klessig 1999). SA modifies the influence of NO by modulating enzymes like ascorbate peroxidase (APX), catalase (CAT) and reduced glutathione (GSH) (Navarre et al., 2000; Prakash et al. 2021). In *Pisum sativum*, SA along with SNP is observed to improve growth and cause enhanced production of antioxidant enzymes and osmoprotectants under salinity stress (Yadu et al. 2017). Exogenously applied NO causes downregulation of genes related to abscisic acid (ABA) biosynthesis causing reduction in ABA contents and, subsequently, decreased salt stress-induced leaf senescence (Nawaz et al. 2017). Cytokinins have also been observed to interact with the NO-associated signaling cascade affecting cell differentiation and development of roots (Freschi 2013). In cotton seedlings, foliar application of NO causes delayed leaf senescence through upregulation of the gene related to CK biosynthesis (isopentenyl transferase, IPT). In *A. thaliana*, interaction of NO and ethylene also appears to maintain ion homeostasis under salinity stress (Wang et al. 2009). Activation of ethylene responsive factors (ERFs) provides salt stress tolerance via decreased accumulation of ROS (Nawaz et al. 2017). Endogenous NO causes reduced accumulation of auxins under salt stress. NO also appears to influence auxin signaling under salt stress, mediating a reduction in the size of root meristem (Liu et al. 2015). NO is also observed to modify signaling and transport of auxin (Sanz et al. 2014). In *Solanum quitoense*, jasmonic acid (JA) and NO caused prevention of *Fusarium oxysporum* infestation. JA has a well-known function of activating defense responses against herbivory and abiotic stresses. NO causes induction of various genes related to JA biosynthesis (Mur et al. 2013). In *Solanum lycopersicum*, exogenous application of NO and JA help to mitigate salt stress-induced oxidative stress by enhancing activities of antioxidant enzymes (Ahmad et al. 2018). Under salinity stress, JA and NO cause accumulation of osmolytes along with Na^+ compartmentalization and prevent ionic toxicity (Davière and Achard 2015). NO inhibits elongation of hypocotyl in dark-grown seedlings of *A. thaliana* (Sanz et al. 2014). Investigations have revealed that gas, light, and NO respond through differential expression of DELLA and PIF proteins (Davière and Achard 2015).

4.9 NO-mediated Modification of Hormones and Fatty Acids Modulate Signaling Pathways or Physiological Responses in Plants

It has been observed in zoological systems that nitration of fatty acids occurs selectively under various circumstances indicating enhanced vulnerability of conjugated fatty acids for nitration as compared with polyenes and dienes (Bonacci et al. 2012). These nitro-fatty acids are generated by redox reactions of nitric oxide ($NO^{\bullet}$) and NO-derived molecules promoting liberation of nitrogen dioxide ($NO_2^{\bullet}$), which actually causes unsaturated fatty acids to be nitrated. The mechanisms forming nitro-fatty acids include via nitrite (NO_2^-), peroxynitrite ($ONOO^-$), or nitrogen dioxide ($NO_2^{\bullet}$) (Freeman et al. 2008). Along with the endogenous presence of nitro-linolenic acid observed in *Arabidopsis*, nitro-fatty acids have also been detected in many other species of plants.

The endogenous presence of nitro-fatty acids is observed as nitro-linolenic acid in *Oryza sativa*, *P. sativum*, and *A. thaliana* (Mata-Pérez et al. 2017). Nitro-linolenic acid level in leaves of *O. sativa* and roots of *P. sativum* were calculated to be 0.748 and 0.072 pmol/g fresh weight (FW). Subcellular contents were worked out as 0.282 and 0.084 pmol/g FW in peroxisomes and mitochondria, respectively. This reveals a variable distribution pattern of nitro-fatty acids in plants as potential signaling molecules in the plant kingdom. Transcriptome-based studies have established the role of nitro-linolenic acid-responsive genes in potential signaling pathways. RNA-seq analysis was carried out for a 9-day-old *Arabidopsis* cell suspension culture supplied with 10 μM and 100 μM concentrations of nitro-fatty acids (Mata-Pérez et al. 2016). Results confirmed that signaling responses of these molecules in plants are in accordance with the various studies conducted on animals.

A strong relatedness has been observed for nitro-linolenic acid under various conditions pertaining to abiotic stresses at all developmental stages (Mata-Pérez et al. 2016). Further analysis of nitro-linolenic acid-modulated genes has revealed that upregulated genes are involved in folding of protein in response to heat stress (Mata-Pérez et al. 2016). The majority of the genes induced after nitro-linolenic acid application were worked out as being in the *HSP90*, *HSP70*, *HSP60*, and *HSP40* families, small heat-shock proteins (*sHSP*), and heat-shock transcription factors (*HSFA7B* and *HSFA2*), revealing engagement of the HSP pathway in protection of the cells. Abiotic stresses generally lead to production of H_2O_2 and ROS causing oxidative stress (Apel et al. 2004; Mata-Pérez et al. 2016). Cytosolic ascorbate peroxidases (APXs) are involved in plant defense upon oxidative stress causing detoxification of the H_2O_2. Moreover, a strong interaction has been observed between the HSFA2 and APX under abiotic stress (Nishizawa et al. 2006). Thus nitro-fatty acids play a role as signaling mediators in the defense of plant against abiotic stress mainly via induction of HSPs.

4.10 Plant Stress Responses and NO

Ongoing environmental changes have resulted in drought and high temperature stresses that are limiting crop growth and yield around the globe. Perception of the stress and subsequent transmission of signals trigger various responses at molecular level, then cellular level, and finally physiological level. Persistent episodes of severe stress lead to the production of ROS and RNS and buildup of oxidative stress leading to oxidation of nucleic acids, proteins, and lipids, resulting in abnormal functions at the cellular level. However, it is reported that induction of NO causes enhanced activities of antioxidant enzymes hence protecting plants against oxidative damage and helping to maintain cellular redox homeostasis. Exogenous application of NO has proved a successful mediator to cut down losses caused by abiotic stresses in plants (Figure 4.2). NO is generally accumulated under various stresses in different plant species and tissues; however, specific insights have been reached using various inhibitors like c-PTIO or L-NAME causing reversal of NO-mediated traits in most of the studies. NO has been reported to have either a direct role in detoxification of ROS by interacting with O_2^- or an indirect role in enhancing performance of the antioxidant machinery. Peroxy nitrite is formed when NO interacts with O_2^- forms and acts as a nitrating agent that regulates activity of proteins under the prevailing stress. NO is also involved in upregulation of the transcription and activity of glutathione reductase (GR) and APX in the ascorbic acid–glutathione (AsA-GSH) pathway in *C. sativus* and *N. tabacum* leaves providing resistance against the prevailing abiotic stress. On the other hand, NO shows rapid reactions with other proteins, thiols, hemes, and oxygen species producing endogenous biochemical signals thereby regulating enzyme

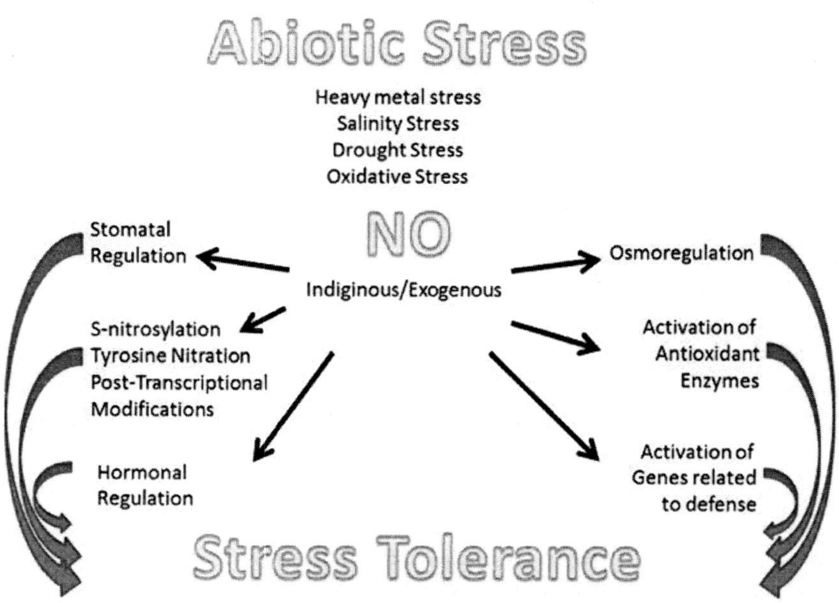

Figure 4.2 Role of nitric oxide in abiotic stress tolerance.

activities directly or indirectly. However, the response observed varies according to the intensity and length of the incident stress and the plant under study. Underlying molecular mechanisms of sensing NO, transduction of signals, and execution of the response remain unclear.

4.11 NO in Connection with Temperature Stress

High temperature treatments trigger an increased accumulation of NO in many plant species rendering NO emission as a vital response causing adaptation of plants to heat stress. However, NO emission is dependent on the extent and duration of high temperature stress. External supply of NO donors also causes a reduction in cellular damage caused by heat stress. Florescent dyes like 5,6-diaminofluorescein diacetate have been used to locate cellular sites of NO synthesis under heat stress. In *N. tabacum*, NO-fluorescence was evident in palisade mesophyll cells, epidermal cells, and trichomes starting in the plastids then in the nucleus, and finally spreading across the cytosol, under the influence of heat stress. Plants proceed toward stepwise adaptation to heat stress starting with perceiving the stress signals and terminating in the expression of targeted genes. After perception of heat stress, stress signaling is triggered by transducers (Ca^{2+}, H_2O_2, and NO) and stress hormones like ABA that initiate synthesis of protein kinases and stimulate downstream gene expression. Differential expression of genes triggers a special cascade of events changing the metabolism of the plant through production of antioxidant enzymes, accumulation and synthesis of osmoprotectants, and translation of HSPs, hence resulting in enabling the plant to survive prevailing heat stress.

Low temperature also promotes alteration in the nitration level and *S*-nitrosoproteome in *P. sativum* and *Brassica juncea* and involvement of NO in rendering tolerance to plants exposed to chilling and freezing stress. NO thus is observed to facilitate osmoprotection, ROS scavenging, regulation of polyamine levels, and maintained hormonal balance. During cold transduction, synthesis of phosphorylated sphingolipids occurs as negative regulation (Thakur and Nayyar 2013).

4.12 NO in Connection with Salinity Stress

Salinity stress is also a major threat for production of crops. NO is observed to enable salt tolerance by causing increased transcription of Na^+/H^+ and H^+-ATPases in *Avicennia marina* thus maintaining homeostasis of ions, causing synthesis of osmoprotectants and initiating operation of antioxidants (APX, guaiacol peroxidase [POX], and SOD). The ameliorative effect of NO against salt stress appears to occur through closure of the stomata thereby protecting against loss of water. Modulation of APX by NO occurs at the posttranslational level where Cys residues are *S*-nitrosated promoting enzyme activity; however, nitration of Tyr residues causes inhibition of enzyme activity. In C4 plants, phosphoenolpyruvate carboxylase is regulated by phosphoenolpyruvate carboxylase-kinase (PEPCK). Under NO treatment, enhanced activity under salinity stress presents a key role of NO linking carbon fixation and mitigation of salt stress. Exogenous application of ABA under salinity stress causes induction of AB15, which, together with AB13, modulates the expression of class I Late Embryogenesis Abundant (LEA) proteins (Em6 and Em1) that are involved in rendering desiccation tolerance in *Arabidopsis* (Shinozaki and Yamaguchi-Shinozaki 1997; Raza et al. 2020).

4.13 NO in Connection with Heavy Metal Stress

Heavy metals (arsenic, aluminum, copper, cadmium, etc.) have the tendency to cause alteration in the normal metabolism and cause toxicity in plants beyond permissible limits. In the soil, accumulation of heavy metals can be due to many factors including industrial activities, mining, and/or application of fertilizers or sewage sludge. Upon entry into the plant cells, they can directly interact with proteins, DNA, and other atoms like sulfur, oxygen, etc. In plants, NO-mediated mitigation of heavy metal stress is either triggered by upregulated antioxidant enzymes and defense proteins, or by regulating the concentration of free metal ions in the cell, which is achieved by preventing buildup to toxic levels due to exclusion of metal ions into the root (Hossain and Komatsu 2013). NO has also been proved to reverse $CuCl_2$-induced inhibition of growth by enhancing plasma membrane-bound H^+-ATPase activity along with enhanced activity of the oxidative defense system through the regulation of the GSH/GSSG (reduced glutathione/oxidized glutathione) ratio (Zhong et al. 2015). A NO-mediated increase was observed in the uptake of minerals and the activity of Rubisco in *N. tabacum* under Cu stress (Khairy et al. 2016).

In acidic soils, Al is observed to affect growth and yield of crops by causing inhibition of mineral and nutrient uptake (Samac and Tesfaye 2003; Yang et al. 2013). However, application of SNP causes alleviation of the adverse effects of Al stress in *Hibiscus moscheutos* (Tian et al. 2007; Wang et al. 2010). NO also causes enhanced Al tolerance by hormonal regulation in *T. aestivum* and *Secale cereale* (He et al. 2012).

In soybean cell suspension cultures, endogenous levels of NO are observed to become elevated under Cd stress (Drazic and Mihailovic 2005; Zawoznik et al. 2007). Under cadmium and copper stress, NO was observed to protect chlorophyll in *H. annuus* and *L. perenne*, respectively (Bidar et al. 2007). NO-mediated Cd tolerance was also observed in *O. sativa* mainly due to restriction of Cd entry caused by NO and associated elevated contents of pectin and hemicellulose in root cell walls (Yemets et al. 2019).

Enhanced activity of NOS produced NO in roots of *T. aestivum* under Zn stress (Wei et al. 2020). Exogenous application of NO was observed to affect Zn translocation from roots to shoot along with higher ascorbate contents in *T. aestivum* (Rizwan et al. 2017).

Decreased H_2O_2 and thiol (nonprotein) contents were observed under NO treatment. NO is reported to cause increased uptake of Fe and Mg, thus improving synthesis of chlorophyll and positively affecting photosynthesis (Graziano et al. 2002).

Oxidative stress induced under Ni stress was alleviated after NO application on *Solanum lycopersicum*, *B. juncea*, and *T. aestivum*. Hence, it is concluded that NO is involved in alleviating metal stress by overcoming oxidative stress through enhanced activities of antioxidants.

4.14 Expression of Other Genes in NO metabolism

Insights into NO-mediated physiological responses highlight cell-specific isoenzymes with characteristic functions. These isoforms constitute a co-expression in various subcellular compartment in different cells (Tuteja 2004). Kant et al. (2008) evaluated expression level of different genes in *Arabidopsis* (sensitive to low-nitrogen stress) and *Thellungiella halophila* (tolerant to low-nitrogen stress) and highlighted the importance of *GS1, GS2, NR2, NRT1.1, NRT2.1*, and *NRT3.1* under low-nitrogen stress. Shi et al. (2010) investigated the performance of rice cultivars under low-nitrogen stress and observed the importance of *OsAMT1* and *OsNRT2*. Luo et al. (2013) evaluated *Populus* species and observed strong sensitivity of genes related to nitrogen metabolism under limited supply of N. Under low-nitrogen stress, Chen et al. (2018) evaluated barley cultivars BI-45 (sensitive cultivar) and BI-04 (tolerant cultivar) for the expression of genes related to N-metabolism. Expression of *HvNiR1*, *HvGS2*, and *HvGLU2* was enhanced in shoot of BI-04. Expression of *HvGLU2* was enhanced in root of BI-04. The glutamine synthetase (GS)/glutamate synthase (GOGAT) pathway plays an important role in primary nitrogen assimilation by making organic nitrogen available for plant while utilizing inorganic nitrogen. Thus, *HvGLU2* and *HvGS2* in roots and shoots appear to be involved in low-nitrogen tolerance. Feraud et al. (2005) described that Fd-GOGAT is involved in the assimilation of photorespiratory and primary ammonium. Certain other reports are also available that document specific expression mediated by NO (Table 4.1).

Table 4.1 Gene expression mediated by nitric oxide (NO).

Phenomenon	Gene(s)/protein(s)	Physiological role	Plant	Reference
Iron homeostasis	*atferl*	NO-dependent ferritin expression	*Arabidopsis thaliana*	Murgia et al. 2004
	rbcL, psbA	Availability of internal metabolically active iron	*Zea mays*	Graziano et al. 2002
Stomatal closure	nial/nia2	NO production via nitrate reductase	*Arabidopsis thaliana*	Desikan et al. (2002)
ABA signaling	Late Embryogenesis Abundant (LEA) protein	Synthesis of abscisic acid (ABA) Stomatal closure	*Nicotiana tabacum*	Gloud et al. 2003
Sexual reproduction	*cat2* *SOD*	Programmed cell death in aleurone layers	*Hordeum vulgare*	Beligni et al. 2002
	noxl or *cuel*	Floral decision Flowering locus C	*Arabidopsis thaliana*	He et al. 2004

(*Continued*)

Table 4.1 (Continued)

Phenomenon	Gene(s)/protein(s)	Physiological role	Plant	Reference
Symbiosis with rhizobia	*LjHb1*	Symbiosis	*Lotus japonicus*	Shimoda et al. 2005
Osmotic stress management	*ABF2*	Alleviate osmotic stress	*Arabidopsis thaliana*	Kim et al. 2004
Transpiration activation	Histone acetyltransferases (HATs)	Helps in transcription	Yeast	Brown et al. 2000
Translation Regulation	1-aminoacyclopropane 1-carboxylate (ACC) synthase	NO-dependent expression of ACC synthase and sHSP	*Nicotiana tabacum*	Lamotte et al. 2004
	Heat-shock proteins (HSPs)	In cell suspension culture		
Protection against ultraviolet-B (UV-B)	*CHS*	Combat effects of UV-B radiation	*Arabidopsis thaliana*	Mackerness 2001

4.15 Engineering Stress Tolerance in Plants Through NO Pathways

Engineering stress tolerance in plants through NO pathways can be accomplished by utilizing application of commercial NO donors, or selective plants having molecular markers related to NO signaling pathways, and/or transgenic plants developed with targeted overexpression of master regulators of NO signaling pathways. Exogenous NO donors with low molecular weight can be utilized in cooperation with ROS (Simontacchi et al. 2015), which facilitates modulated expression of genes (Seabra and Oliveira 2016).

References

Ahmad, P., Abass Ahanger, M., Nasser Alyemeni, M. et al. (2018). Mitigation of sodium chloride toxicity in *Solanum lycopersicum* L. by supplementation of jasmonic acid and nitric oxide. *Journal of Plant Interactions* 13 (1): 64–72.

Apel, K. and Hirt, H. (2004). Reactive oxygen species: metabolism, oxidative stress, and signal transduction. *Annual Review of Plant Biology* 55: 373–399.

Astier, J. and Lindermayr, C. (2012). Nitric oxide-dependent posttranslational modification in plants: an update. *International Journal of Molecular Sciences* 13 (11): 15193–15208.

Astier, J., Kulik, E., Koen, A. et al. (2012). Protein S-nitrosylation: what's going on in plants? *Free Radical Biology & Medicine* 53: 1101–1110.

Begara-Morales, J.C., Chaki, M., Valderrama, R. et al. (2018). Nitric oxide buffering and conditional nitric oxide release in stress response. *Journal of Experimental Botany* 69 (14): 3425–3438.

Beligni, M.V., Fath, A., Bethke, P.C. et al. (2002). Nitric oxide acts as an antioxidant and delays programmed cell death in barley aleurone layers. *Plant Physiology* 129: 1642–1650.

Besson-Bard, A., Pugin, A., and Wendehenne, D. (2008). New insights into nitric oxide signaling in plants. *Annual Review of Plant Biology* 59: 21–39.

Bethke, P.C., Badger, M.R., and Jones, R.L. (2004). Apoplastic synthesis of nitric oxide by plant tissues. *The Plant Cell* 16: 332–341.

Bidar, G., Garçon, G., Pruvot, C. et al. (2007). Behavior of *Trifolium repens* and *Lolium perenne* growing in a heavy metal contaminated field: plant metal concentration and phytotoxicity. *Environmental Pollution* 147 (3): 546–553.

Bonacci, G., Baker, P.R., Salvatore, S.R. et al. (2012). Conjugated linoleic acid is a preferential substrate for fatty acid nitration. *Journal of Biological Chemistry* 287 (53): 44071–44082.

Brown, C.E., Lechner, T., Howe, L. et al. (2000). The many HATs of transcription coactivators. *Trends in Biochemical Sciences* 25 (1): 15–19.

Camargo, A., Llamas, Á., Schnell, R.A. et al. (2007). Nitrate signaling by the regulatory gene NIT2 in *Chlamydomonas*. *The Plant Cell* 19 (11): 3491–3503.

Chaki, M., Alvarez de Morales, P., Ruiz, C. et al. (2015). Ripening of pepper (*Capsicum annuum*) fruit is characterized by an enhancement of protein tyrosine nitration. *Annals of Botany* 116 (4): 637–647.

Chamizo-Ampudia, A., Sanz-Luque, E., Llamas, Á. et al. (2016). A dual system formed by the ARC and NR molybdoenzymes mediates nitrite-dependent NO production in *Chlamydomonas*. *Plant, Cell and Environment* 39 (10): 2097–2107.

Chen, Z., Liu, C., Wang, Y. et al. (2018). Expression analysis of nitrogen metabolism-related genes reveals differences in adaptation to low-nitrogen stress between two different barley cultivars at seedling stage. *International Journal of Genomics* 2018: 8152860.

Clark, D., Durner, J., Navarre, D.A. et al. (2000). Nitric oxide inhibition of tobacco catalase and ascorbate peroxidase. *Molecular Plant–Microbe Interactions* 13 (12): 1380–1384.

Corpas, F.J. and Barroso, J.B. (2013). Nitro-oxidative stress vs oxidative or nitrosative stress in higher plants. *New Phytologist* 199 (3): 633–635.

Corpas, F.J., Barroso, J.B., Carreras, A. et al. (2004). Cellular and subcellular localization of endogenous nitric oxide in young and senescent pea plants. *Plant Physiology* 136 (1): 2722–2733.

Corpas, F.J., Leterrier, M., Valderrama, R. et al. (2011). Nitric oxide imbalance provokes a nitrosative response in plants under abiotic stress. *Plant Science* 181: 604–611.

Correa-Aragunde, N., Graziano, M., and Lamattina, L. (2004). Nitric oxide plays a central role in determining lateral root development in tomato. *Planta* 218 (6): 900–905.

da-Silva, C.J. and Modolo, L.V. (2018). Hydrogen sulfide: a new endogenous player in an old mechanism of plant tolerance to high salinity. *Acta Botanica Brasilica* 32 (1): 150–160.

Del Río, L.A., Corpas, F.J., and Barroso, J.B. (2004). Nitric oxide and nitric oxide synthase activity in plants. *Phytochemistry* 65: 783–792.

Desikan, R., Griffiths, R., Hancock, J. et al. (2002). A new role for an old enzyme: nitrate reductase-mediated nitric oxide generation is required for abscisic acid-induced stomatal closure in *Arabidopsis thaliana*. *Proceedings of the National Academy of Sciences of the United States of America* 99: 16314–16318.

Drazic, G. and Mihailovic, N. (2005). Modification of cadmium toxicity in soybean seedlings by salicylic acid. *Plant Science*168: 511–517.

Durner, J. and Klessig, D.F. (1999). Nitric oxide as a signal in plants. *Current Opinion in Plant Biology* 2 (5): 369–374.

Fan, S.C., Lin, C.S., Hsu, P.K. et al. (2009). The *Arabidopsis* nitrate transporter NRT1. 7, expressed in phloem, is responsible for source-to-sink remobilization of nitrate. *The Plant Cell* 21 (9): 2750–2761.

Feraud, M., Masclaux-Daubresse, C., Ferrario-Méry, S. et al. (2005). Expression of a ferredoxin-dependent glutamate synthase gene in mesophyll and vascular cells and functions of the enzyme in ammonium assimilation in *Nicotiana tabacum* (L.). *Planta* 222 (4): 667–677.

Fernandez, E. and Galván, A. (2008). Nitrate assimilation in *Chlamydomonas*. *Eukaryotic Cell* 7 (4): 555–559.

Foresi, N., Correa-Aragunde, N., Parisi, G. et al. (2010). Characterization of a nitric oxide synthase from the plant kingdom: NO generation from the green alga *Ostreococcus tauri* is light irradiance and growth phase dependent. *The Plant Cell* 22: 3816–3830.

Freeman, B.A., Baker, P.R., Schopfer, F.J. et al. (2008). Nitro-fatty acid formation and signaling. *The Journal of Biological Chemistry* 283 (23): 15515.

Freschi, L. (2013). Nitric oxide and phytohormone interactions: current status and perspectives. *Frontiers in Plant Science* 4: 398.

Gaupels, F., Kuruthukulangarakoola, G.T., and Durner, J. (2011). Upstream and downstream signals of nitric oxide in pathogen defence. *Current Opinion in Plant Biology* 14 (6): 707–714.

Gloud, K.S., Lamotte, O., Klinguer, A. et al. (2003). Nitric oxide production in tobacco leaf cells: a generalized stress response? *Plant, Cell & Environment* 26: 1851–1862.

Graziano, M., Beligni, M.V., and Lamattina, L. (2002). Nitric oxide improves internal iron availability in plants. *Plant Physiology* 130 (4): 1852–1859.

Graziano, M. and Lamattina, L. (2005). Nitric oxide and iron in plants: an emerging and converging story. *Trends in Plant Science* 10 (1): 4–8.

Gupta Kapuganti, J., Igamberdiev, A.U., and Kaiser, W.M. (2010). New insights into the mitochondrial nitric oxide production pathways. *Plant Signaling and Behavior* 5 (8): 999–1001.

Gutteridge, J.M. and Halliwell, B. (2000). Free radicals and antioxidants in the year 2000: a historical look to the future. *Annals of the New York Academy of Sciences* 899 (1): 136–147.

Hasanuzzaman, M. and Fujita, M. (2013). Exogenous sodium nitroprusside alleviates arsenic-induced oxidative stress in wheat (*Triticum aestivum* L.) seedlings by enhancing antioxidant defense and glyoxalase system. *Ecotoxicology* 22 (3): 584–596.

Hasanuzzaman, M., Oku, H., Nahar, K. et al. (2018). Nitric oxide-induced salt stress tolerance in plants ROS metabolism, signaling, and molecular interactions. *Plant Biotechnology Reports* 12: 77–92.

He, H.Y., He, L.F., Gu, M.H. et al. (2012). Nitric oxide improves aluminum tolerance by regulating hormonal equilibrium in the root apices of rye and wheat. *Plant Science* 183: 123–130.

He, Y., Tang, R.H., Hao, Y. et al. (2004). Nitric oxide represses the *Arabidopsis* floral transition. *Science* 305: 1968–1971.

Holzmeister, C., Gaupels, F., Geerlof, A. et al. (2015). Differential inhibition of *Arabidopsis* superoxide dismutases by peroxynitrite-mediated tyrosine nitration. *Journal of Experimental Botany* 66 (3): 989–999.

Hossain, M.S. and Dietz, K.J. (2016). Tuning of redox regulatory mechanisms, reactive oxygen species and redox homeostasis under salinity stress. *Frontiers in Plant Science* 7: 548.

Hossain, Z. and Komatsu, S. (2013). Contribution of proteomic studies towards understanding plant heavy metal stress response. *Frontiers in Plant Science* 3: 310.

Iobbi-Nivol, C. and Leimkühler, S. (2013). Molybdenum enzymes, their maturation and molybdenum cofactor biosynthesis in *Escherichia coli*. *Biochimica et Biophysica Acta (BBA) – Bioenergetics* 1827 (8–9): 1086–1101.

Jeandroz, S., Lamotte, O., Astier, J. et al. (2013). There's more to the picture than meets the eye: nitric oxide cross talk with Ca^{2+} signaling. *Plant Physiology* 163: 459–470.

Joshi, D.S. (1999). Latitudinal variation in locomotor activity rhythm in adult *Drosophila ananassae*. *Canadian Journal of Zoology* 77 (6): 865–870.

Kant, S., Bi, Y.M., Weretilnyk, E. et al. (2008). The *Arabidopsis* halophytic relative *Thellungiella halophila* tolerates nitrogen-limiting conditions by maintaining growth, nitrogen uptake, and assimilation. *Plant Physiology* 147 (3): 1168–1180.

Khairy, A.I.H., Oh, M.J., Lee, S.M. et al. (2016). Nitric oxide overcomes Cd and Cu toxicity in in vitro-grown tobacco plants through increasing contents and activities of RuBisCo and RuBisCo activase. *Biochimie Open* 2: 41–51.

Kim, S., Kang, J.Y., Cho, D.I. et al. (2004). ABF2, an ABRE-binding bZIP factor, is an essential component of glucose signaling and its overexpression affects multiple stress tolerance. *The Plant Journal* 40 (1): 75–87.

Klepper, L. (1979). Nitric oxide (NO) and nitrogen dioxide (NO_2) emissions from herbicide-treated soybean plants. *Atmospheric Environmental* 13: 537–542.

Kneeshaw, S., Gelineau, S., Tada, Y. et al. (2014). Selective protein denitrosylation activity of thioredoxin-h5 modulates plant immunity. *Molecular Cell* 56 (1): 153–162.

Kumar, R., Taware, R., Gaur, V.S. et al. (2009). Influence of nitrogen on the expression of TaDof1 transcription factor in wheat and its relationship with photo synthetic and ammonium assimilating efficiency. *Molecular Biology Reports* 36 (8): 2209–2220.

Lamotte, O., Bertoldo, J.B., Besson-Bard, A. et al. (2015). Protein S-nitrosylation: specificity and identification strategies in plants. *Frontiers in Chemistry* 2: 114.

Lamotte, O., Gould, K., Lecourieux, D. et al. (2004). Analysis of nitric oxide signaling functions in tobacco cells challenged by the elicitor cryptogein. *Plant Physiology* 135: 516–529.

Li, L., Weinberg, C.R., Darden, T.A. et al. (2001). Gene selection for sample classification based on gene expression data: study of sensitivity to choice of parameters of the GA/KNN method. *Bioinformatics* 17 (12): 1131–1142.

Liu, R., Shi, L., Zhu, T. et al. (2018). Crosstalk between nitric oxide and calcium-calmodulin regulate ganoderic acid biosynthesis in *Ganoderma lucidum* under heat stress. *Applied and Environmental Microbiology*. doi: 10.1128/AEM.00043-18.

Liu, W., Li, R.J., Han, T.T. et al. (2015). Salt stress reduces root meristem size by nitric oxide-mediated modulation of auxin accumulation and signaling in *Arabidopsis*. *Plant Physiology* 168 (1): 343–356.

Lozano-Juste, J., Colom-Moreno, R., and León, J. (2011). *In vivo* protein tyrosine nitration in *Arabidopsis thaliana*. *Journal of Experimental Botany* 62 (10): 3501–3517.

Luo, J., Li, H., Liu, T. et al. (2013). Nitrogen metabolism of two contrasting poplar species during acclimation to limiting nitrogen availability. *Journal of Experimental Botany* 64 (14): 4207–4224.

Mackerness, S.A.H., John, C.F., Jordan, B. et al. (2001). Early signaling components in ultraviolet-B responses: distinct roles for different reactive oxygen species and nitric oxide. *FEBS Letters* 489: 237–242.

Mannick, J.B., Schonhoff, C., Papeta, N. et al. (2001). S-Nitrosylation of mitochondrial caspases. *Journal of Cell Biology* 154 (6): 1111–1116.

Mariscal, V., Moulin, P., Orsel, M. et al. (2006). Differential regulation of the *Chlamydomonas* Nar1 gene family by carbon and nitrogen. *Protist* 157: 421–433.

Martinez, A., Peluffo, G., Petruk, A.A. et al. (2014). Structural and molecular basis of the peroxynitrite-mediated nitration and inactivation of *Trypanosoma cruzi* iron-superoxide dismutases (Fe-SODs) A and B: disparate susceptibilities due to the repair of Tyr35 radical by Cys83 in Fe-SODB through intramolecular electron transfer. *Journal of Biological Chemistry* 289 (18): 12760–12778.

Mata-Pérez, C., Sánchez-Calvo, B., Padilla, M.N. et al. (2016). Nitro-fatty acids in plant signaling: nitro-linolenic acid induces the molecular chaperone network in Arabidopsis. *Plant Physiology* 170 (2): 686–701.

Mata-Pérez, C., Sánchez-Calvo, B., Padilla, M.N. et al. (2017). Nitro-fatty acids in plant signaling: new key mediators of nitric oxide metabolism. *Redox Biology* 11: 554–561. doi: 10.1016/j.redox.2017.01.002.

Miller, A.J. and Smith, S.J. (1996). Nitrate transport and compartmentation in cereal root cells. *Journal of Experimental Botany* 47 (7): 843–854.

Mur, L.A., Mandon, J., Persijn, S. et al. (2013). Nitric oxide in plants: an assessment of the current state of knowledge. *AoB Plants* 5: pls052.

Murgia, I., de-Pinto, M.C., Delledonne, M. et al. (2004). Comparative effects of various nitric oxide donors on ferritin regulation, programmed cell death, and cell redox state in plant cells. *Journal of Plant Physiology* 161: 777–783.

Nabi, R.B.S., Tayade, R., Hussain, A. et al. (2019). Nitric oxide regulates plant responses to drought, salinity, and heavy metal stress. *Environmental and Experimental Botany* 161: 120–133.

Navarre, D.A., Wendehenne, D., Durner, J. et al. (2000). Nitric oxide modulates the activity of tobacco aconitase. *Plant Physiology* 122 (2): 573–582.

Nawaz, F., Shabbir, R.N., Shahbaz, M. et al. (2017). Cross talk between nitric oxide and phytohormones regulate plant development during abiotic stresses. In: *Phytohormones: Signaling Mechanisms and Crosstalk in Plant Development and Stress Responses* (Ed. M. El-Esawi), 117–141 Rijeka, Croatia: InTech.

Neill, S.J., Desikan, R., Clarke, A. et al. (2002). Nitric oxide is a novel component of abscisic acid signaling in stomatal guard cells. *Plant Physiology* 128 (1): 13–16.

Noreen, S., Ashraf, M., Hussain, M. et al. (2009). Exogenous application of salicylic acid enhances antioxidative capacity in salt stressed sunflower (*Helianthus annuus* L.) plants. *Pakistan Journal of Botany* 41 (1): 473–479.

Ostroukhova, M. and Ermilova, E. (2019). New insights into NO generation and AOX1 upregulation in *Chlamydomonas*. *Protistology* 13 (1): 19–25.

Palmieri, M.C., Lindermayr, C., Bauwe, H. et al. (2010). Regulation of plant glycine decarboxylase by S-nitrosylation and glutathionylation. *Plant Physiology* 152 (3): 1514–1528.

Parihar, P., Singh, S., Singh, R. et al. (2015). Effect of salinity stress on plants and its tolerance strategies: a review. *Environmental Science and Pollution Research* 22: 4056–4075.

Patel, P., Narayanaswamy, G.K., Kataria, S., and Baghel, L. (2017). Involvement of nitric oxide in enhanced germination and seedling growth of magneto-primed maize seeds. *Plant Signaling & Behavior* 12: e1293217. doi: 10.1080/15592324.2017.1293217.

Patel, R.P., McAndrew, J., Sellak, H. et al. (1999). Biological aspects of reactive nitrogen species. *Biochimica et Biophysica Acta (BBA) – Bioenergetics* 1411 (2–3): 385–400.

Pedroso, M.C., Magalhaes, J.R., and Durzan, D. (2000). A nitric oxide burst precedes apoptosis in angiosperm and gymnosperm callus cells and foliar tissues. *Journal of Experimental Botany* 51 (347): 1027–1036.

Prakash, V., Singh, V.P., Tripathi, D.K., Sharma, S., and Corpas, F.J. (2021). Nitric oxide (NO) and salicylic acid (SA): A framework for their relationship in plant development under abiotic stress. *Plant Biology* 23 (S1): 39–49.

Radi, R. (2004). Nitric oxide, oxidants, and protein tyrosine nitration. *Proceedings of the National Academy of Sciences of the United States of America* 101 (12): 4003–4008.

Raza, A., Charagh, S., Razzaq, A. et al. (2020). Brassicaceae plants response and tolerance to drought stress: physiological and molecular interventions. In: *The Plant Family Brassicaceae* (Ed. M. Hasanuzzaman), 1–43. Singapore: Springer.

Rizwan, M., Ali, S., Hussain, A. et al. (2017). Effect of zinc-lysine on growth, yield and cadmium uptake in wheat (*Triticum aestivum* L.) and health risk assessment. *Chemosphere* 187: 35–42.

Rockel, P. and Kaiser, W.M. (2002). NO production in plants: nitrate reductase versus nitric oxide synthase. In: *Progress in Botany* (Eds. K. Esser, U. Lüttge, W. Beyschlag, and F. Hellwig), vol. 63, 246–257. Berlin and Heidelberg, Germany: Springer-Verlag.

Rubbo, H., Radi, R., Anselmi, D. et al. (2000). Nitric oxide reaction with lipid peroxyl radicals spares α-tocopherol during lipid peroxidation: greater oxidant protection from the pair nitric oxide/α-tocopherol than α-tocopherol/ascorbate. *Journal of Biological Chemistry* 275 (15): 10812–10818.

Samac, D.A. and Tesfaye, M. (2003). Plant improvement for tolerance to aluminum in acid soils – a review. *Plant Cell, Tissue and Organ Culture* 75 (3): 189–207.

Sami, F., Faizan, M., Faraz, A. et al. (2018). Nitric oxide-mediated integrative alterations in plant metabolism to confer abiotic stress tolerance, NO crosstalk with phytohormones and NO-mediated post translational modifications in modulating diverse plant stress. *Nitric Oxide*73: 22–38.

Santisree, P., Bhatnagar-Mathur, P., and Sharma, K.K. (2015). NO to drought-multifunctional role of nitric oxide in plant drought: do we have all the answers? *Plant Science* 239: 44–55.

Sanz, L., Fernández-Marcos, M., Modrego, A. et al. (2014). Nitric oxide plays a role in stem cell niche homeostasis through its interaction with auxin. *Plant Physiology* 166 (4): 1972–1984.

Schrader, N., Fischer, K., Theis, K. et al. (2003). The crystal structure of plant sulfite oxidase provides insights into sulfite oxidation in plants and animals. *Structure* 11 (10): 1251–1263.

Seabra, A.B. and Oliveira, H.C. (2016). How nitric oxide donors can protect plants in a changing environment: what we know so far and perspectives. *AIMS Molecular Science* 3 (4): 692–718.

Sehrawat, A., Abat, J.K., and Deswal, R. (2013). RuBisCO depletion improved proteome coverage of cold responsive S-nitrosylated targets in *Brassica juncea*. *Frontiers in Plant Science* 4: 342.

Shi, W.M., Xu, W.F., Li, S.M. et al. (2010). Responses of two rice cultivars differing in seedling-stage nitrogen use efficiency to growth under low-nitrogen conditions. *Plant and Soil* 326 (1–2): 291–302.

Shimoda, Y., Nagata, M., Suzuki, A. et al. (2005). Symbiotic rhizobium and nitric oxide induce gene expression of non-symbiotic hemoglobin in *Lotus japonicus*. *Plant Cell Physiology* 46: 99–107.

Shinozaki, K. and Yamaguchi-Shinozaki, K. (1997). Gene expression and signal transduction in water-stress response. *Plant Physiology* 115 (2): 327.

Siddiqui, M.H., Alamri, S.A., Al-Khaishany, M.Y. et al. (2017). Exogenous application of nitric oxide and spermidine reduces the negative effects of salt stress on tomato. *Horticulture, Environment and Biotechnology* 58 (6): 537–547.

Simontacchi, M., Galatro, A., Ramos-Artuso, F. et al. (2015). Plant survival in a changing environment: the role of nitric oxide in plant responses to abiotic stress. *Frontiers in Plant Science* 6: 977.

Sun, H., Feng, F., Liu, J. et al. (2017). The interaction between auxin and nitric oxide regulates root growth in response to iron deficiency in rice. *Frontiers in Plant Science* 8: 2169.

Tejada-Jimenez, M., Llamas, A., Galván, A. et al. (2019). Role of nitrate reductase in NO production in photosynthetic eukaryotes. *Plants* 8 (3): 56.

Terrile, M.C., París, R., Calderón-Villalobos, L.I. et al. (2012). Nitric oxide influences auxin signaling through S-nitrosylation of the *Arabidopsis* transport inhibitor response-1 auxin receptor. *The Plant Journal*70 (3): 492–500.

Thakur, P. and Nayyar, H. (2013). Facing the cold stress by plants in the changing environment: sensing, signaling, and defending mechanisms. In: *Plant Acclimation to Environmental Stress* (Eds. N. Tuteja and G.S. Singh), 29–69. New York: Springer.

Tian, Q.Y., Sun, D.H., Zhao, M.G. et al. (2007). Inhibition of nitric oxide synthase (NOS) underlies aluminum-induced inhibition of root elongation in *Hibiscus moscheutos*. *The New Phytologist* 174 (2): 322–331.

Tuteja, N., Chandra, M., Tuteja, R. et al. (2004). Nitric oxide as a unique bioactive signaling messenger in physiology and pathophysiology. *Journal of Biomedicine & Biotechnology* 4: 227–237.

Uchida, A., Jagendorf, A.T., Hibino, T. et al. (2002). Effects of hydrogen peroxide and nitric oxide on both salt and heat stress tolerance in rice. *Plant Science* 163 (3): 515–523.

Wang, H., Liang, X., Wan Q. et al. (2009). Ethylene and nitric oxide are involved in maintaining ion homeostasis in *Arabidopsis* callus under salt stress. *Planta* 230: 293–307.

Wang, H.H., Huang, J.J., and Bi, Y.R. (2010). Nitrate reductase-dependent nitric oxide production is involved in aluminum tolerance in red kidney bean roots. *Plant Science* 179 (3): 281–288.

Wang, J., Krizowski, S., Fischer-Schrader, K. et al. (2015). Sulfite oxidase catalyzes single-electron transfer at molybdenum domain to reduce nitrite to nitric oxide. *Antioxidants and Redox Signaling* 23 (4): 283–294.

Wang, P., Zhu, J.K., and Lang, Z. (2015). Nitric oxide suppresses the inhibitory effect of abscisic acid on seed germination by S-nitrosylation of SnRK2 proteins. *Plant Signaling and Behavior* 10 (6): e1031939.

Wei, L., Zhang, J., Wang, C. et al. (2020). Recent progress in the knowledge on the alleviating effect of nitric oxide on heavy metal stress in plants. *Plant Physiology and Biochemistry* 147: 161–171.

Yadu, S., Dewangan, T.L. Chandrakar, V. et al. (2017). Imperative roles of salicylic acid and nitric oxide in improving salinity tolerance in *Pisum sativum* L. *Physiology and Molecular Biology of Plants* 23 (1) 43–58.

Yanagisawa, S. (2014). Transcription factors involved in controlling the expression of nitrate reductase genes in higher plants. *Plant Science* 229: 167–171.

Yang, Z.B., Rao, I.M., and Horst, W.J. (2013). Interaction of aluminium and drought stress on root growth and crop yield on acid soils. *Plant and Soil* 372 (1): 3–25.

Yemets, A.I., Karpets, Y.V., Kolupaev, Y.E. et al. (2019). Emerging technologies for enhancing ROS/RNS homeostasis. In: *Reactive Oxygen, Nitrogen and Sulfur Species in Plants: Production, Metabolism, Signaling and Defense Mechanisms*, vol 2 (eds. M. Hasanuzzaman, V. Fotopoulos, K. Nahar, and M. Fujita), 873–922. Chichester, UK: John Wiley & Sons Ltd.

Zawoznik, M., Groppa, M.D., Tomaro, M.L. et al. (2007). Endogenous salicylic acid potentiates cadmium-induced oxidative stress in *Arabidopsis thaliana*. *Plant Science* 173: 190–197.

Zelko, I.N., Mariani, T.J., and Folz, R.J. (2002). Superoxide dismutase multigene family: a comparison of the CuZn-SOD (SOD1), Mn-SOD (SOD2), and EC-SOD (SOD3) gene structures, evolution, and expression. *Free Radical Biology and Medicine* 33 (3): 337–349.

Zhang, Y., Wang, L., Liu, Y. et al. (2006). Nitric oxide enhances salt tolerance in maize seedlings through increasing activities of proton-pump and Na^+/H^+ antiport in the tonoplast. *Planta* 224: 545–555.

Zhao, L., Zhang, F., Guo, J. et al. (2004). Nitric oxide functions as a signal in salt resistance in the calluses from two ecotypes of reed. *Plant Physiology* 134: 849–857.

Zhong, L., Wang, L., Xu, L. et al. (2015). The role of NOS-mediated ROS accumulation in an early phase Cu-induced acute cytotoxicity in MCF-7 cells. *Biometals* 28 (1): 113–122.

5

Nitrogen Uptake and Utilization Toward Sustainable Agriculture Development

Role of Nitric Oxide

Mehtab Muhammad Aslam[1], Muhammad Waseem[2], Abah Felix[3], and Iffat Shaheen[4]

[1]*Center for Plant Water-Use and Nutrition Regulation, College of Life Sciences, Joint International Research Laboratory of Water and Nutrient in Cops, Fujian Agriculture and Forestry University, Fuzhou, Fujian, China*
[2]*Horticulture, South China Agricultural University, Guangzhou, China*
[3]*Crop Protection College, Fujian Agriculture and Forestry University, Fuzhou, Fujian, China*
[4]*College of Agriculture, Bahauddin Zakariya University, Multan, Pakistan*

5.1 Introduction

Nitrogen (N_2) is one of the most important plant nutrients that determine plant growth, and its deficiency is one of the main challenges for researchers to untangle (Kumar et al. 2018). It has an important role in plant metabolism, and is a component of nucleic acid, nucleotides, vitamins, pigments, alkaloids, and many other compounds. It is estimated that 13–15% of nucleic acids and proteins are made of nitrogen molecules (Franklin et al. 2017). Therefore, maintaining N_2 metabolism may improve plant performance and yield (Stahl et al. 2016). Plant roots can absorb organic (amino acid, urea, and peptides) or inorganic (ammonium [NH_4^+] or nitrate [NO_3^-]) forms of nitrogen from the soil (Stahl et al. 2016). Inorganic N_2 is universally available in most soils, while organic forms are attributed to plant nutritional demand in a particular ecosystem, e.g. boreal habitats (Jones and Kielland 2002). Generally, N_2 availability is deficient in natural soil, which is influenced by numerous factors such as soil type, temperature, microbiome, and leaching resulting in N_2 depletion (Miller and Cramer 2005). To overcome N_2 scarcity, plants have evolved several morphological and physiological responses to facilitate N_2 use efficiency (Nacry et al. 2013). Therefore, it is worthwhile to identify beneficial plant traits and potential genes that regulate N_2 uptake and utilization efficiency.

Plants can take up diverse N_2 forms, but they fundamentally rely on an inorganic form of N_2 (Franklin et al. 2017). It has been observed that 85–90 million metric tons (t) of N_2 fertilizers are applied to agricultural land each year globally (Masclaux-Daubresse et al. 2010). Nitrogen is the most cost-intensive fertilizer required for plant growth; therefore, it is a serious concern to reduce fertilizer input without affecting plant growth. Thus, there is a rising interest in reducing fertilizer application and breeding plants with improved nitrogen use efficiency (NUE) to enhance plant nutrient acquisition ability. The N_2 uptake capacity of plants is mainly based on soil components, environmental factors, and plant species (Hirel et al. 2007). According to an estimate, 50–70% of N_2 added to the soil is lost (Hodge et al. 2000). However, it is essential to generate improved NUE crops to reduce other environmental problems triggered by water pollution, NO_3^- leaching, and ecosystem saturation.

Nitrogen utilization in plants involves the following steps: N_2 uptake, assimilation, translocation, recycling, and remobilization. NUE is defined as the grain yield/unit of N_2 (constituent N fertilizer) in the soil, or is defined as a physiological index calculated by using fresh/dry matter of plant/nitrogen concentration (Good et al. 2004). In both cases, NUE is a byproduct of nitrogen uptake efficiency (NupE) and nitrogen utilization efficiency (NUtE), considered as a perfect combination of nitrogen assimilation efficiency (NAE) and nitrogen remobilization efficiency (NRE).

Plants growing in a heterogeneous environment can improve N_2 acquisition through modulating root structural architecture, providing a larger surface area to absorb nutrients and water from the soil (Hodge 2010). The mobility of nutrients depends on the plant genotype, soil texture, and other related environmental factors such as temperature and pH (Postma et al. 2014). As a result, plants can take up NO_3^- and NH_4^+ via the roots at different frequencies. Several membrane proteins or transporters are involved in NO_3^- uptake, assimilation, translocation, and remobilization (Table 5.1). For instance, the NO_3^- reductase (NR) enzyme provides organic N_2 and regulates plant NO_3^- metabolism (Chamizo-Ampudia et al. 2017).

Nitric oxide (NO) is a signaling molecule involved in mediating plant responses to several biotic and abiotic stresses via gene expression modulation, altering protein

Table 5.1 Summary of identified nitrate transporters and their functions in plants.

Species name	Gene name	Role in plants	References
Arabidopsis thaliana	*AtNPF1.1*, *AtNPF1.2*	Nitrate allocation to young leaves	(Wang and Tsay 2011; Chiba et al. 2015)
	AtNPF2.3	Root-to-shoot nitrate transport under salinity	(Chiba et al. 2015; Taochy et al. 2015)
	AtNPF2.12	Nitrate transport	(Almagro et al. 2008; Chiba et al. 2015)
	AtNPF2.13	Nitrate remobilization from old to young leaves	(Fan et al. 2009; Nour-Eldin et al. 2012; Chiba et al. 2015)
	AtNPF3.1	Nitrite accumulation in leaves	(Crawford 1995; Sugiura et al. 2007; Pike et al. 2014; Tal et al. 2016)
	AtNPF4.6	Nitrate uptake in roots	(Huang et al. 1999; Kanno et al. 2012; Kanno et al. 2013)
	AtNPF5.5	Nitrogen accumulation	(Léran et al. 2015)
	AtNPF6.2	Nitrate storage in petiole	(Chiu et al. 2004)
	AtNPF6.3	Nitrate uptake in root	(Huang et al. 1996; Guo et al. 2001; Krouk et al. 2010)
	AtNRT1.1 or *CHL1*	Nitrate transporter	(Tsay et al. 1993)

(*Continued*)

Table 5.1 (Continued)

Species name	Gene name	Role in plants	References
Oryza sativa	*OsNPF2.2*	Root-to-shoot nitrate transport	(Li et al. 2015)
	OsNPF2.4	Acquisition and long-distance nitrate transport	(Xia et al. 2015)
	OsNPF7.2	Intracellular nitrate allocation in roots	(Hu et al. 2016)
	OsNPF7.3	Nitrogen allocation	(Ouyang et al. 2010)
	OsNPF8.20	NH_4^+ uptake, nitrogen assimilation, and grain yield	(Fang et al. 2013)
Medicago truncatula	*MtNRT1.3*	increases the ability of the plant to acquire NO_3^- under N-limiting conditions	(Morère-Le Paven et al. 2011)
Brassica napus	*BnNRT1.2*	NO_3^- transporters	(Zhou et al. 1998)
Hordeum vulgare	*HvNRT2.1/-2.4*	NO_3^- inducible	(Trueman et al. 1996)
Nicotiana plumbaginifolia	*NpNRT2*	NO_3^- inducible	(Quesada et al. 1997)

activity, and reactive oxygen species (ROS) interaction (Simontacchi et al. 2015; Prakash et al. 2021). Several studies have provided evidence for the role of NO in enhancing plant tolerance to numerous environmental stresses such as drought (García-Mata and Lamattina 2001), temperature (Sánchez-Vicente and Lorenzo 2021), salinity (Fatima et al. 2021), oxidative (Ahmad et al. 2021), nutrient (González-Moscoso et al. 2021), and ultraviolet-B (UV-B) stresses (Tossi et al. 2009). Therefore, NO has emerged as a key regulator in controlling adaptive plant responses and favoring plant growth. *Arabidopsis thaliana AtNOS* lines synthesized higher levels of NO and displayed increased tolerance to various biotic and abiotic stress. More specifically, *OtNOS* exhibited an enhanced germination rate compared with non-transgenic lines in *A. thaliana* and displayed superior performance under NaCl and drought stress (Foresi et al. 2015), supporting the idea that NO has a role in overcoming abiotic stresses. In this chapter, we discuss how plants acquire nitrogen from the rhizosphere assisted by plant roots, and some key genes involved in regulating N_2 uptake and utilization efficiency, ultimately improving plant growth and yield. For future studies, we propose some crucial genes used to improve NUE through crop genetic manipulation.

5.2 Nitrogen Assimilation and Uptake by Plants

Nitrogen significantly enhances crop yield and also increases food quality, making up a significant part of protein. Application of N_2 in appropriate amounts encourages photosynthetic activities and increases leaf area, blossoms, greenness, longevity, and net assimilation (Leghari et al. 2016; Anas et al. 2020). The amount of N_2 captured by the roots is based primarily on plant adaptability and soil N_2 availability (Tsay et al. 2007;

Masclaux-Daubresse et al. 2010). N_2 availability can fluctuate considerably owing to several environmental factors such as temperature, pH, wind, precipitation, and soil type (Masclaux-Daubresse et al. 2010).

Plants roots take up N_2 from the soil assisted by two NO_3^- transporters that work in a coordinated fashion to acquire NO_3^- from the rhizosphere and allocate it throughout the whole plant body (Tsay et al. 2007). Members of the *NRT1* transporter gene family facilitate a low-affinity transport system (LATS), except for *AtNRT1.1*, which is a dual-affinity transporter and NO_3^- sensor (Ho et al. 2009). When external NO_3^- concentration is reduced a high-affinity transport system (HATS) assists NO_3^- transport (Williams and Miller 2001). However, LATS and HATS transporter genes have been shown to make a significant contribution to N_2 acquisition from the rhizosphere.

Nitrogen is acquired only in inorganic form (NH_4^+ or NO_3^-) under particular soil compositions. Nitrogen assimilation entails converting NO_3^- to NH_4^+, followed by NH_4^+ assimilation into amino acids (De Montaigu et al. 2011). Nitrate reduction takes place in both shoots and roots and is governed by the NR enzyme. NR is a homodimer of its monomer connected with three different groups: a heme, flavin adenine dinucleotide (FAD), and molybdenum cofactor (Moco) (Fernandez and Galvan 2007). Although NR is thought to be located in the cytosol, a study conducted on corn and roots showed an association between nitrite reductase (NiR) and plasma membrane. Thus, NO_3^- assimilation is mainly accomplished by two steps: first, NO_3^- is transported into the cell and acted on by NR enzyme in the cytosol; subsequently it is transported to the chloroplasts where NiR catalyzes its reduction to NH_4^+ (Fernandez and Galvan 2008; Sanz-Luque et al. 2015).

NO has multiple functions in the plant body; its synthesis relies on the availability of internal and external/abiotic factors and particularly on N_2 metabolism (Galatro et al. 2020). NO can be produced inside the plant cell, and it can also be taken up from the environment. Soils are a rich source of NO, where it is synthesized as a result of soil microbial activities. Conversely, symbiotic associations between plant roots and microbes may facilitate NO formulation and may also affect NO synthesis (Galatro et al. 2020). Additionally, NO can also be generated from several enzymatic and non-enzymatic sources (Moreau et al. 2010; Gupta et al. 2011), categorized as oxidation or reduction mechanisms based on the substrate specificity (Gupta et al. 2011). Several studies have provided evidence that NO exhibits a significant effect on organic and inorganic N_2 uptake and translocation to different plant organs, based on NO availability, production, and source (Simon et al. 2009, 2013; Dong et al. 2015).

5.3 Molecular Basis of NUE

NUE is a multifaceted plant trait that involves several steps including assimilation, transport, and signaling. Undeniably, several transgenic methods used to deploy gene expression linked to N_2 uptake, transportation, and remobilization may promote plant growth and yield (Plett et al. 2017). The molecular basis of NUE has been investigated largely to explore the genetic contribution involved in improving crop growth. A number of gene families and many potential genes have been reported to improve crop NUE (Yang et al. 2017; Wang et al. 2018). For instance, NO_3^- and NH_4^+ transporter-related genes significantly improve NUE (Li et al. 2017). Some 70 NO_3^- transporters in

Arabidopsis and >85 transporters in rice have been acknowledged and are considered to increase NUE.

It has been reported that the *NRT1* gene family contained 53 candidate genes in *A. thaliana*, which are differentially expressed among different plant tissues (Tsay et al. 2007), implying a unique contribution of at least some members in alleviating N_2 stress responses. Moreover, nitrogen is captured from the soil and dispersed into aboveground plant tissues assisted by genes such as *AtNRT1.5*, which is responsible for translocating NO_3^- from the root toward the shoot (Lin et al. 2008). Moreover, it has been verified that rice overexpressing *OsNRT1.1* under limited N_2 availability showed enhanced grain yield (32–50% per plant) and NUE (38–54% per plot), and significantly improved seed number (Wang et al. 2018; Bahadur et al. 2019; Zhang et al. 2019). Similarly, rice overexpressing *OsNRT2.1*, *OsNRT2.3b*, and *OsPTR9* enriched NUE, grain yield, and plant overall growth (Huang et al. 2017).

Several studies have evidenced that genetic manipulation of specific enzymes involved in N_2 assimilation serves as a promising approach to improve crop NUE. For instance, glutamine synthetase (GS) converts NH_4^+ to glutamine, which mobilizes inorganic N_2 into an organic form for plants. Ubiquitin-driven expression of *OsGS1* in rice and overexpression of *SbGln* in sorghum may result in enhanced grain yield (Brauer et al. 2011; Urriola and Rathore 2015), indicating the contribution of this enzyme to gaining N_2 from the rhizosphere.

HATS comes into play when NO_3^- availability is limited, assisted by members of the *NRT2* gene family, but their functional characterization remains unexplored among all higher plants. There is increasing interest among researchers in clarifying the role of *NRT2* family members in improving plant growth by encouraging NO_3^- uptake efficiency. One study based on the interaction between *AtNRT2.1* and NAR2 protein, which is part of HATS in *A. thaliana* (Orsel et al. 2006), exhibited that the *AtNRT2.1* mutant lost almost 75% of its NO_3^- uptake compared to the control and had reduced leaf NO_3^- concentration. Consequently, mutant growth was severely impaired with limited NO_3^- availability (Filleur et al. 2001).

NO biosynthesis involves oxidative or reductive pathways (Gupta et al. 2011). The former involves mainly hydroxylamine, polyamine-, and L-arginine-mediated pathways. NO synthase (NOS) is a key enzyme source for nitric oxide metabolism through L-arginine in animals (Crawford 2006). To date the role of the L-arginine pathway in NO production in plants remains to be determined. Nitrate reductase, mitochondrial NR, and plasma membrane-associated NO production are all categorized as reductive reactions. Among them, NR is a well-documented route and pivotal source of NO in plants (Gupta et al. 2011). The molecular mechanism of plant NO has been investigated in the past two decades, which revealed that exogenous and endogenous NO could regulate various metabolic, stress responsive, and hormonal signaling pathways (Astier et al. 2017). It has emerged that this small molecule regulates cyclic guanosine monophosphate (cGMP)-dependent pathways and interacts with ROS) (Mur et al. 2013). Being a signaling molecule, NO typically interacts with phytohormones and related endogenous molecules during plant growth and development (Sanz et al. 2015). At the molecular level, NO has been demonstrated to be responsible for posttranslational modifications of proteins, including *S*-nitrosylation of the thiol group of cysteinyl residues affecting protein subcellular localization (Astier and Lindermayr 2012). More recently, fatty acid nitration by NO has been shown to be a major part of NO signaling in plants (Mata-Pérez et al. 2017). Beligni and Lamattina (2000) showed that NO stimulates seed germination, etiolation, and hypocotyl elongation and emergence. It was also suggested that NO induces adventitious root development in cucumber, whereas 2-4-carboxyphenyl-4,4,5,5-tetramethylimidazoline1-oxyl-3-oxide, an NO scavenger, inhibits its formation (Pagnussat et al. 2002). Previous studies have also suggested the role of NO-responsive RAPID ALKANIZATION FACTOR (*RALF1*, *RALF22*) in root growth and development (Sharma et al. 2016).

5.4 Plant Density to Testify N2 Rate

Nitrogen is required in large quantities by plants (McKague et al. 2005; Leghari et al. 2016) --~1000 μg/kg of dry matter (Leghari et al. 2016). Therefore, it must be provided to plants at the required amount in the rhizosphere region. Plant physiological processes needed to improve agricultural productivity cannot occur without nitrogen application (Leghari et al. 2016). Plants demand a stable quantity of N_2 for healthier growth and development. To boost plant yield and economic productivity, N_2 application rate and plant population density are important parameters to be considered. The optimum application of 120 kg/ha of N_2 improves plant height, number of tillers/hill, dry weight, panicle length, number of filled grains/panicle, straw yield, and harvest index (Malik et al. 2014; Leghari et al. 2016). Boquet et al. reported optimal rice grain yield and profit with the input of 120 kg N/ha at optimal density (Boquet et al. 1994). Moreover, the application of 100 kg/ha of N_2 increased branching, bolls and weight, seed weight, and improved yield in cotton (Boquet et al. 1994). From annuals to perennials, shrubs to trees, the story of N_2 use is the same. However, overuse or underuse of N_2 (Leghari et al. 2016) has a deleterious effect on plant development, yield, and quality.

Plant density is an important factor that influences crop phenology, morphology, and grain yield. Therefore, the optimum utilization of N_2 supplied is affected by plant density. Imran et al. (2015) found that plant density was one factor that determines plant productivity. The higher the plant density, the greater the net biomass accumulation and grain yield per unit area. Depending on the variety, up to a 25% yield increase was recorded at higher plant population density (Imran et al. 2015). Therefore, farmers must be conversant with all cultural, management, and agronomic systems.

Contrarily, overpopulation of plants due to inadequate spacing leads to antagonism for growth factors (Imran et al. 2015), such as light, soil nutrients, and space; thereby causing poor yield, disease susceptibility, and weed pressure. Maintaining proper spacing is a good practice that permits for optimal use of nutrient resources by cotton (Ullah et al. 2019) and other plants. Khan et al. demonstrated that plant density is an important factor for optimizing resources, structure, and increasing the photosynthetic capacity of crops for a good yield, profits for farmers, and CO_2 removal from ambient air. Cotton plants with a population density of 87 000 stands/ha yielded a good result (4546 kg/ha cotton seed and 1682 kg/ha of lint) compared with 15 000 stands/ha, and other densities experimented with in the field (Ullah et al. 2019). This was endorsed by higher biomass accumulation (Anas et al. 2020; de Oliveira Silva et al. 2020). Higher biomass accumulation can only be made possible with adequate spacing when N_2 supply is at its optimum (Anas et al. 2020) among other factors.

Optimum plant density reduces fertilizer application and labor costs, and allows for efficient utilization of irrigation, while output quality and yield are enhanced (Khan et al. 2019; Shah et al. 2021). Although high plant density greater than 10 plants/m^2 poses a disease problem and leads to decreased light interception (Khan et al. 2019), plant population density should be based on plant species (Sangoi 2001), soil type, and nutrient availability. For instance, the current density practices in the Yellow River valley, Yangtze River valley, and northwest region of China are 5.3×10^4–7.5×10^4 plants/ha, 3.0×10^4 plants/ha, and 22.7×10^4 plants/ha, respectively. This is because different soils have different water-holding and ion-exchange capacities. Plant density has a direct correlation with nutrition and other factors such as the phenotype and physiology of plants. Optimum population density results in a high cotton yield (Khan et al. 2019) under adequate N_2 supply (Leghari et al. 2016). No plant grows optimally without optimum N_2 supply – it is an essential macronutrient for healthy plant growth.

Nitrogen deficiency damages crop production; it is the fundamental element of plants. Over- and under population may lead to reduced yield of cotton and other plants, while optimum density increases yield (Rahman et al. 2011), reduces inputs, and maximizes turnover. This can only be made possible with the presence of N_2, which ensures proper plant structure and efficient external and internal metabolic processes. de Oliveira Silva et al. demonstrated that higher planting density produced a significant increase in maize yield and improved N_2 use efficiency by 6.6% and 3.9%, respectively (de Oliveira Silva et al. 2020). Therefore, N_2 is required in a form available for physiological processes to enable high yield at optimum plant population levels. In agricultural management systems, N_2 rate and plant density must be given adequate attention along with other agronomic practices for optimum yield. Recommended N_2 applications vary based on plant species, soil type, and N_2 content.

The primary possibility for NO and N_2 interconnection was that both the routes for generation of NO and assimilation of N_2 share a common enzyme, NO_3^- reductase (Simontacchi et al. 2015). The expression of NR-encoded genes and their corresponding encoded enzymes are influenced by different environmental factors (Yanagisawa 2014). Of those factors, NO_3^- availability in the soil to roots during their germination and development affects NO synthesis in plant tissues (Manoli et al. 2014). Emerging evidence has suggested multiple effects of NO in regulating various aspects of plant development. Seed germination, seedling growth, biomass accumulation, and yield have been reported to increase after exogenous application of NO (Sun et al. 2021). In higher plants, the role of *S*-nitroglutathione reductase in NO metabolism has been recently identified using molecular and genetics approaches. In *Arabidopsis*, *AtGSNOR1* is pivotal for different developmental processes, and the mutant *Atgsnor1* had reduced hypocotyl elongation and apical dominance (Kwon et al. 2012), increased germination rate, changes in leaf shape, decreased photosynthesis, and reduced fruit yield (Hussain et al. 2019).

5.5 NUtE in Plants

NUE is represented by the ratio of yield outcome to available N_2 in plants. The optimum N_2 level is defined as the level that enables the highest yield with the lowest input cost to maximize profits. NUE can be described as the ratio of grain produced to amount of N_2 made available to the plants (Dresbøll and Thorup-Kristensen 2014; Swarbreck et al. 2019), and is quoted as being 33% for cereal crops (Hawkesford and Riche 2020). On the other hand, NUE is a broader term that has several definitions. It encompasses grain yield per unit of N_2 applied (kilogram of yield per kilogram of available nitrogen), total biomass per unit of N_2 uptake, or N_2 uptake/unit of N_2 available or supplied (Dresbøll and Thorup-Kristensen 2014). To achieve NUE and use efficiency in general, plant physiological processes need to be very efficient. Genotypes highly responsive to N_2 are likely the most efficient means to ensure the efficient utilization of N_2 supplied to plants for good yield. Therefore, breeding of N_2-responsive varieties is key. This will not only reduce the supplied quantity of N_2 but also increase crop yield, consequently reducing cost. Wheat breeding enhances NUE through increases in yield, resulting in decreased protein content of the grain because of the dilution process (de Oliveira Silva et al. 2020).

Grain yield and plant N_2 content (de Oliveira Silva et al. 2020) are the direct result of efficient utilization of N_2 supplied to the plant through natural or artificial fertilization. Grain formation and filling require an adequate supply of N_2. Limitation of the N_2 uptake and utilization efficiency caused by physiological and agronomic processes and environmental stress leads to poor yield (de Oliveira Silva et al. 2020). Therefore, the efficient utilization of N_2 is an important component in improving agricultural productivity.

Only one-third of applied N_2 remains in the harvested crops and the rest is lost to the atmosphere and leached down. NUE fluctuates from region to region, soil to soil, and among different crop farming systems. Understanding of genetics and plant–environment management practices and interactions will lead to efficient N_2 utilization in plants. Moreover, N_2 utilization depends, among other things, on environmental factors, plant varieties, and climate scenarios. Genetic variation, environmental conditions and location, and farm agronomic management and their interactions determine crop yield and quality. To improve yields, several studies on plant genetics and environmental interactions have been conducted on wheat crop breeding for improvement and NUE (Dresbøll and Thorup-Kristensen 2014; Hawkesford and Riche 2020).

Varying utilization and application of N_2 fertilizer based on timings, chemical formulation, and quantities influences yield quality while reducing losses and avoiding contravening legislation. The utilization of higher N_2 fertilizer levels leads to increased N_2 input, which promotes greater yields but at lower efficiency. Reduced input of fertilizer to agricultural systems and increased yields are important for future sustainable food production (Dresbøll and Thorup-Kristensen 2014). NUE decreases as N_2 input increases while yield rises progressively with good N_2 management. Overall, the highest NUE is observed in wheat. However, N_2 uptake increases with N_2 rate, which promotes biological yield. Different plants have different N_2 uptake and utilization efficiency levels. Different varieties show different use efficiency even within species irrespective of N_2 rate (Hawkesford and Riche 2020).

Nitrogen assimilation and acquisition from soil is regulated by NO_3^- signaling in soil (Sun et al. 2017). However, NO_3^- indirectly regulates the expression and activity of NR via altering endogenous NO in tissues (Kaiser et al. 2002). NO is a plant signaling key player involved in growth regulation, which suggests that NO is indispensable for improving NUE. Sun et al. (2015) demonstrated that NO generated through the NR route improved N assimilation through lateral root formation and N uptake rates. Similarly, NR was activated in cabbage, suggesting that this differential activation of NR by NO is governed by concentration of NO_3^- in the soil (Du et al. 2008; Jin et al. 2009; Rosales et al. 2011). Altogether, this suggests that NR-mediated NO generation plays a primary role in enhancing N assimilation and acquisition activity representing a strategy of plant adaptation to variable NO_3^- availability and improving NUE.

5.6 Nitrogen Contribution to Sustainable Agriculture Development

Agricultural production is part of and dependent on the nitrogen cycle. All metabolic and biochemical processes in plants that keep them alive and productive are associated with protein, of which N_2 is an essential constituent (Leghari et al. 2016). As a component of the chlorophyll molecule, nitrogen helps in photosynthetic activities (Spiertz 2009) that drive plant growth, development, and yield. Nitrogen makes energy available when and where it is needed by the plant for optimum yield. Water and nutrient uptake are regulated by proteins and enzymes, of which N_2 is the major component. Plants require ample N_2 to drive sufficient yield to feed humans and livestock (Lassaletta et al. 2016). To increase agricultural production to meet the demand of the growing world population N_2 is unavoidable. Therefore, the role of N_2 in crop yield and development of agriculture cannot be overemphasized.

The phenomenon of photosynthesis is also associated with leaf N_2 content (FAO 2017). Insufficient nitrogen is one of the most yield-limiting factors in agricultural production. Several experimental results have shown that in cereal crops including *Oryza sativa* and *Zea mays*, yield, plant height, number of tillers, spikelet per spike, grains per spike, length of spike, and 1000-grain weight are increased by supplying N_2 fertilizers (McKague et al. 2005; Leghari et al. 2016). Leghari et al. (2016) demonstrated that the maximum grain yield of wheat ~3.848 t/ha was produced through the application of 180 kg N_2. All crops need a substantial quantity of N_2 for optimum yield, though the quantity needed depends on variety, environmental factors, and management practices (Leghari et al. 2016). Nitrogen levels significantly influence yield and cotton fibers (Leghari et al. 2016), rice (Lin et al. 2009), and maize grain yield and total starch contents (Chen et al. 2018).

Nitrogen compounds are critical to global crop production particularly for cereals. Optimal N_2 application (100 kg/ha) promotes monopodial and sympodial branches in cotton thus increasing boll/plant and seed yield and weight. Similarly, optimal use of N_2 improves cane weight, vegetative parts, and yield in sugar cane. The output of sugar production in China and India was doubled because of the application of the recommended quantity (300 kg/ha) of N_2 (Wajid et al. 2012). Changing the N_2 optimum application rate increases grain yield by up to 13% in irrigated rice. Higher yields of cereal crops were achieved from selection of N_2-responsive and high-yielding cultivars. The introduction of high-yielding cultivars improves the grain yield per unit land (Spiertz 2009). Efficient use of N_2 sources is a surefire way of ensuring future food security (Rütting et al. 2018). Over 100 million metric tons of wheat are produced each year and the global demand continues to rise (FAO 2017). In the 20th century, scientific and technological innovations enabled the development of synthetic nitrogen fertilizer allowing rapid growth in crop productivity (Roser and Ritchie 2013).

Biomass has a major role in vegetative growth and N_2 application improves the accumulation of biomass. It increases leaf size and greenness (Anas et al. 2020), and application of 100 kg N/ha on *Beta vulgaris* increased the fresh weight of leaves, reduced dry matter, and yield increased steadily with variable levels of N_2 according to Miceli and Miceli (2014). Nitrogen promotes root and plant growth and facilitates the uptake and utilization of other plant nutrients, including potassium and phosphorus. Adequate N_2 utilization can support plant performance and crop output (Anas et al. 2020). On the other hand, livestock manure returned to the soil for plant use plays an important role in maintaining plant growth and productivity. NO_3^- and NH_4^+ are the primary sources of N_2 in the soil, and are critical for biochemical and physiological functions in plants governing growth and development (Ribaudo et al. 2011). The availability of these N_2 sources significantly boosts crop yield and quality (Zhang et al. 2020).

To meet sustainable agriculture targets on our non-expanding land, N_2 has a vital role. Synthetic N_2 not only increased crop yields but also enabled an expansion of livestock production (Roser and Ritchie 2013). Therefore, the N_2 requirement is the most central feature for plant production, with deficiency resulting in stunted plant growth and early leaf senescence, and, subsequently, poor crop production and quality.

Agricultural productivity largely depends on nutrient availability, particularly N, which is critical for the subterranean environment. The phenomenon of exchange of NO between plants and the environment is relatively scarce. The capacity of plants to exchange NO was first observed in soybean (Corpas and Barroso 2014), but the genetic and physiological reasons for emissions were not clear (Chen et al. 2012). Years later, some authors considered the possibility that plants may have adapted such a mechanism to remove excessive NO_3^- and reduce N toxicity (Saxe 1986; Marvasi 2017). NO has been demonstrated to play a pivotal role in all physiological processes in a wide range of plant species including in the cell cycle itself. Some of these processes include senescence (Procházková and Wilhelmová 2011), flower development (Zafra et al. 2010), and fruit ripening (Leshem and Pinchasov 2000; Zafra et al. 2010). These reports show the beneficial effects of NO and how it mitigates the adverse effects of distinct environmental stimuli including heavy metals, drought, salinity, delays in flowering and fruit ripening, fruit shield life, fruit and flower quality, seed germination, and UV radiation. Taken together, many advances have been made in understanding of the role of NO in crop plants. There are still great challenges ahead to understand how small NO molecules and their derivatives function at a molecular and genetic level and the detailed mechanisms of interaction of NO with other signaling molecules.

References

Ahmad, A., Khan, W.U., Shah, A.A. et al. (2021). Synergistic effects of nitric oxide and silicon on promoting plant growth, oxidative stress tolerance and reduction of arsenic uptake in *Brassica juncea*. *Chemosphere* 262: 128384.

Almagro, A., Lin, S.H., and Tsay, Y.F. (2008). Characterization of the *Arabidopsis* nitrate transporter NRT1.6 reveals a role of nitrate in early embryo development. *The Plant Cell* 20: 3289–3299.

Anas, M., Liao, F., Verma, K. et al. (2020). Fate of nitrogen in agriculture and environment: agronomic, eco-physiological and molecular approaches to improve nitrogen use efficiency. *Biological Research* 53: 1–20.

Astier, J., Gross, I., and Durner, J. (2017). Nitric oxide production in plants: an update. *Journal of Experimental Botany* 69: 3401–3411. doi: 10.1093/jxb/erx420.

Astier, J. and Lindermayr, C. (2012). Nitric oxide-dependent posttranslational modification in plants: an update. *International Journal of Molecular Sciences* 13: 15193–15208.

Bahadur, A., Batool, A., Nasir, F.et al. (2019). Mechanistic insights into arbuscular mycorrhizal fungi-mediated drought stress tolerance in plants. *International Journal of Molecular Sciences* 20: 4199.

Beligni, M.V. and Lamattina, L. (2000). Nitric oxide stimulates seed germination and de-etiolation, and inhibits hypocotyl elongation, three light-inducible responses in plants. *Planta* 210: 215–221.

Boquet, D.J., Moser, E.B., and Breitenbeck, G.A. (1994). Boll weight and within-plant yield distribution in field-grown cotton given different levels of nitrogen. *Agronomy Journal* 86: 20–26.

Brauer, E.K., Rochon, A., Bi, Y.M. et al. (2011). Reappraisal of nitrogen use efficiency in rice overexpressing glutamine synthetase1. *Physiologia Plantarum* 141: 361–372.

Chamizo-Ampudia, A., Sanz-Luque, E., Llamas, A. et al. (2017). Nitrate reductase regulates plant nitric oxide homeostasis. *Trends in Plant Science* 22: 163–174.

Chen, G., Cao, H., Liang, J. et al. (2018). Factors affecting nitrogen use efficiency and grain yield of summer maize on smallholder farms in the North China Plain. *Sustainability* 10: 363.

Chen, J., Wu, F.H., Liu, T.W. et al. (2012). Emissions of nitric oxide from 79 plant species in response to simulated nitrogen deposition. *Environmental Pollution* 160: 192–200. doi: 10.1016/j.envpol.2011.09.007.

Chiba, Y., Shimizu, T., Miyakawa, S. et al. (2015). Identification of *Arabidopsis thaliana* NRT1/PTR FAMILY (NPF) proteins capable of transporting plant hormones. *Journal of Plant Research* 128: 679–686.

Chiu, C.-C., Lin, C.-S., Hsia, A.-P. et al. (2004). Mutation of a nitrate transporter, AtNRT1:4, results in a reduced petiole nitrate content and altered leaf development. *Plant and Cell Physiology* 45: 1139–1148.

Corpas, F.J. and Barroso, J.B. (2014). Functional implications of peroxisomal nitric oxide (NO) in plants. *Frontiers in Plant Science* 5: 97. doi: 10.3389/fpls.2014.00097.

Crawford, N.M. (1995). Nitrate: nutrient and signal for plant growth. *The Plant Cell* 7: 859.

Crawford, N.M. (2006). Mechanisms for nitric oxide synthesis in plants. *Journal of Experimental Botany* 57: 471–478.

De Montaigu, A., Sanz-Luque, E., Macias, M.I. et al. (2011). Transcriptional regulation of CDP1 and CYG56 is required for proper NH4+ sensing in *Chlamydomonas*. *Journal of Experimental Botany* 62: 1425–1437.

de Oliveira Silva, A., Ciampitti, I.A., Slafer, G.A. et al. (2020). Nitrogen utilization efficiency in wheat: a global perspective. *European Journal of Agronomy* 114: 126008.

Dong, F., Simon, J., Rienks, M. et al. (2015). Effects of rhizospheric nitric oxide (NO) on N uptake in *Fagus sylvatica* seedlings depend on soil CO_2 concentration, soil N availability and N source. *Tree Physiology* 35: 910–920.

Dresbøll, D.B. and Thorup-Kristensen, K. (2014). Will breeding for nitrogen use efficient crops lead to nitrogen use efficient cropping systems?: a simulation study of G× E× M interactions. *Euphytica* 199: 97–117.

Du, S., Zhang, Y., Lin, X. et al. (2008). Regulation of nitrate reductase by nitric oxide in Chinese cabbage pakchoi (*Brassica chinensis* L.). *Plant, Cell and Environment* 31: 195–204.

Fan, S.-C., Lin, C.-S., Hsu, P.-K. et al. (2009). The *Arabidopsis* nitrate transporter NRT1.7, expressed in phloem, is responsible for source-to-sink remobilization of nitrate. *The Plant Cell* 21: 2750–2761.

Fang, Z., Xia, K., Yang, X. et al. (2013). Altered expression of the PTR/NRT 1 homologue Os PTR 9 affects nitrogen utilization efficiency, growth and grain yield in rice. *Plant Biotechnology Journal* 11: 446–458.

Fatima, A., Husain, T., Suhel, M. et al.(2021). Implication of nitric oxide under salinity stress: the possible interaction with other signaling molecules. *Journal of Plant Growth Regulation* 1–15.

Fernandez, E. and Galvan, A. (2007). Inorganic nitrogen assimilation in *Chlamydomonas. Journal of Experimental Botany* 58: 2279–2287.

Fernandez, E. and Galvan, A. (2008). Nitrate assimilation in *Chlamydomonas. Eukaryotic Cell* 7: 555–559.

Filleur, S., Dorbe, M.-F., Cerezo, M. et al. (2001). An *Arabidopsis* T-DNA mutant affected in Nrt2 genes is impaired in nitrate uptake. *FEBS Letters* 489: 220–224.

Food and Agriculture Organization of the United Nations (FAO). (2017). *World Fertilizer Trends and Outlook to 2020*. Rome, Italy: FAO.

Foresi, N., Mayta, M.L., Lodeyro, A.F. et al. (2015). Expression of the tetrahydrofolate-dependent nitric oxide synthase from the green alga *Ostreococcus tauri* increases tolerance to abiotic stresses and influences stomatal development in *Arabidopsis. The Plant Journal* 82: 806–821.

Franklin, O., Cambui, C.A., Gruffman, L. et al. (2017). The carbon bonus of organic nitrogen enhances nitrogen use efficiency of plants. *Plant, Cell and Environment* 40: 25–35.

Galatro, A., Ramos-Artuso, F., Luquet, M. et al. (2020). An update on nitric oxide production and role under phosphorus scarcity in plants. *Frontiers in Plant Science* 11: 413.

García-Mata, C. and Lamattina, L. (2001). Nitric oxide induces stomatal closure and enhances the adaptive plant responses against drought stress. *Plant Physiology* 126: 1196–1204.

González-Moscoso, M., González-García, Y., Martínez-Villegas, N.V. et al. (2021). Nitric oxide modified growth, nutrient uptake and the antioxidant defense system in tomato seedlings stressed with arsenic. *Theoretical and Experimental Plant Physiology* 1–19.

Good, A.G., Shrawat, A.K., and Muench, D.G. (2004). Can less yield more? Is reducing nutrient input into the environment compatible with maintaining crop production? *Trends in Plant Science* 9: 597–605.

Guo, F.-Q., Wang, R., Chen, M. et al. (2001). The *Arabidopsis* dual-affinity nitrate transporter gene AtNRT1.1 (CHL1) is activated and functions in nascent organ development during vegetative and reproductive growth. *The Plant Cell* 13: 1761–1777.

Gupta, K.J., Fernie, A.R., Kaiser, W.M. et al. (2011). On the origins of nitric oxide. *Trends in Plant Science* 16: 160–168.

Hawkesford, M.J. and Riche, A.B. (2020). Impacts of G x E x M on nitrogen use efficiency in wheat and future prospects. *Frontiers in Plant Science* 11: 1157.

Hirel, B., Le Gouis, J., Ney, B. et al. (2007). The challenge of improving nitrogen use efficiency in crop plants: towards a more central role for genetic variability and quantitative genetics within integrated approaches. *Journal of Experimental Botany* 58: 2369–2387.

Ho, C.-H., Lin, S.-H., Hu, H.-C. et al. (2009). CHL1 functions as a nitrate sensor in plants. *Cell* 138: 1184–1194.

Hodge, A. (2010). Roots: the acquisition of water and nutrients from the heterogeneous soil environment. In: *Progress in Botany* 71, 307–337. Berlin, Heidelberg, Germany: Springer.

Hodge, A., Robinson, D., and Fitter, A. (2000). Are microorganisms more effective than plants at competing for nitrogen? *Trends in Plant Science* 5: 304–308.

Hu, R., Qiu, D., Chen, Y. et al. (2016). Knock-down of a tonoplast localized low-affinity nitrate transporter OsNPF7.2 affects rice growth under high nitrate supply. *Frontiers in Plant Science* 7: 1529.

Huang, N.-C., Chiang, C.-S., Crawford, N.M. et al. (1996). CHL1 encodes a component of the low-affinity nitrate uptake system in *Arabidopsis* and shows cell type-specific expression in roots. *The Plant Cell* 8: 2183–2191.

Huang, N.-C., Liu, K.-H., Lo, H.-J. et al. (1999). Cloning and functional characterization of an *Arabidopsis* nitrate transporter gene that encodes a constitutive component of low-affinity uptake. *The Plant Cell* 11: 1381–1392.

Huang, S., Zhao, C., Zhang, Y. et al. (2017). Nitrogen use efficiency in rice. In: *Nitrogen in Agriculture – Updates* (eds. A. Fahad and S. Fahad), 187–208. London: InTech.

Hussain, A., Yun, B.-W., Kim, J.H. et al. (2019). Novel and conserved functions of S-nitrosoglutathione reductase in tomato. *Journal of Experimental Botany* 70: 4877–4886.

Imran, S., Arif, M., Khan, A. et al. (2015). Effect of nitrogen levels and plant population on yield and yield components of maize. *Advances in Crop Science and Technology* 1–7.

Jin, C.W., Du, S.T., Zhang, Y.S. et al. (2009). Differential regulatory role of nitric oxide in mediating nitrate reductase activity in roots of tomato (*Solanum lycocarpum*). *Annals of Botany* 104: 9–17.

Jones, D.L. and Kielland, K. (2002). Soil amino acid turnover dominates the nitrogen flux in permafrost-dominated taiga forest soils. *Soil Biology and Biochemistry* 34: 209–219.

Kaiser, W.M., Weiner, H., Kandlbinder, A. et al. (2002). Modulation of nitrate reductase: some new insights, an unusual case and a potentially important side reaction. *Journal of Experimental Botany* 53: 875–882.

Kanno, Y., Hanada, A., Chiba, Y. et al. (2012). Identification of an abscisic acid transporter by functional screening using the receptor complex as a sensor. *Proceedings of the National Academy of Sciences* 109: 9653–9658.

Kanno, Y., Kamiya, Y., and Seo, M. (2013). Nitrate does not compete with abscisic acid as a substrate of AtNPF4.6/NRT1.2/AIT1 in *Arabidopsis*. *Plant Signaling and Behavior* 8: e26624.

Khan, N., Han, Y., Wang, Z. et al. (2019). Role of proper management of nitrogen in cotton growth and development. *International Journal of Biosciences* 14: 5.

Krouk, G., Lacombe, B., Bielach, A. et al. (2010). Nitrate-regulated auxin transport by NRT1.1 defines a mechanism for nutrient sensing in plants. *Developmental Cell* 18: 927–937.

Kumar, S., Bhavya, P.S., Ramesh, R. et al. (2018). Nitrogen uptake potential under different temperature-salinity conditions: implications for nitrogen cycling under climate change scenarios. *Marine Environmental Research* 141: 196–204.

Kwon, E., Feechan, A., Yun, B.-W. et al. (2012). AtGSNOR1 function is required for multiple developmental programs in *Arabidopsis*. *Planta* 236: 887–900.

Lassaletta, L., Billen, G., Garnier, J. et al. (2016). Nitrogen use in the global food system: past trends and future trajectories of agronomic performance, pollution, trade, and dietary demand. *Environmental Research Letters* 11: 095007.

Leghari, S.J., Wahocho, N., Laghari, G.M. et al. (2016). Role of nitrogen for plant growth and development: a review. *Advances in Environmental Biology* 10: 209–219.

Léran, S., Garg, B., Boursiac, Y. et al. (2015). AtNPF5.5, a nitrate transporter affecting nitrogen accumulation in *Arabidopsis* embryo. *Scientific Reports* 5: 1–7.

Leshem, Y.Y. and Pinchasov, Y. (2000). Non-invasive photoacoustic spectroscopic determination of relative endogenous nitric oxide and ethylene content stoichiometry during the ripening of strawberries *Fragaria anannasa* (Duch.) and avocados *Persea americana* (Mill.). *Journal of Experimental Botany* 51: 1471–1473.

Li, H., Hu, B., and Chu, C. (2017). Nitrogen use efficiency in crops: lessons from *Arabidopsis* and rice. *Journal of Experimental Botany* 68: 2477–2488.

Li, Y., Ouyang, J., Wang, Y. et al. (2015). Disruption of the rice nitrate transporter OsNPF2.2 hinders root-to-shoot nitrate transport and vascular development. *Scientific Reports* 5: 1–10.

Lin, S.-H., Kuo, H.F., Canivenc, G. et al. (2008). Mutation of the *Arabidopsis* NRT1.5 nitrate transporter causes defective root-to-shoot nitrate transport. *The Plant Cell* 20: 2514–2528.

Lin, X.-Q., Zhu, D.-F., Chen, H.-Z. et al. (2009). Effects of plant density and nitrogen application rate on grain yield and nitrogen uptake of super hybrid rice. *Rice Science* 16: 138–142.

Malik, T.H., Lal, S.B., Wani, N.R. et al. (2014). Effect of different levels of nitrogen on growth and yield attributes of different varieties of basmati rice (*Oryza sativa* L.). *International Journal of Scientific and Technology Research* 3: 444–448.

Manoli, A., Begheldo, M., Genre, A. et al. (2014). NO homeostasis is a key regulator of early nitrate perception and root elongation in maize. *Journal of Experimental Botany* 65: 185–200.

Marvasi, M. (2017). Potential use and perspectives of nitric oxide donors in agriculture. *Journal of the Science of Food and Agriculture* 97: 1065–1072. doi: 10.1002/jsfa.8117.

Masclaux-Daubresse, C., Daniel-Vedele, F., Dechorgnat, J. et al. (2010). Nitrogen uptake, assimilation and remobilization in plants: challenges for sustainable and productive agriculture. *Annals of Botany* 105: 1141–1157.

Mata-Pérez, C., Sánchez-Calvo, B., Padilla, M.N. et al. (2017). Nitro-fatty acids in plant signaling: new key mediators of nitric oxide metabolism. *Redox Biology* 11: 554–561.

McKague, K., Reid, K., and Simpson, H. (2005). *Environmental Impacts of Nitrogen Use in Agriculture*. Ontario: Ministry of Agriculture, Food and Rural Affairs.

Miceli, A. and Miceli, C. (2014). Effect of nitrogen fertilization on the quality of Swiss chard at harvest and during storage as minimally processed produce. *Journal of Food Quality* 37: 125–134.

Miller, A.J. and Cramer, M.D. (2005). Root nitrogen acquisition and assimilation. *Plant and Soil* 274: 1–36.

Moreau, M., Lindermayr, C., Durner, J. et al. (2010). NO synthesis and signaling in plants – where do we stand? *Physiologia Plantarum* 138: 372–383.

Morère-Le Paven, M.-C., Viau, L., Hamon, A. et al. (2011). Characterization of a dual-affinity nitrate transporter MtNRT1.3 in the model legume *Medicago truncatula*. *Journal of Experimental Botany* 62: 5595–5605.

Mur, L.A.J., Mandon, J., Persijn, S. et al. (2013). Nitric oxide in plants: an assessment of the current state of knowledge. *AoB Plants* 5: pls052.

Nacry, P., Bouguyon, E., and Gojon, A. (2013). Nitrogen acquisition by roots: physiological and developmental mechanisms ensuring plant adaptation to a fluctuating resource. *Plant and Soil* 370: 1–29.

Nour-Eldin, H.H., Andersen, T.G., Burow, M. et al. (2012). NRT/PTR transporters are essential for translocation of glucosinolate defence compounds to seeds. *Nature* 488: 531–534.

Orsel, M., Chopin, F., Leleu, O. et al. (2006). Characterization of a two-component high-affinity nitrate uptake system in *Arabidopsis*. Physiology and protein-protein interaction. *Plant Physiology* 142: 1304–1317.

Ouyang, J., Cai, Z., Xia, K. et al. (2010). Identification and analysis of eight peptide transporter homologs in rice. *Plant Science* 179: 374–382.

Pagnussat, G.C., Simontacchi, M., Puntarulo, S. et al. (2002). Nitric oxide is required for root organogenesis. *Plant Physiology* 129: 954–956.

Pike, S., Gao, F., Kim, M.J. et al. (2014). Members of the NPF3 transporter subfamily encode pathogen-inducible nitrate/nitrite transporters in grapevine and *Arabidopsis*. *Plant and Cell Physiology* 55: 162–170.

Plett, D., Garnett, T., and Okamoto, M. (2017). Molecular genetics to discover and improve nitrogen use efficiency in crop plants. In: *Plant Macronutrient Use Efficiency* (eds. M.A. Hossain, T. Kamiya, D.J. Burritt, L.-S.P. Tran, and T. Fujiwara), 93–122. London: Academic Press.

Postma, J.A., Dathe, A., and Lynch, J.P. (2014). The optimal lateral root branching density for maize depends on nitrogen and phosphorus availability. *Plant Physiology* 166: 590–602.

Prakash, V., Singh, V.P., Tripathi, D.K. et al. (2021). Nitric oxide (NO) and salicylic acid (SA): a framework for their relationship in plant development under abiotic stress. *Plant Biology* 23: 39–49.

Procházková, D. and Wilhelmová, N. (2011). Nitric oxide, reactive nitrogen species and associated enzymes during plant senescence. *Nitric Oxide* 24: 61–65. doi: 10.1016/j.niox.2011.01.005.

Quesada, A., Krapp, A., Trueman, L.J. et al. (1997). PCR-identification of a *Nicotiana plumbaginifolia* cDNA homologous to the high-affinity nitrate transporters of the crnA family. *Plant Molecular Biology* 34: 265–274.

Rahman, M., Hossain, M., and Bell, R.W. (2011). Plant density effects on growth, yield and yield components of two soybean varieties under equidistant planting arrangement. *Asian Journal of Plant Sciences* 10: 278–286.

Ribaudo, M., Delgado, J., Hansen, L. et al. (2011). Nitrogen in agricultural systems: implications for conservation policy. USDA-ERS Economic Research Report No. 127. https://ssrn.com/abstract=2115532 (accessed 22 November 2021).

Rosales, E.P., Iannone, M.F., Groppa, M.D. et al. (2011). Nitric oxide inhibits nitrate reductase activity in wheat leaves. *Plant Physiology and Biochemistry* 49: 124–130.

Roser, M. and Ritchie, H. (2013). Fertilizers. Our World in Data. https://ourworldindata.org/fertilizers (accessed 22 November 2021).

Rütting, T., Aronsson, H., and Delin, S. (2018). *Efficient Use of Nitrogen in Agriculture*. Cham, Switzerland: Springer.

Sánchez-Vicente, I. and Lorenzo, O. (2021). Nitric oxide regulation of temperature acclimation: a molecular genetic perspective. *Journal of Experimental Botany* 72: 5789–5794.

Sangoi, L. (2001). Understanding plant density effects on maize growth and development: an important issue to maximize grain yield. *Ciência Rural* 31: 159–168.

Sanz-Luque, E., Chamizo-Ampudia, A., Llamas, A. et al. (2015). Understanding nitrate assimilation and its regulation in microalgae. *Frontiers in Plant Science* 6: 899.

Sanz, L., Albertos, P., Mateos, I. et al. (2015). Nitric oxide (NO) and phytohormones crosstalk during early plant development. *Journal of Experimental Botany* 66: 2857–2868.

Saxe, H. (1986). Effects of NO, NO_2 and CO_2 on net photosynthesis, dark respiration and transpiration of pot plants. *New Phytologist* 103: 185–197. doi: 10.1111/j.1469-8137.1986.tb00607.x.

Shah, A.N. Wu, Y., Tanveer, M.et al. (2021). Interactive effect of nitrogen fertilizer and plant density on photosynthetic and agronomical traits of cotton at different growth stages. *Saudi Journal of Biological Sciences* 28: 3578–3584.

Sharma, A., Hussain, A., Mun, B.-G. et al. (2016). Comprehensive analysis of plant rapid alkalization factor (RALF) genes. *Plant Physiology and Biochemistry* 106: 82–90.

Simon, J., Dong, F., Buegger, F. et al. (2013). Rhizospheric NO affects N uptake and metabolism in Scots pine (*Pinus sylvestris* L.) seedlings depending on soil N availability and N source. *Plant, Cell and Environment* 36: 1019–1026.

Simon, J., Stoelken, G., Rienks, M. et al. (2009). Rhizospheric NO interacts with the acquisition of reduced N sources by the roots of European beech (*Fagus sylvatica* L.). *FEBS Letters* 583: 2907–2910.

Simontacchi, M., Galatro, A., Ramos-Artuso, F. et al. (2015). Plant survival in a changing environment: the role of nitric oxide in plant responses to abiotic stress. *Frontiers in Plant Science* 6: 977. doi: 10.3389/fpls.2015.00977.

Spiertz, J.H.J. (2009). Nitrogen, sustainable agriculture and food security: a review. In: *Sustainable Agriculture* (eds. E. Lichtfouse, M. Navarrete, P. Debaeke, S. Véronique, and C. Alberola) 635–651. Dordrecht, the Netherlands: Springer

Stahl, A., Friedt, W., Wittkop, B. et al. (2016). Complementary diversity for nitrogen uptake and utilisation efficiency reveals broad potential for increased sustainability of oilseed rape production. *Plant and Soil* 400: 245–262.

Sugiura, M., Georgescu, M.N., and Takahashi, M. (2007). A nitrite transporter associated with nitrite uptake by higher plant chloroplasts. *Plant and Cell Physiology* 48: 1022–1035.

Sun, C., Zhang, Y., Liu, L. et al. (2021). Molecular functions of nitric oxide and its potential applications in horticultural crops. *Horticulture Research* 8: 71. doi: 10.1038/s41438-021-00500-7.

Sun, H., Li, J., Song, W. et al. (2015a). Nitric oxide generated by nitrate reductase increases nitrogen uptake capacity by inducing lateral root formation and inorganic nitrogen uptake under partial nitrate nutrition in rice. *Journal of Experimental Botany* 66: 2449–2459. doi: 10.1093/jxb/erv030.

Sun, H., Tao, J., Zhao, Q. et al. (2017). Multiple roles of nitric oxide in root development and nitrogen uptake. *Plant Signaling and Behavior* 12: e1274480. doi: 10.1080/15592324.2016.1274480.

Swarbreck, S.M., Wang, M., Wang, Y. et al. (2019). A roadmap for lowering crop nitrogen requirement. *Trends in Plant Science* 24: 892–904.

Tal, I., Zhang, Y., Jørgensen, M. et al. (2016). The *Arabidopsis* NPF3 protein is a GA transporter. *Nature Communications* 7: 1–11.

Taochy, C., Gaillard, I., Ipotesi, E. et al. (2015). The *Arabidopsis* root stele transporter NPF2.3 contributes to nitrate translocation to shoots under salt stress. *The Plant Journal* 83: 466–479.

Tossi, V., Lamattina, L., and Cassia, R. (2009). An increase in the concentration of abscisic acid is critical for nitric oxide-mediated plant adaptive responses to UV-B irradiation. *New Phytologist* 181: 871–879.

Trueman, L.J., Richardson, A., and Forde, B.G. (1996). Molecular cloning of higher plant homologues of the high-affinity nitrate transporters of *Chlamydomonas reinhardtii* and *Aspergillus nidulans*. *Gene* 175: 223–231.

Tsay, Y.-F., Chiu, C.-C., Tsai, C.-B. et al. (2007). Nitrate transporters and peptide transporters. *FEBS Letters* 581: 2290–2300.

Tsay, Y.-F., Schroeder, J.I., Feldmann, K.A. et al. (1993). The herbicide sensitivity gene CHL1 of *Arabidopsis* encodes a nitrate-inducible nitrate transporter. *Cell* 72: 705–713.

Ullah, A., Akbar, A., Luo, Q. et al. (2019). Microbiome diversity in cotton rhizosphere under normal and drought conditions. *Microbial Ecology* 77: 429–439.

Urriola, J. and Rathore, K.S. (2015). Overexpression of a glutamine synthetase gene affects growth and development in sorghum. *Transgenic Research* 24: 397–407.

Wajid, N., Ahmad, A., Bano, A. et al. (2012). Effect of nitrogen on yield and oil quality of sunflower (*Helianthus annuus* L.) hybrids under sub humid conditions of Pakistan. *American Journal of Plant Sciences* 3: 243–251.

Wang, W., Hu, B., Yuan, D. et al. (2018). Expression of the nitrate transporter gene OsNRT1.1A/OsNPF6.3 confers high yield and early maturation in rice. *The Plant Cell* 30: 638–651.

Wang, Y.-Y. and Tsay, Y.-F. (2011). Arabidopsis nitrate transporter NRT1.9 is important in phloem nitrate transport. *The Plant Cell* 23: 1945–1957.

Williams, L.E. and Miller, A.J. (2001). Transporters responsible for the uptake and partitioning of nitrogenous solutes. *Annual Review of Plant Biology* 52: 659–688.

Xia, X., Fan, X., Wei, J. et al. (2015). Rice nitrate transporter OsNPF2.4 functions in low-affinity acquisition and long-distance transport. *Journal of Experimental Botany* 66: 317–331.

Yanagisawa, S. (2014). Transcription factors involved in controlling the expression of nitrate reductase genes in higher plants. *Plant Science* 229: 167–171.

Yang, X., Xia, X., Zhang, Z. et al. (2017). QTL mapping by whole genome re-sequencing and analysis of candidate genes for nitrogen use efficiency in rice. *Frontiers in Plant Science* 8: 1634.

Zafra, A., Rodríguez-García, M.I., and Alché Jde, D. (2010). Cellular localization of ROS and NO in olive reproductive tissues during flower development. *BMC Plant Biology* 10: 36. doi: 10.1186/1471-2229-10-36.

Zhang, J., Liu, Y.X., Zhang, N. et al. (2019). NRT1.1B is associated with root microbiota composition and nitrogen use in field-grown rice. *Nature Biotechnology* 37: 676–684.

Zhang, J., Tong, T., Potcho, P.M. et al. (2020). Nitrogen effects on yield, quality and physiological characteristics of giant rice. *Agronomy* 10: 1816.

Zhou, J.-J., Theodoulou, F.L., Muldin, I. et al. (1998). Cloning and functional characterization of a *Brassica napus* transporter that is able to transport nitrate and histidine. *Journal of Biological Chemistry* 273: 12017–12023.

6

Regulation of Plant Growth by Microbe-Assisted Nitric Oxide Production

Sagar Bag, Anupam Mondal, and Avishek Banik

Laboratory of Microbial Interaction, School of Biotechnology, Presidency University, Kolkata, West Bengal, India

6.1 Introduction

Agricultural farm workers face considerable yield losses due to a variety of infectious diseases. Different environmental stresses caused by microorganisms on agricultural plants diminish crop productivity and crop quality. Crop-associated diseases are caused by numerous stresses and these plant diseases must be handled in order to sustain the amount of food supplied by farmers across the entire globe (Khan et al. 2021). Plants are normally stressed in the regular biological system because of a diverse range of biotic factors (microbes and infectious agents) and abiotic factors (salt, heat, heavy metals, and cold). These stresses can influence ordinary development, advancement, and yield, thus influencing worldwide agricultural productivity. A fundamental research topic that has arisen in the context of environmental modifications and nutritional sustainability is how certain plant species detect and respond to such signals. Because of both abiotic and biotic stress conditions, plants go through several changes via regulation of gene expression and other physiological modifications. Plants often respond to abiotic stress conditions by generating reactive oxygen species (ROS). ROS are conceivably poisonous, being either beneficial or detrimental to plant tissues that rely mainly upon fixation (Miller et al. 2010). ROS disrupt cell redox equilibrium and generate oxidative damage in plant species. Furthermore, overmanufacture of ROS can harm different cell lines and result in necrotic damage throughout plants (Petrov et al. 2015). When ROS are produced, NO interacts with numerous signaling additives as a way to maintain a constant ROS concentration (Vranová et al. 2002). Generally, exogenous or endogenous NO plays a crucial role in ROS detoxification and enhances antioxidant responses by modifying several protein functions and gene transcription in conditions where ROS production is enormously increased (Nabi et al. 2019). The importance of NO as a signal molecule in several plant species was realized in the last decade of the twentieth century and it has since been linked to a wide range of activities. NO is a reactive and crucial gaseous molecule, which is found in the troposphere region and is involved in a series of environmental chemical methods that have extensive effects on ozone concentration in the troposphere region, formation of acid rain, aerosol manufacturing, and the nitrogen cycle (Fowler et al. 2009). Despite the fact that NO was initially thought of as a major air contaminant, it has subsequently been demonstrated to be associated with different signaling activities and numerous physiological events like leaf senescence, seed germination, and environmental biotic and abiotic stress responses (Mur et al. 2013). NO also possesses many significant properties, acting as a signaling messenger molecule during stress conditions (the presence of small redox molecules, free radicals), and it is easily diffused through the plasma membrane. NO is also engaged in a wide range of biologically significant activities in plant species, including organogenesis, stress tolerance, and immunology (Wurm and Lindermayr 2021). NO additionally has a function in the mitochondrial electron transport chain modulating ROS to activate defense mechanisms in plants via improved antioxidant production when subjected to a variety of abiotic stress conditions (Zottini et al. 2002). Cellular gene expression and protein functions are altered by NO when it interacts indirectly and directly with a diverse range of targets and phenotypic responses are also altered. Stress responses that are mediated by NO and the fundamental components have also been widely analyzed by many researchers (Yu et al. 2012); however, with different outcomes found because of the utilization of various approaches (Begara-Morales et al. 2014; Ahmad et al. 2016). In this chapter, we will discuss the biological generation of NO in plant species and the microbial regulation of NO production under different stress conditions (biotic and abiotic). We will also offer an update on NO signaling in plants and interrelations with microorganisms, and current information available regarding this discussion.

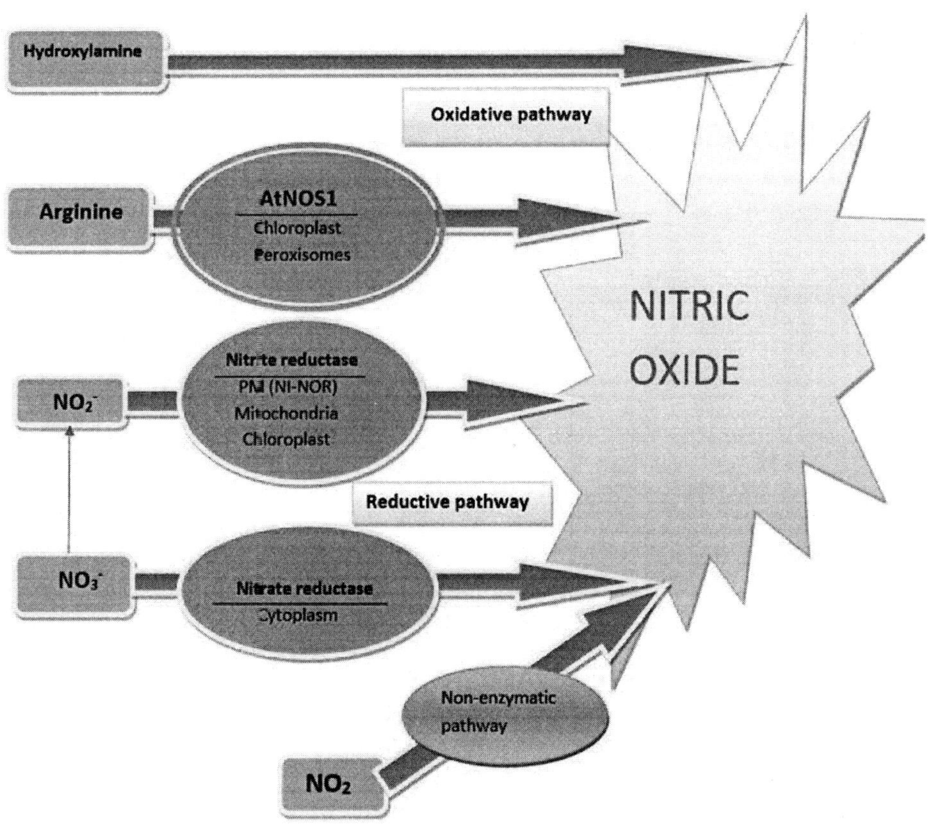

Figure 6.1 Sources and different pathways related to NO production in plants.

6.2 Sources and Production of NO

The origin of NO in plants is far from a clear picture, and several probable locations for its incorporation and synthesis have been proposed (Mur et al. 2013). NO is found in the soil, plants, and bacterial species, and this valuable molecule helps to trigger many biochemical pathways during plant–microbial interactions. Agricultural soils are the primary source of NO production (Figure 6.1). The level of NO produced from soil globally is in the range of 8.9 Tg N per year (IPCC 2007). Its production depends on various factors like moisture content, temperature, nitrogen content, and soil pH.

6.2.1 Biosynthesis of NO in Plants

In plants, NO synthesis seems to be a more difficult process because no enzyme has been recognized with similar functions. In plants, NO biosynthesis occurs by reductive and oxidative pathways. Oxidative pathways are carried out by oxidation of polyamines, arginine, and hydroxylamine. Reductive pathways involve the reduction of cytochrome c oxidase/reductase, xanthine oxidoreductase, and nitrate reductase (NR). In plants, NO production occurs in the peroxisome, chloroplast, cytoplasm, apoplast, mitochondria, and plasma membrane. In chloroplast, NO synthesis occurs by nitrite reduction (NiR) where NO_2^- is transformed into NO. Electron transport system (ETS) based NiR takes place in mitochondria using donor molecule reduced nicotinamide adenine dinucleotide phosphate (NADPH) and substrate nitrite to generate NO (Nabi et al. 2019). Synthesis of NO in plants takes place by two important pathways, i.e. the oxidative pathway and the reductive pathway.

6.2.1.1 Oxidative Pathway

6.2.1.1.1 NOS Enzyme

NO is derived from L-arginine in reactions catalyzed by nitric oxide synthase (NOS) in mammals (Stuehr et al. 2004; Daff 2010). In this process, co-substrate oxygen, electron donor NADPH, and some cofactors such as the reduced form of flavin adenine dinucleotide (FADH), tetrahydopterin, and flavin mononucleotide (FMN) are required. In animals, three isoforms of NOS have been found: inducible (iNOS), endothelial (eNOS), and neuronal (nNOS) (Nabi et al. 2019). nNOS and eNOS are expressed constitutively in animal cells, whereas the iNOS form is produced under different stress conditions to increase biosynthesis of NO. However, there is a vigorous controversy about the occurrence of standard NOS in plants. A few works have recognized a NOS-like activity in plants, utilizing different cofactors and inhibitors from animal NOS (Astier et al. 2018). Despite the analysis and sequencing of numerous plant genomes, there is no significant evidence of real NOS in higher plant species (Jeandroz et al. 2016). Recently, in *Synechococcus* PCC 7335 an additional globin domain of NOS at the *N*-terminal region has been identified (Correa-Aragunde et al. 2018). The structure and arrangement of plant NOS are entirely distinct from those of the animal NOS and the NOS-like actions are thought to be involved with a variety of physiological and stress-associated activities in higher plants (Begara-Morales et al. 2020).

6.2.1.1.2 Polyamines (PAs)

PAs have been suggested to be capable of generating NO from the basic amino acid, arginine (Tun et al. 2006). Principal PAs are spermine, spermidine, and putrescine, which are produced from L-arginine. PAs are present in all living organisms and have a function in the regulation of several plant-related processes. PAs tend to participate in different stress-mediated responses and environmental alterations (Wimalasekera, Tebartz et al. 2011).

6.2.1.2 Reductive Pathway

This is perhaps the best characterized mechanism for NO synthesis in plants. In this reductive pathway, NO_2^- can be transformed to NO by several enzymatic pathways (NR, nitrite:NO reductase [NiNOR], NOS-like enzymes, and xanthine oxidoreductase [XOR]) and nonenzymatic pathways.

6.2.1.2.1 NR Pathway

This pathway for NO biosynthesis is the best-known enzymatic pathway. NR has both membrane-bound and cytosol free forms. Its main function is the transformation of NO_3^- to NO_2^- (Moreau et al. 2010). However, several in vitro and in vivo investigations have revealed that NO generation by NR is associated with low O_2 levels and intracellular pH as the most critical factors for NR activity (Prochazkova et al. 2014). A number of studies have been conducted to investigate the role of NR in the formation of NO during flowering, the growth of lateral roots, and most significantly during different environmental stresses (Moreau et al. 2010).

$$NO_3^- + 2H^+ + e^- = NO_2^- + H_2O$$

$$NO_2^- + 2H^+ + e^- = NO + H_2O$$

6.2.1.2.2 NiNOR Pathway

In *Nicotiana tabacum*, nitrite is converted to NO through an enzymatic pathway involving membranous NiNOR (Prochazkova et al. 2014). Its molecular mass is greater than NR (Hayat et al. 2010). Cytochrome c is required as an electron donor for the synthesis of NO by NiNOR (Wilson et al. 2008).

6.2.1.2.3 NOS-like Enzymes and XOR-Mediated Pathway

L-arginine is converted into L-citrulline and NO by NOS-like enzymes. These NOS-like enzymes require oxygen and NADPH for this deamination reaction (Rümer et al. 2009). A gene, *AtNOS1*, was identified from the genome of *Arabidopsis. thaliana*, which coded for NOS-like proteins, participating in hormonal signaling and growth of plants (Guo et al. 2003). XOR is a Mo-containing, ubiquitous enzyme. There are two isoforms of this enzyme: xanthine oxidase and xanthine dehydrogenase, both of which are convertible to one another (Prochazkova et al. 2014). XOR accelerates NO production under hypoxic conditions, offering an alternative pathway for NO synthesis whenever NOS-like enzymes are not accessible (Cantu-Medellin and Kelley 2013). Furthermore, xanthine dehydrogenase controls the formation and scavenging of ROS in plant species under varied environmental stresses (Hofmann 2016).

6.2.1.3 Other Enzymatic Pathways

Several investigations have demonstrated that NO may also be produced via the activity of polyamine oxidase, cytochrome P450, horseradish peroxidase, and catalase enzymes, and enzymatic generation of NO can also mediated by hemoglobin (Wimalasekera, Villar et al. 2011). For example, the oxidation reaction of NOHA (*N*-hydroxy arginine) to generate NO is catalyzed by cytochrome P450, which requires oxygen and NADH as cofactors (del Río et al. 2004).

6.2.1.4 Nonenzymatic Pathways

In several plant species, NO is also produced by multiple nonenzymatic pathways (Astier et al. 2018). In the ascorbate (AsA) mediated pathway, NO_2^- is reduced and forms DHA (dehydroascorbic acid) and NO (del Río et al. 2004). In addition, carotenoids may convert NO_2 to NO via a nonenzymatic process similar to that reported in spinach (Cooney et al. 1994).

6.2.2 NO Production by Microbes

Because NO is now established as a crucial signaling agent in plant studies, an important task is to illustrate the different procedures of NO production by microbes. Most of the NO is discharged as a by-product of energy metabolic processes performed by bacterial species. A few bacteria additionally make NO for cellular signaling functions. In soil, different biological transformation methods like nitrification and denitrification are carried out by nitrifying bacteria (*Nitrosomonas* sp., *Nitrosococcus* sp., *Nitrobactor* sp., etc.) and denitrifying bacteria (*Pseudomonas* sp.).Those transformation processes are the major origin of NO. NO and N_2O are formed as by-products in an autotrophic aerobic nitrification method through the transformation of ammonia to nitrate. In the absence of oxygen (anaerobic conditions), those reactions are reversed and produce NO. In denitrification, nitrite is converted into N_2 with nitrous oxide, and NO as reaction intermediate products. This procedure is carried out by nitrous oxide reductase, nitric oxide reductase, nitrite reductase, and NR. The denitrification process can happen both in the presence of O_2 and also in the absence of O_2, and NO production occurs best under anaerobic conditions (Figure 6.2).

Additional procedures related to production of NO are dissimilatory nitrate reduction to ammonium (DNRA), anammox, chemo-denitrification, co-denitrification, and NOS (Medinets et al. 2015). Chemo-denitrification is regarded as a critical step in the generation of NO. Nonenzymatically, nitrite/nitrate is transformed to N_2 by this mechanism. Chemo-denitrification generally requires reduced metals, ammonium ions, and high carbon and moisture content of the soil. DNRA is also noted as a crucial route for utilization of nitrate in terrestrial ecosystem, wetlands, and coastal regions (Giblin et al. 2013).

6.3 NO: A Signaling Molecule

NO functions as a vital signaling molecule in various physiological events during nutrient assimilation and development of plants. Much evidence has suggested the involvement of NO in the growth of plant roots (Fernández-Marcos et al. 2011; Wang et al. 2013). A study has suggested that during lateral root formation, NO and the plant hormone auxin are

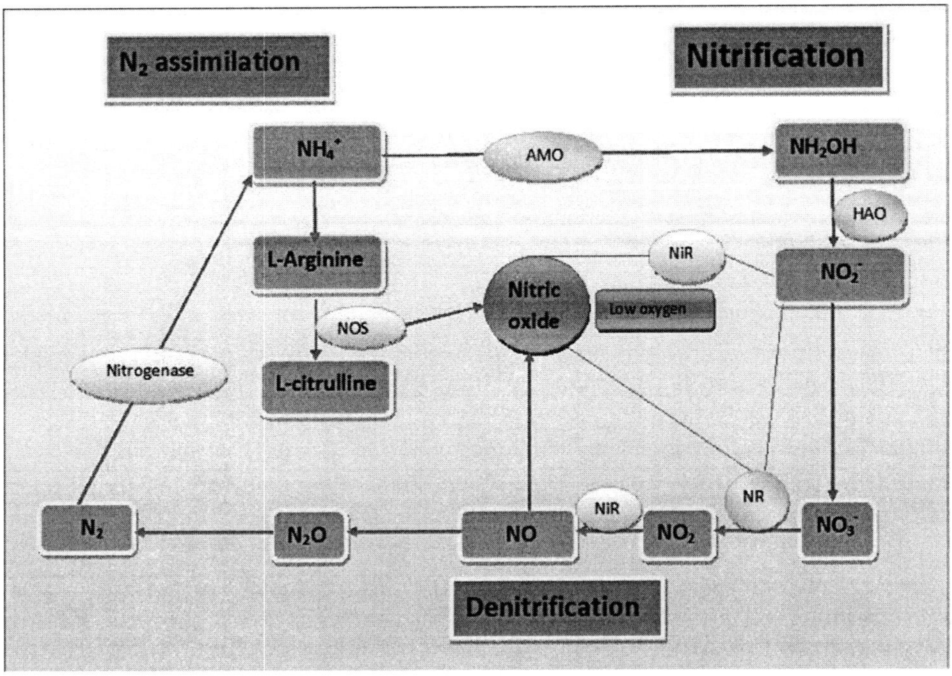

Figure 6.2 Production of NO by microbes via different pathways (NiR, nitrite reductase; HAO, hydroxylamine oxidoreductase; NR, nitrate reductase; AMO, ammonia monooxygenas; NOS, nitric oxide synthase).

complexed together in a signaling pathway in tomato (Correa-Aragunde et al. 2004). Furthermore, owing to its neutral properties, small size, and the easily diffusible characteristics of reactive nitrogen species (RNS), NO is considered an effective signaling agent with an ability to interact with phytohormones, such as auxin and abscisic acid, thus restricting programmed cell death and reducing harmful consequences of stress (Nabi et al. 2019). NO and its associated compounds mainly target chloroplast and mitochondrial protein complexes (Palmieri et al. 2010); thiols, amine-containing nucleic acids, lipoproteins, and lipid radical in the membrane (Rubbo 2000); and the TIR1 receptor of auxin (Terrile et al. 2012). NO is an uncharged and diatomic gaseous molecule with a single unpaired electron. At higher concentrations, NO is very toxic. However, at lower concentration, in eukaryotes, NO is a crucial signaling molecule (Bian et al. 2008). NO performs as a modulator via stimulating guanylate cyclases, cycloxygenases, and ADP ribosyl transferases proteins. Nitric oxide can also repress the action of H^+-ATPase, aconitase, lipoxygenase, ribonucleotide reductase, etc. In plants, the occurrence and action of NOS has been described in numerous systems by blot analysis using immunoglobulin against NOS of animal cells and by observing the transformation of labeled arginine to citrulline (Tuteja et al. 2004). NO is both water soluble and lipid soluble, but NO is very reactive and simply converted into other N_2 oxides by oxidation. NO reacts with superoxide and produces peroxynitrite and also reacts with other components like transition metals and thiol groups to produce *S*-nitrosothiols. Several studies have revealed that the signaling molecule NO activates different gene defense mechanisms and also acts to regulate many developmental processes in plants. However, in plants, it was reported that a NAD(P)H based reduction reaction occurs for the transformation of NO^{3-} to NO by NR (Tuteja et al. 2004). Following microbial infection, the NO synthesis machinery, NOS, and other NO-producing enzymes are activated, according to extensive investigations on microbial signaling, with an increasing concentration of NO leading to the switching on of guanylate cyclase activity. NO promotes leaf extension, root growth, seed germination, and inhibits fruit maturation and leaf senescence. Many genes such as phenylalanine ammonia lyase, pathogen-related genes, and chalcone synthase associated with defense mechanisms are also activated by NO (Romero-Puertas and Delledonne 2003) (Romero-Puertas et al., 2004). Exogenous NO elevates growth of different plant parts, protecting them from oxidative destruction by adjusting ROS metabolism. In conjugation with ROS, NO is also important for acquired systemic resistance by modulating the Nonexpresser of PR genes1 (NPR1) (TGA1) is a member of bZIP transcription factor system (Mittler et al. 2011). Additionally, *S*-nitrosoglutathione reductase (GSNOR) and *S*-nitrosoglutathione (GSNO), a substrate of GSNOR, perform a vital function in different environmental abiotic and biotic stress responses (Wünsche et al. 2011).

6.4 NO in Response to Abiotic Stress

Several abiotic stresses like temperature, flood, metal toxicity, and soil problems (alkalinity and salinity) create a serious risk to sustainable production of agricultural crops (Banik et al. 2018). These stress conditions are also reported to influence the spread and distribution of insects, weeds, and pathogens (Peters et al. 2014). They also affect the interactions between plant and pest by modifying defense responses and physiology of plants (Scherm and Coakley 2003; Banik et al. 2019a).

6.4.1 Heavy Metal Stress

Heavy metals are a group of metalloids and metals that are harmful to plant survival at lower concentrations. A few heavy metal micronutrients are essential for normal healthy development and growth in plant species. Most of the heavy metals are not essential for any physiological function in plants (Ghosh et al. 2016; Naskar et al. 2016). At higher concentrations, accumulated heavy metals inhibit growth, create toxic effects in plants, and change metabolic activity. Heavy metal stress can also have negative impacts on photosynthesis, antioxidant activity, cell differentiation, and cell division (Banik et al. 2018; Purushanahalli Shivagangaiah et al. 2021). Heavy metal uptake and transport in plant species are controlled by several substances like metal ion exchangers and crosslinking agents. These specialized transporters are found in the cell membrane of the root and are necessary for heavy metal ion assimilation (Yan et al. 2020; Chandra and Banik 2021). Studies have shown that in *Arabidopsis*, root length is altered (about 38%) by heavy metal stress (Corpas et al. 2017). Several studies have investigated the impact of several heavy metal stresses in plants and the recovery mechanisms of plants (He and Chen 2014; Kolbert et al. 2017). Investigations have also reported that NO is associated with different important mechanisms like relaxation of the cell wall, safeguarding the lipid bilayer, enhancing expansion of the cell wall, and improving plant health in general (Seabra and Oliveira

2016). Endogenous NO is accumulated in response to heavy metals and varies with specificity, condition, and type of metals. In *Arabidopsis*, overproduction of NO due to lead stress influences various reactions that are found in the peroxisomes and lateral and primary root lengths (Kolbert et al. 2017). Studies have found contradictory consequences of endogenous nitric oxide production and metal toxicity in various plant species. For example, in the roots of rose mallow seedlings, there was a sudden decrease in endogenous nitric oxide levels after aluminum stress (Tian et al. 2007). In contrast, endogenous NO levels quickly increased in soybean after 72 h under heavy metal stress (Kopyra et al. 2006). In mung bean and rice, arsenic-induced toxicity is minimized by exogenous NO application by decreasing malondialdehyde (MDA) levels and ROS (Singh et al. 2009; Ismail 2012). Cadmium exhibited harmful effects on the roots of plants with loss of productivity and reduced nutrient uptake (Hédiji et al. 2015; Banik et al. 2018). During cadmium stress, NO is thought to induce apoptosis via activation of stress-induced protein kinase and mitogen-activated protein kinase (MAPK) (Ye et al. 2013).

6.4.2 Salt Stress

An overall rise in soil salinity disrupts agricultural production and growth of plants (Zhu 2003). Arable (over 20%) land and irrigated land (approximately 50%) are influenced by the effects of excessive salinity (Shrivastava and Kumar 2015). Osmotic balance and ion homeostasis of plants are altered by salt stress. Furthermore, various physiological and biochemical processes of plants are inhibited by higher saline conditions (Munns and Tester 2008). Salt stress can have a substantial impact on the organization of the soil microbial populations, metabolic activities, and the neighboring micro-environment (Wang et al. 2021). Studies have shown that in maize NO releases chitosan nanoparticles (CSNPs), which reduce the harmful effect of salinity (Oliveira et al. 2016). In addition, several studies also reported an expansion in the dry weight after application of NO in seashore mallow seedlings and maize under saline stressed conditions (Guo et al. 2009). Sodium nitroprusside (SNP), an important NO donor, has been found to reduce osmotic stresses and enhances the growth of seedling under stressed (salt) conditions in different plant species including cucumber, rice, and lupin (Fan et al. 2013). Moreover, SNP-treated plants created more biomass in stressed conditions demonstrating a positive function of NO under salt pressure (Adamu et al. 2018). Nabi et al. investigated the function of NO in saline stress alleviation in various rice genotypes. They found a remarkable rise in the expression of *OsGR1*, *OsP5CS2*, and *OsHIPP38* in SNP-treated seedlings of rice genotypes under salt stress (Nabi et al. 2019). During stress conditions, NO is involved in decreasing ROS levels by increasing enzymatic antioxidant activities and activates the expression of defense genes stimulated by cyclic GMP (cGMP). Protein posttranslational alterations like *S*-nitrosylation and tyrosine nitration are mediated through the action of NO and enhance the activity of enzymes related to stress tolerance (Khan et al. 2015).

6.4.3 Drought Stress

Drought is one of the main constraints on food production because it restricts the yield of crops. Drought has become one of the most critical plant stresses, triggering a variety of biochemical, morphological physiological, and molecular modifications in the plant. Plant growth, development, and crop production are negatively influenced by drought stress.

Drought's consequences in agriculture are exacerbated by the loss of water supplies and the rising food demand caused by the world's rapid population expansion. The unpredictability of drought is due to a number of variables, including an unequal and irregular rainfall distribution, evaporation and transpiration, and water storage capacity across the rhizosphere region (Seleiman et al. 2021). Plants have many strategies to minimize drought stress such as altering the permeability of the cuticle, forming a waxy coating or spongy material, modification of the root system to increase water supply, and sunken stomata to reduce evaporation of water. Many of these modifications due to insufficient water conditions require the involvement of NO and often a symbiotic relationship with root-associated microbes that leads to root architecture modifications (Creus et al. 2005; Banik et al. 2019b). In addition, NO has been identified as a key element of drought tolerance of a variety of plant species by improving antioxidant mechanisms and metabolism of ROS, osmolytes, and proline (Shi et al. 2014). NO may potentially operate as a mediator to prevent loss of water through abscisic acid (ABA)-induced stomatal responses via different signaling mechanisms, for example, MAPK, cGMP, and Ca^{2+}. Also at the molecular scale, under drought stress, plants show DNA methylation activities and undergo genome modifications. These cellular modifications are affected by NO as can be seen with the exogenous implementation of SNP (Fan et al. 2012).

6.4.4 Heat Stress

As global warming continues, it is becoming increasingly crucial to investigate how plant species are adapting to such climatic changes. High temperatures have a deleterious effect on plant growth and development. Heat stress caused by global warming will have an impact on plant health. Excessive heat has an impact on plant reproduction and phases of floral meristem formation and progression (Jagadish et al. 2021). Heat stress occurs due to excessive generation of ROS (Kotak et al. 2007). In various plants, there is a rise in NO concentration due to the treatment of the plant species with high temperatures (Yu et al. 2014). During heat stress, NO accumulation is mainly induced by nitrite-dependent and NOS pathways. Under heat stressed conditions, NO concentrations are also modulated by several plant hormones. NO might potentially be produced by polyamines and also by GSNO reductase (GSNOR) by regulating the action of GSNO (Parankusam et al. 2017).

6.5 Impression of Microorganisms on Plant Growth Under Abiotic Stress

Beneficial microbes can protect plant species from different environmental stresses by utilizing features such as phosphate solubilization, phytohormones, biological N_2 fixation, siderophore synthesis, and synthesis of numerous volatile compounds (Rastogi et al. 2021). Plant growth-promoting microbes (PGPMs) such as fungi and bacteria stimulate plant development and enhance tolerance against diverse environmental stresses. Through a number of suggested processes, PGPMs stimulate plant development in heavy metal-polluted environments. A few PGPM species of *Rhizobium* and *Pseudomonas* perform critical roles in maintaining plant health under environmental stressful conditions (Sepehri et al. 2021).

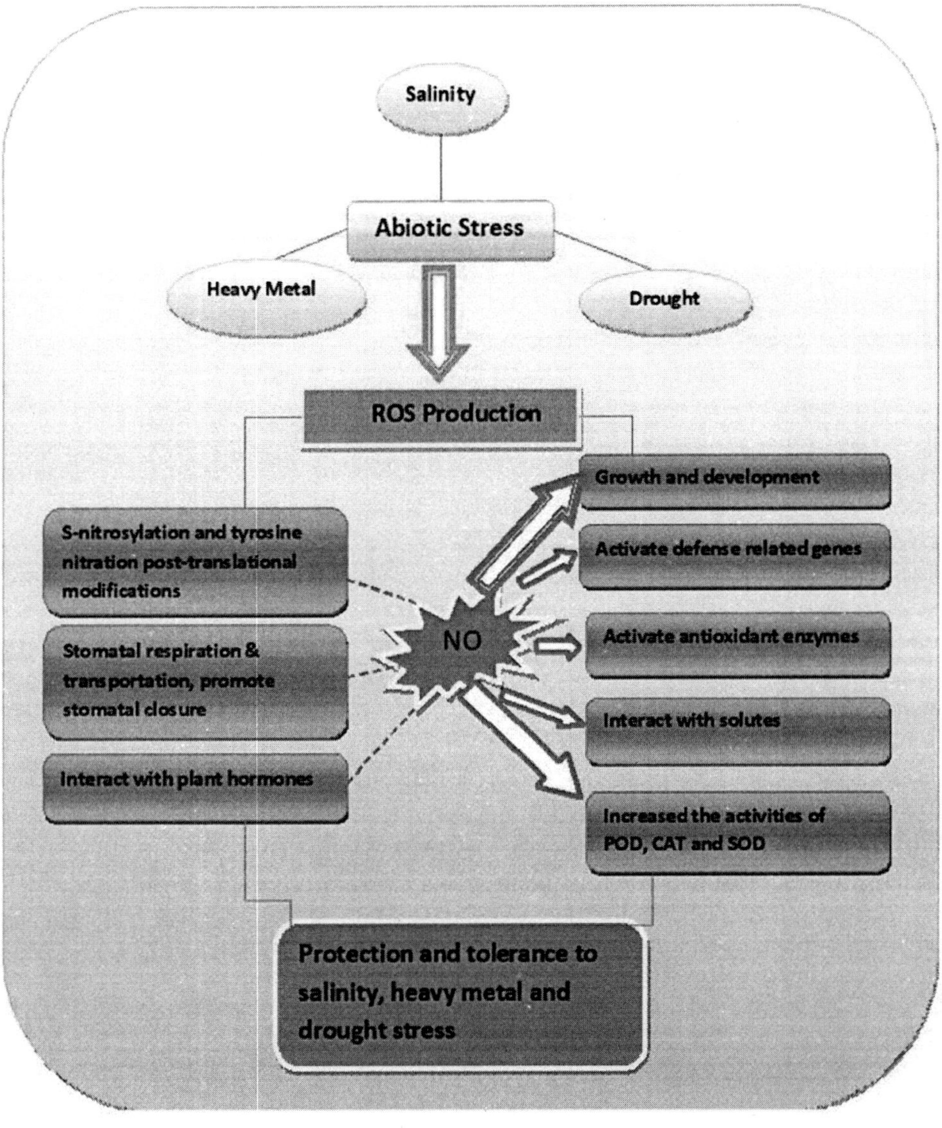

Figure 6.3 A simplified model of the function of NO in plants against various abiotic stress responses.

6.6 NO in Response to Biotic Stress

As with abiotic stresses, plants often face many biotic stresses sequentially, such as infections by pathogens to cause various disease symptoms. Biotic stress can also occur as a result of the damage done by living microorganisms such as fungus, weeds, harmful and beneficial insects, bacteria, and wild or domesticated crop plants (Newton et al. 2011). Plants react to biotic stress via a defensive mechanism. This defensive mechanism is categorized as a systematic and innate response. ROS is produced during infection and pathogen spread is restricted by oxidative bursts (Atkinson and Urwin 2012). Plants also increase cell lignification in response to the pathogen attack. This system prevents the penetration of parasites and reduces the susceptibility of the host (Madani et al. 2019). NO is produced in response to pathogenic microbes through the nitrite or L-arginine mediated pathways (through cGMP) and participates in different developmental functions, possibly activating several virulence factors. Many infectious pathogens lead to the accumulation of NO and may also increase the *S*-nitrosylation status of the membrane channels or receptors of the host and facilitate the entry of NO. When a pathogenic agent invades plant parts, it damages many compartments, which leads to uncontrolled NO production. Host plants also initiate several defense mechanisms. High levels of NO also trigger nitrosative stress, simultaneously increasing the nitration status of host protein, lipid, and DNA, and facilitating tissue destruction by the pathogen (Arasimowicz-Jelonek et al. 2014) (Figure 6.4).

6.6.1 NO and Disease Resistance

To react to several infectious pathogens and unfavorable environmental circumstances, plants depend on a complex web of signal transduction routes, which result in physiological and genomic reprogramming (Valeri et al. 2020). Microbe infections may lead to a sudden increase in the NO level during plant–pathogenic associations, suggesting the participation of NO in resistance to different diseases (Malik et al. 2011; Yun 2011). Several studies have focused on the role of nitrite-dependent NO biosynthesis (NR-mediated) throughout plant–microbial interactions, and contributing NO-mediated immune responses are heavily dependent on the nitrogen nutrition of plants (Sun et al. 2020). In *A. thaliana*, *nia1* and *nia2* (NR mutant) were highly hypersensitive and defective as a result of *Pseudomonas syringae* infections, but NO treatment alleviated symptoms and restored transcription of numerous active genes involved in defensive mechanisms (Vitor et al. 2013). NO has also been implicated in a wide range of interactions between microorganisms and plants and their subsequent immune response, such as transcription regulation of several genes related to pathogenesis (Bellin et al. 2013). Enhanced NO concentration was associated with effector-mediated and pathogen-associated molecular patterns (PAMP) -mediated immunity resulting in hypersensitive expressions and apoptosis (Martínez-Medina et al. 2019a, 2019b). The proposed signaling actions of NO are precisely controlled by tyrosine nitration and *S*-nitrosation (posttranslational alterations) of certain specific proteins, which are related to different physiological systems (Yu et al. 2014). This approach is assisted by use of mutated plant lines with modified GSNOR levels displaying defective responses of the immune system, and by proteomic profiling during hypersensitive reactions showing widespread *S*-nitrosation of particular target proteins, ROS generation, signaling of hormones, and regulation of apoptosis (Romero-Puertas et al. 2008). In addition, NO has also been reported to regulate transcription of several genes associated with immune responses by modulating the function of specific transcriptional factors (SRG1, TGA1, NPR1) related to defense mechanisms (Cui et al. 2018). Particularly for SRG1 (Zn finger protein), a negative feedback circuit is triggered by *S*-nitrosation, which can inhibit the defensive responses (Cui et al. 2018) (Figure 6.4).

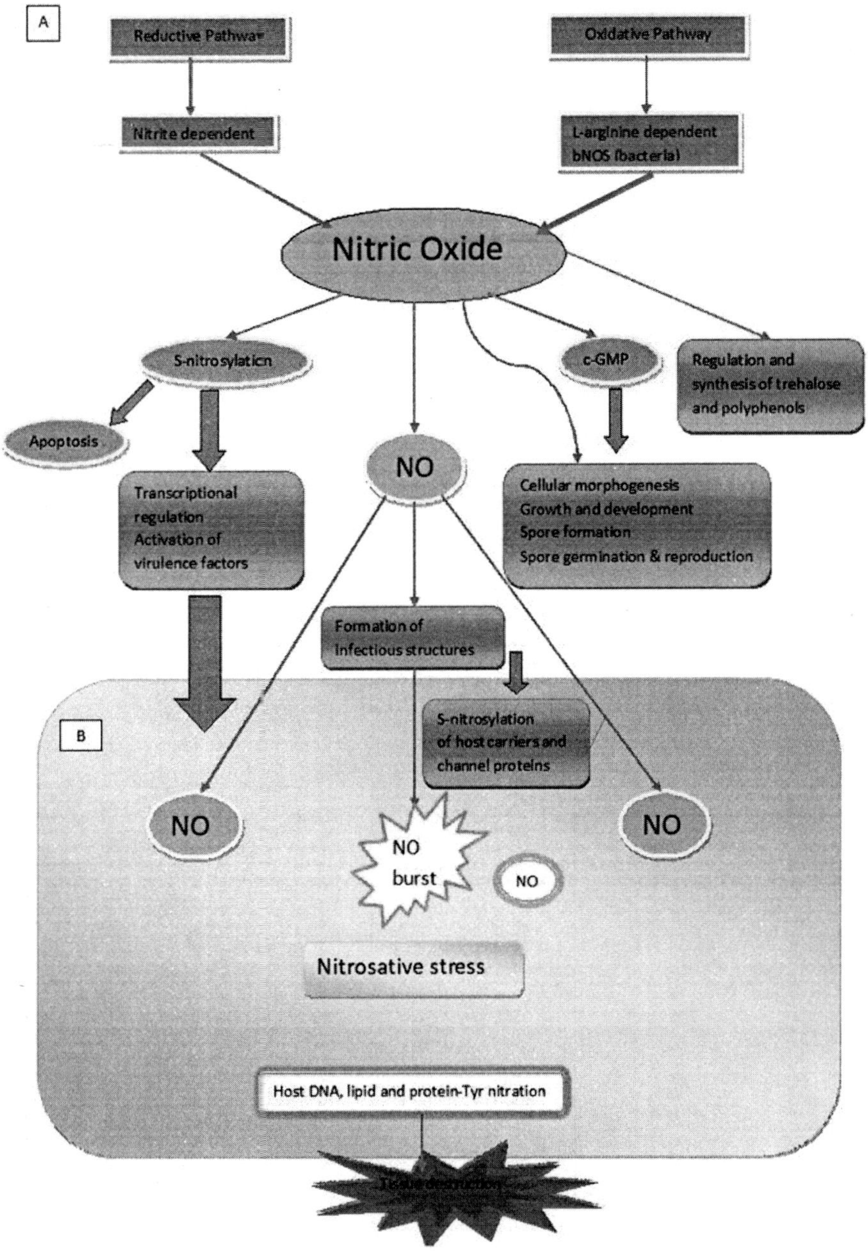

Figure 6.4 Mode of action of NO during plant–pathogen interaction. (A) Pathogenic system. (B) Plant–pathogen association.

6.6.2 ROS and NO During Pathogen Defense

NO is an essential signaling molecule in plants because it is rapidly generated, elicits specific actions inside cells, and is swiftly scavenged if no longer necessary (Manrique-Gil et al. 2021). Following pathogenic infection (during symbiotic association or herbivore attacks), ROS and NO are formed. During microbial infection, respiratory burst oxidase homolog D (RBOHD, a primary source of ROS (e.g. O^{2-}) may be blocked by the *S*-nitrosylation procedure. The interaction between NO and glutathione produces GSNO (*S*-nitrosoglutathione), which is then degraded by the action of GSNOR to produce NH_3 and Oxidized glutathione (GSSG). Defense signal pathways, which are especially dependent on the synthesis of ethylene and jasmonic acid, are influenced by NO. Protein modifications (posttranslational) are directly mediated by the action of ROS and NO or sometimes interact together to produce peroxynitrite, a reactive compound for protein posttranslational modifications probably degraded by peroxyredoxine II E (PrxII E). H_2O_2 can be obtained from O^{2-} by superoxide dismutase(SOD) and is further degraded by the action of ascorbate peroxidases (APX) to produce H_2O. Such an association is involved between ROS and NO pathways for the establishment of symbiotic relationships and defense mechanisms of plant by modulating the expression of defense genes (Martínez-Medina et al. 2019a, 2019b) (Figure 6.5).

6.7 NO and Symbiotic Interactions in Roots

Apart from its role in pathogenic interactions, NO has also been implicated in the regulation and formation of symbiotic relationships between plant species and microorganisms. N_2 fixation is a significant source of naturally accessible N_2 in agriculture and natural habitats. This symbiotic relationship leads to the development of root nodules containing a bacteroid (symbiotic type of rhizobia) that converts N_2 to NH_3. The activity and the organization of the complex symbiotic association depends on well-organized biochemical and molecular crosstalk among the symbionts, where NO has been shown to perform a number of important and diverse physiological activities (Banik et al. 2016; Berger et al. 2019). During symbiotic association, NO is generated by both bacteria and plants. In plant nodules, the best described NO-producing method is a two-step phase regulated by the mitochondrial Electron transport chain (ETC) and the activity of Nitrate reductase (NR) (Horchani et al. 2011). Furthermore, the administration

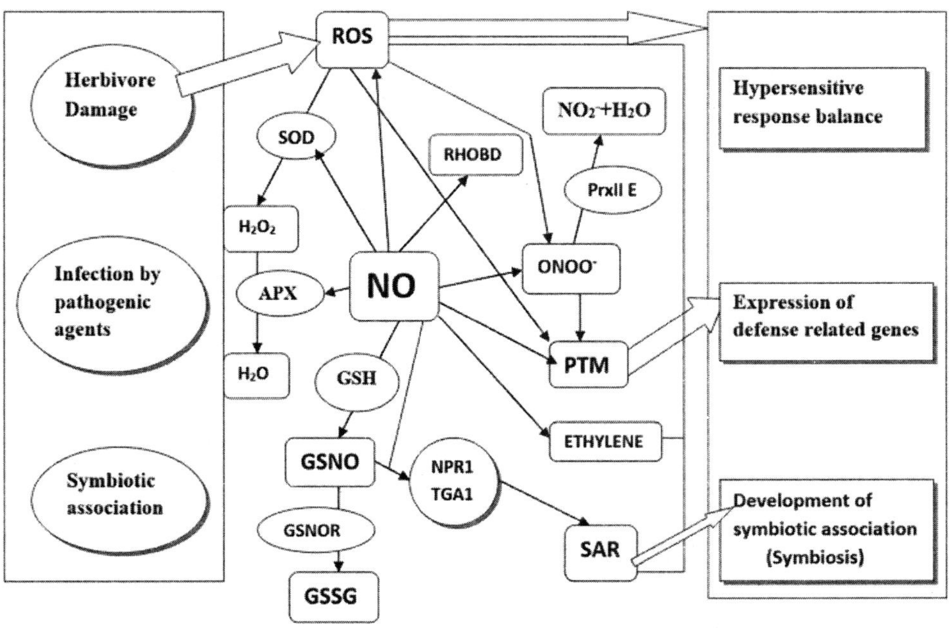

Figure 6.5 ROS and NO signaling during biotic stress.

of NG-Monomethyl-L-Arginine (L-NMMA), a powerful NOS enzyme inhibitor, to nodules revealed the possible role of NO generation from arginine in *Medicago truncatula*, *Sinorhizobium meliloti* and *Lupinus albus–Bradyrhizobium* sp. symbiotic pairs (Baudouin et al. 2006). During rhizobium symbiosis, NO in nodules has been suggested to fulfill several biochemical, physiological, and molecular functions ranging from energy equilibrium to senescence and toxicity (Berger et al. 2019). The clear line between the harmful and beneficial effects of NO during symbiosis increases the necessity for temporal and spatial regulation of production of NO in nodules, and illustrates the significance of its tighter regulations for an efficient symbiotic relationship (Berger et al. 2019). It is commonly acknowledged that NO is a signaling molecule that may readily pass through biological membranes, thus it can regulate indirectly or directly the actions of extracellular and cellular proteins. It has been suggested that not only specifically microbes generate nitric oxide, but also several enzymes from a large variety of microorganisms could affect concentrations of NO (Medinets et al. 2015) Table 6.1.

Table 6.1 Influences of microbes on plants under abiotic stress.

Stress	Microbes	Plant	Effects	References
Heavy metal (iron)	*Bacillus megaterium*	*Oryza sativa*	Increase plant growth and development	Terré et al. 2007
Heavy metal (zinc)	*Pseudomonas aeruginosa*	*Triticum aestivum*	Increase dry weight of plant and enhances nutritional uptake of N and P	Islam et al. 2014
Heavy metal (cadmium)	*Pseudomonas putida*	Mustard	Enhance shoot and root length, increased indoleacetic acid production, increased germination rate	Pal et al. 2019
Heavy metal (copper)	*Azospirillum brasilense*	Cucumber	Enhance nutritional uptake, improved root growth and development	Marastoni et al. 2019
Salt or salinity	*Trichoderma harzianum*	Mustard	Enhance antioxidant activity	Ahmad et al. 2015
Salt or salinity	Arbuscular mycorrhizal fungi	*Jatropha curcas*	Increase plant growth and enhance antioxidant activity	Kumar et al. 2010
Heat	*Azospirillum brasilense*	Wheat	Decrease generation of ROS Preactivation of transcription factors	Abd El-Daim et al. 2014
Drought and heat	*Curvularia protuberata*	*Solanum lycopersicum*	Enhance colonization of roots	de Zelicourt et al. 2013
Salt and heat	*Novosphingobium* sp.	Citrus	Enhance root growth	Vives-Peris et al. 2018
Drought	*Pseudomonas chlororaphis*	Tomato	Enhance antioxidant activity and generation of abscisic acid and indoleacetic acid	Brilli et al. 2019

6.8 ROS and NO in Plant Responses to Abiotic and Biotic Stresses

Plants are continually challenged in their natural habitat by abiotic and biotic stresses, such as drought, exposure to heavy metals, thermal stress, pathogens, and salinity. Plant responses to different sources of environmental stress include generation of RNS and ROS. In ROS production, O_2 (molecular oxygen) functions as an electron acceptor thereby triggering the aggregation of OH^-, O^{2-}, H_2O_2, and $^1O_2$ (singlet oxygen). H_2O_2 (the most stable form of ROS) is produced spontaneously by the reaction between $^1O_2$ and O_2 (Foyer and Noctor 2009). Because RNS and ROS are both powerful oxidizing agents with a great ability to destroy cellular structures, their synthesis in stress responses was thought to be particularly poisonous and damaging (Demidchik 2015). However, recent research has revealed that reactive nitrogen species (RNS) and reactive oxygen species (ROS) play a role in preventing electron transport system (ETS) deregulation or disruption at the subcellular and cellular levels (Groß et al. 2013). These crucial signaling molecules activate several physiological responses of plants via the stimulation of secondary messenger molecules, and increase enzymatic activities and activation of gene transcriptional factors (Mengel et al. 2013; Lamotte et al. 2015). For these particular reasons, the levels of RNS and ROS in cells must be strictly regulated via the scavenging process (Foyer and Noctor 2009). This is especially important since ROS and NO interact with antioxidant mechanisms in a wide range of cellular systems, including plant responses to biotic and abiotic stresses (Joudoi et al. 2013). Many studies have shown that the activities of RNS and ROS in plant species are carefully controlled by one another. RNS have been demonstrated to be engaged in the facilitation of posttranslational alterations of different enzymes like SOD (Singh et al. 2021). Furthermore, the latest reports have suggested that there is a close connection between ROS and NO, indicating a NO-mediated defense system in plants against different stress conditions. NO is a small and relatively stable molecule but can interact quickly with ROS and other radicals by generating RNS such as NO^+, NO^-, N_2O_3, and $ONOO^-$ (Hill et al. 2010). The present interpretation of the relationships of ROS, NO, and the antioxidant mechanism has previously been evaluated with an emphasis on origins, processing sites, and association with signaling compounds in plants (Del Castello et al. 2019). In *Stylosanthes guianensis*, NO triggers ABA-induced stimulation of CAT, ascorbate peroxidase, and SOD (Zhou et al. 2005). In addition, in *M. truncatula*, glutathione synthesis (a major antioxidant pathway) is modulated by NO (Innocenti et al. 2007). All of these data demonstrate a significant three-way relationship between ROS, NO, and antioxidant activity in plants.

6.9 Conclusion

Abiotic and biotic stressors are the most serious hazards to plant growth and development. These unfavorable stress conditions, such as heavy metal contamination, drought, salt, and disease infections, typically diminish agricultural yield significantly. All of these stresses may lead to excessive generation of ROS, which produces molecular and physiological damage in the nucleic acids, proteins, and lipids of the cell. NO has been recognized as an essential and key player that regulates different pathways related to cell signaling. It is now widely agreed that the signaling molecule NO has the capacity to freely diffuse across cellular membranes and it can also regulate or alter the functions of major cellular and extracellular proteins. A large variety of microbes can affect the concentration of NO by secreting extracellular enzymes. To maintain all the essential and primary functions and activities of the cell, NO maintains the inner atmosphere of the plant cell and modulates the gene expression and cell homeostasis. Intrinsically NO generation by different biosynthetic pathways, involving molecular and physiological activities in response to particular stimuli, continues to be a subject of curiosity. Multidimensional investigations will be useful for additional explorations with respect to the function of NO in plant biology. Future research into the complicated process of NO signaling in plants will likely provide promising results.

Acknowledgments

The Science and Engineering Research Board, Department of Science and Technology, Government of India, has financially supported this chapter through a Start-up Research Grant, File No. SRG/2020/000586 to Avishek Banik.

References

Abd El-Daim, I.A., Bejai, S., and Meijer, J. (2014). Improved heat stress tolerance of wheat seedlings by bacterial seed treatment. *Plant and Soil* 379 (1–2): 337–350.

Adamu, T.A., Mun, B.G., Lee, S.U. et al. (2018). Exogenously applied nitric oxide enhances salt tolerance in rice (*Oryza sativa* L.) at seedling stage. *Agronomy* 8 (12): 276.

Ahmad, P., Hashem, A., Abd-Allah, E.F. et al. (2015). Role of *Trichoderma harzianum* in mitigating NaCl stress in Indian mustard (*Brassica juncea* L) through antioxidative defense system. *Frontiers in Plant Science* 6: 868.

Arasimowicz-Jelonek, M. and Floryszak-Wieczorek, J. (2014). Nitric oxide: an effective weapon of the plant or the pathogen? *Molecular Plant Pathology* 15 (4): 406–416.

Astier, J., Gross, I., and Durner, J. (2018). Nitric oxide production in plants: an update. *Journal of Experimental Botany* 69 (14): 3401–3411.

Atkinson, N.J. and Urwin, P.E. (2012). The interaction of plant biotic and abiotic stresses: from genes to the field. *Journal of Experimental Botany* 63 (10): 3523–3543.

Banik, A., Chattopadhyay, A., Ganguly, S. et al.(2019a). Characterization of a tea pest specific *Bacillus thuringiensis* and identification of its toxin by MALDI-TOF mass spectrometry. *Industrial Crops and Products* 137: 549–556.

Banik, A., Dash, G.K., Swain, P. et al. (2019b). Application of rice (*Oryza sativa* L.) root endophytic diazotrophic *Azotobacter* sp. strain Avi2 (MCC 3432) can increase rice yield under green house and field condition. *Microbiological Research* 219: 56–65.

Banik, A., Mukhopadhaya, S.K., and Dangar, T.K. (2016). Characterization of N_2-fixing plant growth promoting endophytic and epiphytic bacterial community of Indian cultivated and wild rice (*Oryza* spp.) genotypes. *Planta* 243 (3): 799–812.

Banik, A., Pandya, P., Patel, B. et al. (2018). Characterization of halotolerant, pigmented, plant growth promoting bacteria of groundnut rhizosphere and its in-vitro evaluation of plant-microbe protocooperation to withstand salinity and metal stress. *Science of the Total Environment* 630: 231–242.

Baudouin, E., Pieuchot, L., Engler, G. et al. (2006). Nitric oxide is formed in *Medicago truncatula-Sinorhizobium meliloti* functional nodules. *Molecular Plant-Microbe Interactions* 19 (9): 970–975.

Begara-Morales, J.C., Chaki, M., Valderrama, R. et al. (2020). Nitric oxide under abiotic stress conditions. In: *Plant Life Under Changing Environment* (eds. D.K. Tripathi, and V.P. Singh, Chauhan D, Sharma S, Prasad S, Dubey NK, Ramawat N), 735–754. Academic Press.

Begara-Morales, J.C., Sánchez-Calvo, B., Chaki, M. et al. (2014). Dual regulation of cytosolic ascorbate peroxidase (APX) by tyrosine nitration and S-nitrosylation. *Journal of Experimental Botany* 65 (2): 527–538.

Bellin, D., Asai, S., Delledonne, M. et al. (2013). Nitric oxide as a mediator for defense responses. *Molecular Plant-Microbe Interactions* 26 (3): 271–277.

Berger, A., Boscari, A., Frendo, P. et al. (2019). Nitric oxide signaling, metabolism and toxicity in nitrogen-fixing symbiosis. *Journal of Experimental Botany* 70 (17): 4505–4520.

Bian, K., Doursout, M.F., and Murad, F. (2008). Vascular system: role of nitric oxide in cardiovascular diseases. *The Journal of Clinical Hypertension* 10 (4): 304–310.

Brilli, F., Pollastri, S., Raio, A. et al. (2019). Root colonization by *Pseudomonas chlororaphis* primes tomato (*Lycopersicum esculentum*) plants for enhanced tolerance to water stress. *Journal of Plant Physiology* 232: 82–93.

Cantu-Medellin, N. and Kelley, E.E. (2013). Xanthine oxidoreductase-catalyzed reduction of nitrite to nitric oxide: insights regarding where, when and how. *Nitric Oxide* 34: 19–26.

Chandra, R. and Banik, A. (2021). Detoxification and bioconversion of arsenic and chromium. In: *Nanobiotechnology, Microbes and Plant Assisted Synthesis of Nanoparticles, Mechanisms and Applications* (eds. S. Ghosh, and T. Webster), 253–270. doi: 10.1016/B978-0-12-822878-4.00016-X.

Cooney, R.V., Harwood, P.J., Custer, L.J. et al. (1994). Light-mediated conversion of nitrogen dioxide to nitric oxide by carotenoids. *Environmental Health Perspectives* 102 (5): 460–462.

Corpas, F.J., Barroso, J.B., Palma, J.M. et al. (2017). Plant peroxisomes: a nitro-oxidative cocktail. *Redox Biology* 11: 535–542.

Correa-Aragunde, N., Foresi, N., Del Castello, F. et al.(2018). A singular nitric oxide synthase with a globin domain found in *Synechococcus* PCC 7335 mobilizes N from arginine to nitrate. *Scientific Reports* 8 (1): 1–11.

Creus, C.M., Graziano, M., Casanovas, E.M. et al. (2005). Nitric oxide is involved in the *Azospirillum brasilense*-induced lateral root formation in tomato. *Planta* 221 (2): 297–303.

Cui, B., Pan, Q., Clarke, D. et al. (2018). S-nitrosylation of the zinc finger protein SRG1 regulates plant immunity. *Nature Communications* 9 (1): 1–12.

Daff, S. (2010). NO synthase: structures and mechanisms. *Nitric Oxide* 23 (1): 1–11.

de Zelicourt, A., Al-Yousif, M., and Hirt, H. (2013). Rhizosphere microbes as essential partners for plant stress tolerance. *Molecular Plant* 6 (2): 242–245.

Del Castello, F., Nejamkin, A., Cassia, R. et al. (2019). The era of nitric oxide in plant biology: twenty years tying up loose ends. *Nitric Oxide* 85: 17–27.

del Río, L.A., Corpas, F.J., and Barroso, J.B. (2004). Nitric oxide and nitric oxide synthase activity in plants. *Phytochemistry* 65 (7): 783–792.

Demidchik, V. (2015). Mechanisms of oxidative stress in plants: from classical chemistry to cell biology. *Environmental and Experimental Botany* 109: 212–228.

Fan, H., Li, T., Guan, L. et al. (2012). Effects of exogenous nitric oxide on antioxidation and DNA methylation of *Dendrobium huoshanense* grown under drought stress. *Plant Cell, Tissue and Organ Culture (PCTOC)* 109 (2): 307–314.

Fan, H.F., Du, C.X., and Guo, S.R. (2013). Nitric oxide enhances salt tolerance in cucumber seedlings by regulating free polyamine content. *Environmental and Experimental Botany* 86: 52–59.

Fernández-Marcos, M., Sanz, L., Lewis, D.R. et al. (2011). Nitric oxide causes root apical meristem defects and growth inhibition while reducing PIN-FORMED 1 (PIN1)-dependent acropetalauxin transport. *Proceedings of the National Academy of Sciences* 108 (45): 18506–18511.

Fowler, D., Pilegaard, K., Sutton, M.A. et al. (2009). Atmospheric composition change: ecosystems–atmosphere interactions. *Atmospheric Environment* 43 (33): 5193–5267.

Foyer, C.H. and Noctor, G. (2009). Redox regulation in photosynthetic organisms: signaling, acclimation, and practical implications. *Antioxidants & Redox Signaling* 11 (4): 861–905.

Ghosh, A., Adhikari, S., Ta, S. et al. (2016). Tuning of azine derivatives for selective recognition of Ag+ with the in vitro tracking of endophytic bacteria in rice root tissue. *Dalton Transactions* 45 (48): 19491–19499.

Giblin, A.E., Tobias, C.R., Song, B. et al. (2013). The importance of dissimilatory nitrate reduction to ammonium (DNRA) in the nitrogen cycle of coastal ecosystems. *Oceanography* 26 (3): 124–131.

Groß, F., Durner, J., and Gaupels, F. (2013). Nitric oxide, antioxidants and prooxidants in plant defence responses. *Frontiers in Plant Science* 4: 419.

Guo, F.Q., Okamoto, M., and Crawford, N.M. (2003). Identification of a plant nitric oxide synthase gene involved in hormonal signaling. *Science* 302 (5642): 100–103.

Hayat, S., Hasan, S.A., Mori, M. et al. (2010). Nitric oxide: chemistry, biosynthesis, and physiological role. *Nitric Oxide in Plant Physiology* 69 (14): 1–15.

He, J. and Chen, J.P. (2014). A comprehensive review on biosorption of heavy metals by algal biomass: materials, performances, chemistry, and modeling simulation tools. *Bioresource Technology* 160: 67–78.

Hill, B.G., Dranka, B.P., Bailey, S.M. et al. (2010). What part of NO don't you understand? Some answers to the cardinal questions in nitric oxide biology. *Journal of Biological Chemistry* 285 (26): 19699–19704.

Hofmann, N.R. (2016) Opposing functions for plant xanthine dehydrogenase in response to powdery mildew infection: production and scavenging of reactive oxygen species. *The Plant Cell*, 28 (5), 1001 https://doi.org/10.1105/tpc.16.00381

Horchani, F., Prévot, M., Boscari, A. et al. (2011). Both plant and bacterial nitrate reductases contribute to nitric oxide production in *Medicago truncatula* nitrogen-fixing nodules. *Plant Physiology* 155 (2): 1023–1036.

Innocenti, G., Pucciariello, C., Le Gleuher, M. et al. (2007). Glutathione synthesis is regulated by nitric oxide in *Medicago truncatula* roots. *Planta* 225 (6): 1597–1602.

Islam, F., Yasmeen, T., Ali, Q. et al. (2014). Influence of *Pseudomonas aeruginosa* as PGPR on oxidative stress tolerance in wheat under Zn stress. *Ecotoxicology and Environmental Safety* 104: 285–293.

Ismail, G.S.M. (2012). Protective role of nitric oxide against arsenic-induced damages in germinating mung bean seeds. *Acta Physiologiae Plantarum* 34 (4): 1303–1311.

Jagadish, S.K., Way, D.A., and Sharkey, T.D. (2021). Plant heat stress: concepts directing future research. *Plant, Cell & Environment.* 44 (7): 1992–2005

Jeandroz, S., Wipf, D., Stuehr, D.J. et al. (2016). Occurrence, structure, and evolution of nitric oxide synthase-like proteins in the plant kingdom. *Science Signaling* 9 (417): re2–re2.

Joudoi, T., Shichiri, Y., Kamizono, N. et al. (2013). Nitrated cyclic GMP modulates guard cell signaling in Arabidopsis. *The Plant Cell* 25 (2): 558–571.

Khan, M., Khan, A.U., Hasan, M.A. et al. (2021). Agro-nanotechnology as an emerging field: a novel sustainable approach for improving plant growth by reducing biotic stress. *Applied Sciences* 11 (5): 2282.

Khan, M.N., Mobin, M., Mohammad, F. et al. (eds.) (2015). *Nitric Oxide Action in Abiotic Stress Responses in Plants*, 51–52. Basel, Switzerland: Springer International Publishing.

Kolbert, Z., Feigl, G., Bordé, Á. et al. (2017). Protein tyrosine nitration in plants: present knowledge, computational prediction and future perspectives. *Plant Physiology and Biochemistry* 113: 56–63.

Kopyra, M., Stachoń-Wilk, M., and Gwóźdź, E.A. (2006). Effects of exogenous nitric oxide on the antioxidant capacity of cadmium-treated soybean cell suspension. *Acta Physiologiae Plantarum* 28 (6): 525–536.

Kotak, S., Larkindale, J., Lee, U. et al. (2007). Complexity of the heat stress response in plants. *Current Opinion in Plant Biology* 10 (3): 310–316.

Kumar, A., Sharma, S., and Mishra, S. (2010). Influence of arbuscular mycorrhizal (AM) fungi and salinity on seedling growth, solute accumulation, and mycorrhizal dependency of *Jatropha curcas* L. *Journal of Plant Growth Regulation* 29 (3): 297–306.

Lamotte, O., Bertoldo, J.B., Besson-Bard, A. et al. (2015). Protein S-nitrosylation: specificity and identification strategies in plants. *Frontiers in Chemistry* 2: 114.

Madani, B., Mirshekari, A., and Imahori, Y. (2019). Physiological responses to stress. In: *Postharvest Physiology and Biochemistry of Fruits and Vegetables* (eds. E.Yahia, and A. Carrillo-Lopez), 405–423. Woodhead Publishing.

Malik, S.I., Hussain, A., Yun, B.W. et al. (2011). GSNOR-mediated de-nitrosylation in the plant defence response. *Plant Science* 181 (5): 540–544.

Manrique-Gil, I., Sánchez-Vicente, I., Torres-Quezada, I. et al. (2021). Nitric oxide function during oxygen deprivation in physiological and stress processes. *Journal of Experimental Botany* 72 (3): 904–916.

Marastoni, L., Pii, Y., Maver, M. et al. (2019). Role of *Azospirillum brasilense* in triggering different Fe chelate reductase enzymes in cucumber plants subjected to both nutrient deficiency and toxicity. *Plant Physiology and Biochemistry* 136: 118–126.

Martínez-Medina, A., Pescador, L., Fernández, I. et al. (2019a). Nitric oxide and phytoglobin PHYTOGB1 are regulatory elements in the *Solanum lycopersicum–Rhizophagus irregularis* mycorrhizal symbiosis. *New Phytologist* 223 (3): 1560–1574.

Martínez-Medina, A., Pescador, L., Terrón-Camero, L.C. et al. (2019b). Nitric oxide in plant–fungal interactions. *Journal of Experimental Botany* 70 (17): 4489–4503.

Medinets, S., Skiba, U., Rennenberg, H. et al. (2015). A review of soil NO transformation: associated processes and possible physiological significance on organisms. *Soil Biology & Biochemistry* 80: 92–117.

Mengel, A., Chaki, M., Shekariesfahlan, A. et al. (2013). Effect of nitric oxide on gene transcription–S-nitrosylation of nuclear proteins. *Frontiers in Plant Science* 4: 293.

Miller, G., Suzuki, N., Ciftci-Yilmaz, S. et al. (2010). Reactive oxygen species homeostasis and signalling during drought and salinity stresses. *Plant, Cell & Environment* 33 (4): 453–467.

Mittler, R., Vanderauwera, S., Suzuki, N. et al. (2011). ROS signaling: the new wave? *Trends in Plant Science* 16 (6): 300–309.

Moreau, M., Lindermayr, C., Durner, J. et al.(2010). NO synthesis and signaling in plants–where do we stand? *Physiologia Plantarum* 138 (4): 372–383.

Munns, R. and Tester, M. (2008). Mechanisms of salinity tolerance. *Annual Review of Plant Biology* 59: 651–681.

Mur, L.A., Mandon, J., Persijn, S. et al. (2013). Nitric oxide in plants: an assessment of the current state of knowledge. *AoB Plants* 5, pls052, https://doi.org/10.1093/aobpla/pls052

Nabi, R.B.S., Tayade, R., Hussain, A. et al. (2019). Nitric oxide regulates plant responses to drought, salinity, and heavy metal stress. *Environmental and Experimental Botany* 161: 120–133.

Naskar, B., Modak, R., Sikdar, Y. et al. (2016). A simple Schiff base molecular logic gate for detection of Zn^{2+} in water and its bio-imaging application in plant system. *Journal of Photochemistry and Photobiology A: Chemistry* 321: 99–109.

Newton, A.C., Johnson, S.N., and Gregory, P.J. (2011). Implications of climate change for diseases, crop yields and food security. *Euphytica* 179 (1): 3–18.

Oliveira, H.C., Gomes, B.C., Pelegrino, M.T. et al. (2016). Nitric oxide-releasing chitosan nanoparticles alleviate the effects of salt stress in maize plants. *Nitric Oxide* 61: 10–19.

Pal, A.K., Mandal, S., and Sengupta, C. (2019). Exploitation of IAA producing PGPR on mustard (*Brassica nigra* L.) seedling growth under cadmium stress condition in comparison with exogenous IAA application. *Plant Science Today* 6 (1): 22–30.

Palmieri, M.C., Lindermayr, C., Bauwe, H. et al. (2010). Regulation of plant glycine decarboxylase by S-nitrosylation and glutathionylation. *Plant Physiology* 152 (3): 1514–1528.

Parankusam, S., Adimulam, S.S., Bhatnagar-Mathur, P. et al. (2017). Nitric oxide (NO) in plant heat stress tolerance: current knowledge and perspectives. *Frontiers in Plant Science* 8: 1582.

Prochazkova, D., Haisel, D., and Pavlikova, D. (2014). Nitric oxide biosynthesis in plants – the short overview. *Plant, Soil and Environment* 60 (3): 129–134.

Purushanahalli Shivagangaiah, C., Sanyal, D., Dasgupta, S. et al. (2021). Phycoremediation and photosynthetic toxicity assessment of lead (Pb^{2+}) by two freshwater microalgae *Scenedesmus acutus* and *Chlorella pyrenoidosa*. *Physiologia Plantarum*. doi:10.1111/ppl.13368.

Rastogi, S., Ratna, S., Said, O.B., and Kumar, R. (2021). Physiological and molecular aspects of retrieving environmental stress in plants by microbial interactions. In: *Microbes and Signaling Biomolecules Against Plant Stress* (eds. A. Sharma), 107–125. Singapore: Springer.

Romero-Puertas, M. and Delledonne, M. (2003). Nitric oxide signaling in plant-pathogen interactions. *IUBMB Life* 55 (10–11): 579–583.

Romero-Puertas, M., Perazzolli, M., Zago, E.D., and Delledonne, M. (2004). Nitric oxide signalling functions in plant–pathogen interactions. *Cellular Microbiology* (2004) 6 (9): 795–803.

Romero-Puertas, M.C., Campostrini, N., Mattè, A. et al. (2008). Proteomic analysis of S-nitrosylated proteins in *Arabidopsis thaliana* undergoing hypersensitive response. *Proteomics* 8 (7): 1459–1469.

Rubbo, H., Radi, R., Anselmi, D. et al. (2000). Nitric oxide reaction with lipid peroxyl radicals spares α-tocopherol during lipid peroxidation greater oxidant protection from the pair nitric oxide/α-tocopherol than α-tocopherol/ascorbate. *Journal of Biological Chemistry* 275 (15): 10812–10818.

Rümer, S., Gupta Kapuganti, J., and Kaiser, W.M. (2009). Oxidation of hydroxylamines to NO by plant cells. *Plant Signaling & Behavior* 4 (9): 853–855.

Seabra, A.B. and Oliveira, H.C. (2016). How nitric oxide donors can protect plants in a changing environment: what we know so far and perspectives. *AIMS Molecular Science* 3 (4): 692.

Seleiman, M.F., Al-Suhaibani, N., Ali, N. et al. (2021). Drought stress impacts on plants and different approaches to alleviate its adverse effects. *Plants* 10 (2): 259.

Sepehri, M. and Khatabi, B. (2021). Combination of siderophore-producing bacteria and *Piriformospora indica* provides an efficient approach to improve cadmium tolerance in alfalfa. *Microbial Ecology* 81 (3): 717–730.

Shi, H., Ye, T., Zhu, J.K. et al. (2014). Constitutive production of nitric oxide leads to enhanced drought stress resistance and extensive transcriptional reprogramming in Arabidopsis. *Journal of Experimental Botany* 65 (15): 4119–4131.

Shrivastava, P. and Kumar, R. (2015). Soil salinity: a serious environmental issue and plant growth promoting bacteria as one of the tools for its alleviation. *Saudi Journal of Biological Sciences* 22 (2): 123–131.

Singh, H.P., Kaur, S., Batish, D.R. et al. (2009). Nitric oxide alleviates arsenic toxicity by reducing oxidative damage in the roots of *Oryza sativa* (rice). *Nitric Oxide* 20 (4): 289–297.

Singh, K., Shukla, I., Tiwari, A.K. et al. (2021). Physiological, biochemical, and molecular mechanism of nitric oxide-mediated abiotic stress tolerance. In: *Plant Growth Regulators: Signalling under Stress Conditions* (eds. T. Aftab, and KR. Hakeem), Springer Nature Switzerland, 217.

Stuehr, D.J., Santolini, J., Wang, Z.Q. et al. (2004). Update on mechanism and catalytic regulation in the NO synthases. *Journal of Biological Chemistry* 279 (35): 36167–36170.

Sun, Y., Wang, M., Mur, L.A.J. et al. (2020). Unravelling the roles of nitrogen nutrition in plant disease defences. *International Journal of Molecular Sciences* 21 (2): 572.

Terré, S., Asch, F., Padham, J. et al. (2007). Influence of root zone bacteria on root iron plaque formation in rice subjected to iron toxicity. In: *Utilisation of Diversity in Land Use Systems: Sustainable and Organic Approaches to Meet Human Needs* (ed. E. Tielkes), 446. Göttingen: Cuvillier Verlag.

Terrile, M.C., París, R., Calderón-Villalobos, L.I. et al. (2012). Nitric oxide influences auxin signaling through S-nitrosylation of the Arabidopsis TRANSPORT INHIBITOR RESPONSE 1 auxin receptor. *The Plant Journal* 70 (3): 492–500.

Tian, Q.Y., Sun, D.H., Zhao, M.G. et al. (2007). Inhibition of nitric oxide synthase (NOS) underlies aluminum-induced inhibition of root elongation in *Hibiscus moscheutos*. *New Phytologist* 174 (2): 322–331.

Tun, N.N., Santa-Catarina, C., Begum, T. et al. (2006). Polyamines induce rapid biosynthesis of nitric oxide (NO) in *Arabidopsis thaliana* seedlings. *Plant and Cell Physiology* 47 (3): 346–354.

Tuteja, N., Chandra, M., Tuteja, R. et al. (2004). Nitric oxide as a unique bioactive signaling messenger in physiology and pathophysiology. *Journal of Biomedicine and Biotechnology* 2004 (4): 227.

Valeri, M.C., Novi, G., Weits, D.A. et al. (2020). *Botrytis cinerea* induces local hypoxia in *Arabidopsis* leaves. *New Phytologist* 229 (1): 173–185.

Vitor, S.C., Duarte, G.T., Saviani, E.E. et al. (2013). Nitrate reductase is required for the transcriptional modulation and bactericidal activity of nitric oxide during the defense response of *Arabidopsis thaliana* against *Pseudomonas syringae*. *Planta* 238 (3): 475–486.

Vives-Peris, V., Molina, L., Segura, A. et al. (2018). Root exudates from citrus plants subjected to abiotic stress conditions have a positive effect on rhizobacteria. *Journal of Plant Physiology* 228: 208–217.

Vranová, E., Atichartpongkul, S., Villarroel, R. et al. (2002). Comprehensive analysis of gene expression in *Nicotiana tabacum* leaves acclimated to oxidative stress. *Proceedings of the National Academy of Sciences* 99 (16): 10870–10875.

Wang, H., Xiao, W., Niu, Y. et al. (2013). Nitric oxide enhances development of lateral roots in tomato (*Solanum lycopersicum* L.) under elevated carbon dioxide. *Planta* 237 (1): 137–144.

Wang, M., Zhao, S., Wang, L. et al. (2021). Salt stress-induced changes in microbial community structures and metabolic processes result in increased soil cadmium availability. *Science of the Total Environment* 782: 147125.

Wilson, I.D., Neill, S.J., and Hancock, J.T. (2008). Nitric oxide synthesis and signalling in plants. *Plant, Cell and Environment* 31 (5): 622–631.

Wimalasekera, R., Tebartz, F., and Scherer, G.F. (2011). Polyamines, polyamine oxidases and nitric oxide in development, abiotic and biotic stresses. *Plant Science* 181 (5): 593–603.

Wimalasekera, R., Villar, C., Begum, T. et al. (2011). COPPER AMINE OXIDASE1 (CuAO1) of *Arabidopsis thaliana* contributes to abscisic acid- and polyamine-induced nitric oxide biosynthesis and abscisic acid signal transduction. *Molecular Plant* 4 (4): 663–678.

Wünsche, H., Baldwin, I.T., and Wu, J. (2011). S-Nitrosoglutathione reductase (GSNOR) mediates the biosynthesis of jasmonic acid and ethylene induced by feeding of the insect herbivore *Manduca sexta* and is important for jasmonate-elicited responses in *Nicotiana attenuata*. *Journal of Experimental Botany* 62 (13): 4605–4616.

Wurm, C.J. and Lindermayr, C. (2021). Nitric oxide signaling in the plant nucleus: the function of nitric oxide in chromatin modulation and transcription. *Journal of Experimental Botany* 72 (3): 808–818.

Yan, A., Wang, Y., Tan, S.N. et al. (2020). Phytoremediation: a promising approach for revegetation of heavy metal-polluted land. *Frontiers in Plant Science* 11: 359.

Ye, Y., Li, Z., and Xing, D. (2013). Nitric oxide promotes MPK6-mediated caspase-3-like activation in cadmium-induced *Arabidopsis thaliana* programmed cell death. *Plant, Cell and Environment* 36 (1): 1–15.

Yu, M., Lamattina, L., Spoel, S.H. et al. (2014). Nitric oxide function in plant biology: a redox cue in deconvolution. *New Phytologist* 202 (4): 1142–1156.

Yun, B.W., Feechan, A., Yin, M. et al. (2011). S-nitrosylation of NADPH oxidase regulates cell death in plant immunity. *Nature* 478 (7368): 264–268.

Zhou, B., Guo, Z., Xing, J. et al. (2005). Nitric oxide is involved in abscisic acid-induced antioxidant activities in *Stylosanthes guianensis*. *Journal of Experimental Botany* 56 (422): 3223–3228.

Zhu, J.K. (2003). Regulation of ion homeostasis under salt stress. *Current Opinion in Plant Biology* 6 (5): 441–445.

7

Nitric Oxide and Reactive Oxygen Species Interaction for Stress Signaling

Ester Badiani[1], Stefania Pasqualini[2], Mario Ciaffi[1], Anna Rita Paolacci[1], Agostino Sorgonà[3], and Maurizio Badiani[3]

[1]*Dipartimento per la Innovazione nei Sistemi Biologici, Agroalimentari e Forestali, Università della Tuscia, Viterbo, Italy*
[2]*Dipartimento di Chimica, Biologia e Biotecnologie, Università di Perugia, Perugia, Italy*
[3]*Dipartimento di Agraria, Università Mediterranea di Reggio Calabria, Reggio Calabria, Italy*

7.1 Introduction

Plants' interactions with the environment involve the participation and reciprocal cross-talk among several molecular routes. One of the most common plant responses to environmental challenges is the deployment of reactive nitrogen species (RNS) and of reactive oxygen species (ROS). Both are important in signaling and, by activating signaling chains, regulate several different developmental processes, through both transcriptional gene activation and influencing enzyme activity at the posttranscriptional level (Farnese et al. 2016; Patel et al. 2019).

ROS derive from the univalent reduction of molecular oxygen. They comprise the superoxide radical anion ($\bullet O_2^-$), hydrogen peroxide (H_2O_2), and the hydroxyl radical ($\bullet OH$). ROS have distinct effects on biological systems, having in common an elevated chemical reactivity and a much reduced half-life: the hydroxyl radical is among the most chemically reactive and indiscriminate species known, whereas $\bullet O_2^-$ and hydrogen peroxide have a more selective and specific chemical reactivity (del Río 2015).

As well as ROS, H_2S and its derivatives, denoted as reactive sulfur species, are important signaling molecules, not just coming into play in plants' responses to environmental stress. However, either alone or together with NO, H_2S regulates the main events during normal growth and development, but also during the plant growth developmental program, including photosynthesis, seed germination, root development, stomatal behavior, and fruit ripening (Chen et al. 2015a, 2015b; Mukherjee and Corpas 2020).

RNS describes both short-lived free radicals and nonradical species, which are able to either oxidize or reduce appropriate reaction partners or undergo transformation into radicals. Apart from NO, and mainly derived from it, other RNS include: *S*-nitrosothiols (SNOs), peroxynitrite (ONOO−), dinitrosyl-iron complexes, and higher oxides of nitrogen (NOx; Umbreen et al. 2018). NO is a gaseous, small, readily diffusible molecule reacting with several different compounds, including other radicals, such as alkyl- and lipid peroxyl radicals. It is also able to behave as an antioxidant, by interrupting spontaneous, self-fueling radical chain reactions leading to the peroxidation of lipids, so acting as a defensive agent against the cytotoxic effects caused by ROS in plants (Correa-Aragunde et al. 2015; Arora et al. 2016).

Because they can impair structures and functions in the plant cell whenever redox homeostasis is overcome, ROS and RNS generation was regarded at first as an undesirable and unavoidable consequence of life in the presence of O_2 (Demidchik 2015; Lushchak 2015). Further studies in the past few decades, however, have made it clear that these molecules are also important as signal transducers in various plant processes, providing they are kept under control by the antioxidant machinery of the aerobic cell (del Río 2015; Farnese et al. 2016).

ROS and NO, either alone or in combination are important players in many different processes of the plant cell, such as reactions to physical/chemical stress and plant immunity (Joudoi et al. 2013; Montilla-Bascon et al. 2017; Sahay and Gupta 2017; Begara-Morales et al. 2018; Hasanuzzaman et al. 2018; Sami et al. 2018; Sun et al. 2018; Corpas et al. 2019; Kaya et al. 2020; Kushwaha et al. 2019; Liu et al. 2019; Munawar et al. 2019; Saddhe et al. 2019; Sharma et al. 2019; Jedelska et al. 2020), symbiotic associations (Fukudome et al. 2016; Hichri et al. 2016; Martinez-Medina et al. 2019), promotion of high microbial diversity in the rhizospheric environment (Ma et al. 2020), organogenesis, ion homeostasis, hormonal regulation, germination, flowering, and fruit ripening/long-term storage (Freschi 2013; Mur et al. 2013b; Parí et al. 2013; Zhou and Zhang 2014; Sanz et al. 2015; Astier et al. 2018; Corpas et al. 2018; Moni et al. 2018; Del Castello et al. 2019; Kolbert et al. 2019; Steelheart et al. 2019; Raya-Gonzalez et al. 2019; Jafari and Daneshvar 2020, Khan et al. 2020; Ren et al. 2020).

It is mostly the environmental context that sets the route and the extent of the changes triggered by RNS and ROS. Because of the high complexity of these processes, much remains to do to unravel the signaling mechanisms triggered by ROS and NO, either alone or in combination, and their interactions with the plethora of other signaling molecules in plants, as affected by the overall balance among pro-oxidants and antioxidants. Many studies in the past few years have been focused on these issues, whose interest for plant scientists is still steadily growing. In this chapter, the most recent of such studies are sketchily summarized, concentrating on the adaptive responses of plants to those environmental changes able to affect the generation of ROS and RNS and their reciprocal interactions.

7.2 Plant Metabolism of ROS and Nitric Oxide: A Summary

In plants, many sources of environmental stress are able to rapidly alter the cell redox balance, both by increasing the production of ROS and NO and by inactivating the enzymes involved in antioxidant defense (del Río 2015). This increases the formation of reactive species, which is needed to activate defense/compensation responses, leading to the upregulation of antioxidant systems (Shi et al. 2014) or to the launching of a cell death program (PCD), which is important for adaptation to oxidative stress as well as for deploying resistance mechanisms (Yun et al. 2011).

ROS and NO syntheses take place at different locations in the plant cell, and their localization in time and space is important for signal generation and their intracellular spread (Groß et al. 2013). Depending on the compartment of origin, ROS and NO have distinct roles, deriving from local interactions with specific reaction partners in each organelle (Shapiguzov et al. 2012; Mur et al. 2013a); in the presence of hypoxia, for example, an important role is assigned to NO generated at the plasma membrane level, whereas stress due to heavy metals evokes NO generation in the chloroplasts and mitochondria (Kumar and Trivedi 2016).

Reactions and dynamics governing ROS production have been investigated in plant cells in response to a variety of environmental circumstances and developmental processes (see reviews by Krasensky-Wrzaczek and Kangasjärvi 2018, Ozgur et al. 2018, and Turkan et al. 2018). The plasmalemma has been found to be among the most important sites for ROS production, because of the activity of proteins belonging to the respiratory burst oxidase homolog family (RBOH), commonly denoted as reduced nicotinamide adenine dinucleotide phosphate (NADPH) oxidases. These enzymes are integral to the membrane and catalyze the transfer electrons taken from NADPH in the cytoplasm to molecular oxygen in the apoplast, forming the superoxide radical anion and causing extracellular ROS accumulation (Das and Roychoudhury 2014). Many different stresses activate the transcription and activity of NADPH oxidases, producing a so-called oxidative burst, i.e. localized ROS overproduction (Jajic et al. 2015; Wang X. et al. 2015). Other cell wall-associated oxidases and peroxidases also participate in apoplastic ROS overproduction, even if their participation in the stress response is not fully understood (Das and Roychoudhury 2014). Apart from triggering specific signaling events, in cooperation with other local signals, the oxidative burst mediated by RBOH can spread into surrounding cells, causing a far-reaching signaling cascade denoted as the ROS wave. Each cell receiving the ROS wave will activate their own ROS-generating RBOH, so propagating the wave systemically, from the primary source to the whole plant, a picture referred to as systemic acquired acclimation (Mittler and Blumwald 2015). Therefore, not just in the plant cells that were the first to perceive the stressor, but in principle in all the remaining cells in the plant, gene expression and metabolism are altered to counteract the adverse stimulus. Although the ROS wave is needed for inducing acclimation to stress, the resulting response is not always stress-specific; therefore, the main functional significance of the wave could be to alert the plant cells to produce stress-specific signals and responses (Gilroy et al. 2014).

Apart from the apoplast, mitochondria and chloroplasts can also generate ROS, as a result of the possible leakage of electrons during the flow along the electron transport chains and the consequent partial reduction of molecular oxygen (Gupta and Igamberdiev 2015). In addition peroxisomes can generate ROS as a consequence of their metabolic activity, e.g. during photorespiration, the β-oxidation of fatty acids, and the glyoxylate cycle (del Río 2015). ROS pools produced in different compartments do not necessarily remain isolated from each other, but can instead communicate with each other, contributing to metabolic regulation. An example, it has been shown that the ROS signal produced by the apoplastic oxidative burst can be communicated to the chloroplasts, generating there a secondary ROS wave (Shapiguzov et al. 2012).

Such signal communication probably requires components from the cytosol, and implies that ROS have to cross membrane systems – for example by converting $\bullet O_2^-$ to H_2O_2 in the apoplast, by the action of an apoplast-localized superoxide dismutase (SOD), which would allow H_2O_2 to enter the cell via aquaporins – or ROS signals may be detected by proteins in apoplast receptors associated with membrane systems (de Dios Barajas-Lopez et al. 2013). In addition, ROS originating from the chloroplasts or from the peroxisome can bring the signal to the nucleus, where they can regulate the expression of many defense-related genes (de Dios Barajas-Lopez et al. 2013; Sandalio and Romero-Puertas 2015; Locato et al. 2018).

The equilibrium between ROS formation and removal primarily depends on the action of antioxidants, which are involved in the elimination of such reactive species through both nonenzymatic and enzymatic mechanisms. Among the enzymes involved in ROS scavenging, SOD is particularly important in that it removes the ROS that is formed first upon exposure to several stressors, namely $\bullet O_2^-$. Other ROS-scavenging enzymes include glutathione peroxidase (GPX), ascorbate peroxidase (APX), and catalase (CAT), which converts hydrogen peroxide to H_2O (Lazaro et al. 2013). Either alone or as the electron-donating substrates of some of these enzymes, ascorbate, glutathione, and tocopherol, i.e. nonenzymatic antioxidants, also have a decisive role in keeping the balance between ROS formation and removal in plant cells. To allow the occurrence of the oxidative burst and the consequent generation of the ROS signal(s) to take place, the cellular levels of such nonenzymatic antioxidant molecules might be kept low at first (del Río 2015), but once the ROS signal has been generated, their levels become upregulated and act in a cooperative manner, e.g. through the so-called ascorbate–glutathione cycle (Viehweger 2014).

Unlike ROS, the generation of NO in plant cells is less well understood, and this still constitutes one of the major obstacles in understanding its roles as a signal molecule (Farnese et al. 2016; Kolbert et al. 2019). NO is intrinsically produced in the plant cells as a result of several catabolic transformations starting from of nitrogen-containing molecules. In addition, it can be taken up by plants from the external environment, including air and soil.

NO has two biosynthetic pathways, one of which is oxidative in nature, via the oxidation of polyamines (PAs) or L-arginine, whereas the other is a reductive one, in which the activity of nitrate reductase (NR) is heavily involved.

NO synthesis by the oxidative pathway usually involves polyamines or L-arginine as well as hydroxylamine. L-arginine is oxidized to produce NO plus citrulline by the action of peroxidase (POD), starting from N-omega hydroxy-L-arginine (NOHA) and H_2O_2. In addition, cytochrome P450 and hemoglobin can oxidize NOHA leading to NO formation, in the presence of NADPH and by using the oxidizing power of oxygen (Astier et al. 2016; Bhuyan et al. 2020). Moreover, arginase can convert L-arginine to urea and L-ornithine and urea via the enzyme arginase, which leads to PAs and hence to NO, after PA oxidation (Peng et al. 2016; da Silva et al. 2018). If hydroxylamine is provided exogenously, NO can be generated via the oxidative pathway, but the enzyme responsible for such conversion is not known (Hasanuzzaman et al. 2018). In the presence of hypoxic conditions, cytochrome c oxidase and/or reductase are reputed to oxidize NADH/NADPH to produce NO. In addition, NO can be formed from the oxidation of salicylhydroxamate, whereas POD can produce NO by using hydroxyurea and H_2O_2 as the substrates (Hasanuzzaman et al. 2018).

It is thought that NR in the cytosol, whose primary role is the assimilation of reductive N, is essential for reductive NO biosynthesis in plants (Horchani et al. 2011) and this has been observed under a variety of physiological circumstances, either in the presence of adversities, such as bacterial infection (Mur et al. 2013a), hypoxia, cold, drought (reviewed by Farnese et al. 2016), or during normal development, as occurs when the lateral roots are formed and during floral transition (Mur et al. 2013a). In ordinary growth circumstances, however, NR usually plays its inherent role, i.e. reducing nitrate to nitrite, and only in the presence of specific circumstances, such as anaerobiosis or high concentrations of nitrite, significant amounts of NR-derived nitric oxide can be measured (Gupta et al. 2011; Mur et al. 2013a).

As well as NR, enzymes participating in the mitochondrial electron transport chain can be involved in the reductive pathway of NO (Leon and Costa-Broseta 2020), as can xanthine oxidoreductase and NO_2-dependent NO reductase (Sahay and Gupta 2017). The reductive pathway of NO synthesis normally prevails under normal partial pressure of O_2, whereas under conditions of microaerobiosis or anoxia the oxidative pathway is stimulated (Leon and Costa-Broseta 2020; Zhou et al. 2021a, 2021b).

Among those putatively involved in NO synthesis, one enzyme that conceivably has been among the most intensively studied is nitric oxide synthase (NOS). Previous literature surveys concerning the presence of a typical NOS enzyme in plants gave negative results (Hancock and Neill 2014; Foresi et al. 2017). More convincingly, Jeandroz et al. (2016) reported on a genomic search of canonical NOS sequences within a large dataset of sequenced transcriptomes from land plants, which gave no result. This led those authors to conclude that the molecular mechanisms for producing NO are probably different among plants and animals. As a consequence, it has been recommended to avoid using terms such as "NOS" or "NOS-like" referring to higher plants, for which instead the expression "nitric oxide-generating" enzymes and proteins should be used (Santolini et al. 2017; Astier et al. 2018; Hancock and Neill 2019).

Aside from the enzymatic pathways presented above, nonenzymatic routes also exist for the production of NO in plant cells. During atmospheric N_2 fixation by nitrification and denitrification, NO can be released. The reaction between two molecules of HNO_2 can produce NO and NO_2, which is further converted to NO and superoxide radical anion. Acidic pH can favor nonenzymatic NO production, while NO_2 is converted to NO and NO_3. Upon exposure to light, β-carotene reduces NO_2 to NO. Reducing electrons for converting NO_2 to NO can be also donated by ascorbic acid, which becomes oxidized to dehydroascorbic acid (DHA). In addition, NO is produced nonenzymatically from *S*-nitrosoglutathione (GSNO), and by deoxygenated heme-proteins too, which under anoxic conditions convert NO_2 to NO (Sami et al. 2018; Bhuyan et al. 2020).

As seen earlier for ROS, the formation of NO also occurs in various organelles in the plant cell, such as peroxisomes, chloroplasts, the endoplasmic reticulum, mitochondria, plasma membrane, apoplast, as well as in the cytosol (Sahay and Gupta 2017). The mechanism of NO formation in each cell compartment depends on the local predisposition to carry out reductive or oxidative NO synthesis (see earlier). At the plasma membrane level, reductive synthesis is preferred, because a comparatively lower pH favors the conversion of NO_2 to NO and O_2; reductive NO biosynthesis is also favored in the apoplast, from which the root-specific NO is released by using the local abundance of nitrite substrates. In chloroplasts, mitochondria, and peroxisomes, either of the two NO formation routes is favored, and a switch between the two can also occur, dictated by the local availability of substrates and by the environment. In the cytosol, an NADPH-dependent conversion of nitrate to nitrite occurs, which is further reduced to NO at the expense of polyamine oxidation. In the cytoplasm, NO is always produced as an intermediate during the conversion of reduced glutathione (GSH) to GSNO (Nabi et al. 2019; Zhou et al. 2021a).

NO given exogenously is absorbed by the plant cell as nitrate, by the action of nitrate transporters, and then nitrate is regenerated as NO by NR, at the expense of NADPH (Zhou et al. 2021a).

As with many other metabolites, not to say all of them, cellular concentration of NO has to be maintained at definite values in space and time, which is accomplished by scavenging and metabolic activities initiated in turn by specific enzymatic activities and the associated substrates. Similarly to other organisms, plants have their own NO scavengers, namely nonsymbiotic hemoglobins known as phytoglobins (Hill et al. 2016), which, in the presence of hypoxia, are particularly efficient in converting NO to nitrate, being oxidized to metphytoglobin (Gupta and Igamberdiev 2011). These are then reduced back to phytoglobins at the expense of ascorbate, by the action of a specific reductase (Gupta and Igamberdiev 2011; Gupta et al. 2020a; Zhou et al. 2021b).

Endogenous levels of NO are also controlled by its reaction with GSH, to form GSNO, which is possibly used to supply back NO employed in protein nitrosylation, thus contributing to the cellular homeostasis of NO (Leon and Costa-Broseta 2020; Kalinina and Novichkova 2021). *S*-nitrosoglutathione can be transported in the phloem, thus acting as a stable and soluble source for the transport of the NO signal over long distances, which may be important for plant acclimation to stress (Arasimowicz-Jelonek et al. 2014). *S*-nitrosoglutathione also contributes to setting the cellular level of NO by inhibiting the pathways of N_2 assimilation (Fungillo et al. 2014). The turnover of GSNO also depends on the activity of GSNO reductase (GSNOR), which deaminates *S*-nitrosoglutathione to glutathione disulfide (GSSG) and ammonia. As a consequence, GSNOR can control the cellular levels of GSNO, which in turn has important consequences for maintaining NO homeostasis and its transitory signaling activity in cells (Corpas 2011; Malik et al. 2011; Lindermayr 2018). The expression of GSNOR is responsive to environmental stress and is influenced by the ROS/NO ratio (Cheng et al. 2015; Wang et al. 2015; Yang et al. 2015; Farnese et al. 2016), thus constituting a pivotal point in the crosstalk between the two signaling routes. A further oxidation of GSNO produces ammonia and GSSG by the action of a specific reductase (Leon and Costa-Broseta 2020). Upon denitrosylation of previously nitrosylated proteins, NO is released back and this affects NO cellular homeostasis (Leon and Costa-Broseta 2020). This occurs when the S-nitrosylation-dependent signaling is altered upon attack of pathogens, which activate responses at the immunological level. Protein *S*-nitrosylation is thought to be regulated by thioredoxins, another interaction providing its own contribution to NO homeostasis and participating in plant immunity (Mata-Perez and Spoel 2019; Leon and Costa-Broseta 2020).

Dehydrogenases of the NAD(P)H dehydrogenase-type are seen as additional regulators of NO degradation. They convert NO into peroxynitrite via reaction with superoxide anion, their activity being stimulated by calcium and hindered under anoxia by SOD (Gupta et al. 2020a).

NO can be also scavenged when it reacts with O_2, yielding nitrate and nitrite, or it reacts with lipid peroxyl radicals, forming nitro-fatty acids (Hancock 2012; Rubbo 2013; Leon and Costa-Broseta 2020). Hemoglobins can contribute to scavenging NO by reducing themselves to Fe(II)-hemoglobin, and then regaining oxidation at the expense of NO (Chamizo-Ampudia et al. 2017). In the presence of O_2, NO is metabolized to nitrous anhydride (N_2O_3)

and NO_2. In an aqueous milieu, they are then transformed to nitrite and nitrate, taking part in the response to nitrooxidative stress (Corpas and Palma 2018).

7.3 ROS and NO in Action at the Molecular Level

The action molecular mechanisms of ROS and NO have been widely studied in many different organisms, including plants (Green et al. 2014; Lamotte et al. 2015; Morales et al. 2015). The effects of NO and related chemical species depend on their ability to chemically modify proteins, according to three possible distinct mechanisms: the nitrosylation of metals, the nitration of tyrosinyl residues in proteins, and *S*-nitrosylation (Lamotte et al. 2015). When nitrosylation of metals occurs, NO binds to transition metals in metalloproteins; NO binds to the iron at the center of heme, the prosthetic group of several enzymes involved in ROS-scavenging enzymes, like POD, CAT, and SOD, not to say guanylate cyclase, among others, and this results in their inhibition or activation. The possibly opposite effects of NO on proteins bearing a heme group, namely inhibition or activation, are thought to depend on their oxidation state. Because NO resembles O_2, as far as paramagnetic properties, hydrophobicity, charge, and size are concerned, it can be predicted that any metalloenzyme containing a heme group reactive toward O_2 will be potentially prone to inhibition by NO (Arora et al. 2016).

During tyrosine nitration, a nitro group is added to tyrosinyl residues in proteins. This is mainly brought about by peroxynitrite, which is formed from the reaction between NO and the superoxide radical anion. For some time tyrosine nitration was regarded as a consequence of nitrosative stress, but is assumed nowadays to also behave as a signal (Mengel et al. 2013).

Finally, *S*-nitrosylation, or more rigorously *S*-nitrosation, during which NO binds to the cysteinyl residue of a given protein, appears to be the main avenue for transducing the NO signal. *Trans*-nitrosation is an alternative manner for *S*-nitrosation to occur, in which there is an exchange of NO from one *S*-nitrosylated residue to another thiol group, mediated by nitrosothiols, including GSNO (Lamotte et al. 2015). In both cases, *S*-nitrosation causes NO to become reversibly bound to a S atom, resulting in the production of *S*-nitrosothiol (-SNO). Since *S*-nitrosation does not depend on enzymatic catalysis, its extent is dictated by the reactivity between the nitrosating agent and the local redox potential. *S*-nitrosation is selectively brought about on specific cysteinyl residues and targeted proteins become posttranslationally modified (PTM), which increases specificity and flexibility in the adaptive responses to a changing cellular environment (Gong et al. 2014; Arora and Bhatla 2015; Arora et al. 2016; Hancock 2020). So far, *S*-nitrosation-dependent reversible PTMs have been found in more than 200 proteins, allowing the identification of possible new functions for many proteins during ad hoc proteomic studies (Minguez et al. 2012; Sevilla et al. 2015).

S-nitrosation has a remarkable contribution to plant adaptation to abiotic stress. The stress caused by excess salinity, for example, appears to be able to induce *S*-nitrosation of respiratory, photorespiratory, and antioxidant enzymes (Camejo et al. 2013), whereas upon exposure to cold a preferential *S*-nitrosation of the enzymes involved in carbon metabolism was observed (Puyaubert et al. 2014), which could suggest a major role for *S*-nitrosation in maintaining the balance between energy-capturing and energy-processing reactions under stress conditions.

Apart from influencing the posttranscriptional expression of plant enzymes, *S*-nitrosation may also impact the transcriptional level, deriving *S*-nitrosation of transcription factors (TFs), which affects their binding to *cis*-acting elements on DNA and/or their cellular localization. The myeloblastosis (MYB) family of TF, which play important roles in the onset of tolerance to both abiotic and biotic stress, can be inactivated by *S*-nitrosation, which could suggest that the they would need to be downregulated after the initial stress response (Tavares et al. 2014).

S-nitrosation can also impact nuclear proteins, including histone deacetylases, deeply involved in the histone acetylation/deacetylation cycle, which contributes to the reversible transitioning from the more condensed and less accessible status of chromatin (heterochromatin) to its less compact and more accessible counterpart (euchromatin; Mengel et al. 2013; Chaki et al. 2015; Ageeva-Kieferle et al. 2021). Upon *S*-nitrosation, at least in animal systems, histone deacetylase detaches from chromatin, and this promotes acetylation, making DNA more accessible for replication/transcription (Farnese et al. 2016). By the way, this suggests the intriguing possibility that NO could have regulatory roles in the epigenetic control of plant processes (Floryszak-Wieczorek et al. 2012; Chaki et al. 2015; Ageeva-Kieferle et al. 2021).

Both *S*-nitrosation and denitrosation, involving the removal of NO from cysteinyl residues, have to occur in order to allow the reversibility of the subsequent effects in terms of enzyme activities, protein–protein interactions, and as signaling (Sevilla et al. 2015). NO-dependent *S*-nitrosation of the cysteinyl residue of GSH leads to the synthesis of GSNO, which, as mentioned before, can reversibly store NO, and thus participate in the control of its biological activity, e.g. by donating NO for *S*-nitrosation-dependent proteins PTM, as well as acting as a long-distance signal on its own (as mentioned earlier). The aforementioned nitrosoglutathione reductase (GSNOR), first described in plants by Sakamoto et al. (2002), by regulating the endogenous levels of *S*-nitrosoglutathione and subsequently the global extent of *in planta S*-nitrosation, is now recognized as a central regulator of NO signaling in plants (Leterrier et al. 2011; Jahnová et al. 2019). A direct reversion mechanism leading to *S*-denitrosation, thought to be specific for a subset of plant proteins, has been envisaged for thioredoxin (Trx)h5 (Kneeshaw et al. 2014). Therefore, at least two different strategies might exist in plants to arrest NO-mediated signaling through *S*-nitrosation: one indirect, mediated by the action of GSNOR on GSNO (Malik et al. 2011; Kolbert et al. 2019); and the other direct, in which Trxh5 selectively denitrosates proteins (Kneeshaw et al. 2014).

Similar to NO, ROS signals are also transmitted via PTM of proteins (see earlier), cysteinyl residues again being involved. However, unlike via *S*-nitrosation operated by NO, ROS appear to bring about several different types of oxidative PTM (Ox-PTM), reversible or not, which include disulfide bond formation, sulfhydration, and *S*-glutathionylation (Akter et al. 2015). Each protein involved can be subjected to different modes of Ox-PTMs, and it has been proposed that each of these Ox-PTM types may have a different role (Couturier et al. 2013). The initial reaction of ROS-induced PTM is the reversible oxidation of a cysteinyl residue, which yields sulfenic acid (R-SOH). Being remarkably unstable, sulfenic acid will readily react with free protein thiols, forming -*S*-*S*- bonds or causing the covalent attachment of glutathione, increasing the extent of *S*-glutathionylation; such a reaction plays an important role in signaling and protects the protein from the deleterious action of $\bullet O_2^-$. Deglutathionylation and reduction of disulfide bonds, respectively regulated by thioredoxin and glutaredoxins, terminate the *ab initio* Ox-PTM signal (Waszczak et al. 2014, 2015).

Although mostly studied in animals and in bacteria, one Ox-PTM in particular, namely *S*-glutathionylation, is reputed to be a key player in the abiotic stress response in plants, being able to impact the functional state of enzymatic proteins and TFs, as well as the chromatin structure (Zagorchev et al. 2013; Dietz 2014).

Even though much experimental evidence has been accumulating over recent years concerning the impact of Ox-PTMs and *S*-nitrosation on metabolism and gene expression in plants, a unified picture concerning signals mediated by ROS, NO, and their crosstalk is still far from being proposed.

7.4 Crosstalk Between Reactive Nitrogen, Oxygen, and Sulfur Species

Every time exposure to a stress causes the production of NO, it is never produced alone, and this implies that the accumulation of other signals will occur in parallel, including ROS and hydrogen sulfide. In this respect, the case of plant glyceraldehyde 3-phosphate dehydrogenase (GAPDH) is paradigmatic: this enzyme can be *S*-nitrosated, becoming able to interact in a specific manner with partial gene sequence coding NADP-dependent malate dehydrogenase in nuclear DNA. On the other hand, GAPDH modification can also occur via oxidation brought about by H_2O_2 (Hancock and Neill 2019) and moreover it can be *S*-persulfidated by hydrogen sulfide (Aroca et al. 2017), which causes its translocation to the nucleus. This exemplifies that signals generated by reactive species can compete for the same protein target (Hancock et al. 2017), and therefore that the NO-dependent signal will be the dominant one is not guaranteed. In this context, methods for identifying thiol modifications could be helpful for understanding signal overlapping and crosstalk (Williams et al. 2015, 2019). Every time NO and $\bullet O_2^-$ are formed together in the plant cell, the synthesis of peroxynitrite can result, which has signaling capacity on its own (Speckmann et al. 2016).

S-nitrosation can mediate the crosstalk between nitric oxide and other signaling routes, included those in which ROS are involved. This is exemplified by the case of NADPH oxidase, a fundamental ROS-generating enzyme (see earlier), which is subjected to modification caused by NO (Yu et al. 2012a). As seen earlier, tyrosine nitration is another PTM in which NO is involved (Kolbert et al. 2017; Jiménez et al. 2019), again possibly resulting in functional changes. By nitrating the tyrosinyl residues of target proteins, peroxynitrite can regulate their function. For example, $ONOO^-$ can influence the activities of SOD enzymes, thus indirectly regulating the formation of other signaling molecules, such as ROS (Holzmeister et al. 2015). As in the case of *S*-nitrosation, tyrosine nitration brought about by NO can cause PTM of the target proteins, leading to either decreased or increased activities when the target proteins are enzymes.

Because NO and ROS are formed in parallel in the cell, their interactions and the complexity of their chemistry have to be considered. A fairly well-studied interaction between

NO and ROS, in particular $\bullet O_2^-$, leads to the formation of $ONOO^-$. This has two potentially remarkable consequences. First, such a reaction scavenges both $\bullet O_2^-$ and NO from the cell milieu, lowering in this manner the effective concentrations and hence the signaling capacity of both of them. Furthermore, a different signaling molecule is generated, namely $ONOO^-$ (Speckmann et al. 2016), able to trigger an array of effects potentially different from those that would have resulted from ROS or NO per se (Hancock and Neill 2019).

It should also be considered that the formation of NO will often occur in an antioxidant-rich cellular environment. As mentioned earlier, NO may influence the expression of antioxidant enzymes, as occurs when SOD activity is altered in the presence of $ONOO^-$ (Holzmeister et al. 2015; Romero-Puertas and Sandalio 2016), thereby reducing the removal of $\bullet O_2^-$ and the production of a key signaling molecule such as hydrogen peroxide. Likewise, NO can influence CAT activity (Bauer 2015), thereby reducing the removal of H_2O_2 and possibly stabilizing and sustaining the signal brought about by ROS. On the other hand, there are many reports suggesting that NO can activate ROS-scavenging enzymes, such as SOD and CAT (reviewed in Arora et al. 2016; Klein et al. 2018), with the participation of H_2S signaling (da Silva et al. 2018). This apparent contradiction could stem from the fact that interactions between different ROS-scavenging enzymes (and their different isoenzymes) and NO take place at different levels, so the intracellular equilibrium among reactive oxygen species and nitric oxide is key in establishing the sign and the extent of modulation of the enzyme activities: as discussed earlier, in fact, on the one hand, NO-dependent nitrosylation can modulate, not only posttranslationally but even at the genetic level, the activities of ROS-scavenging enzymes, all of which are metalloproteins; on the other, and reciprocally, endogenous NO availability can be regulated by ROS-scavenging enzymes. In this respect, the case of SOD is paradigmatic: SOD isoforms can not only be in competition with NO for the common substrate/reaction partner, namely superoxide radical anion, but also catalyze the release of NO from GSNO. Indeed, providing hydrogen peroxide is present, GSNO is rapidly acted upon by Cu/ZnSOD to yield oxidized glutathione and NO (Arora et al. 2016). Hence, NO-based PTMs may modulate the activities of enzymatic antioxidants (Yang et al. 2015; Kolbert and Feigl 2017; Gupta et al. 2020b). In general, NO has been proposed to control to a remarkable degree the cellular levels of ROS, and, as a consequence, their downstream signaling potential, by strengthening antioxidant defense, reducing lipid peroxidation, and balancing the redox status of the cell to mitigate concentration peaks of ROS production following oxidative stress (Fancy et al. 2017; Hasanuzzaman et al. 2018; Su et al. 2018; Jiménez et al. 2019; Sharma et al. 2019).

A considerable part of the burden of antioxidant defense in the plant cell is borne by nonenzymatic antioxidants. Significantly, NO and GSH can generate GSNO, and this can lower the GSH/GSSG ratio in the plant cell, thereby altering the intracellular redox balance, and even more so if the reaction is restricted within the boundaries of a given cell compartment. Furthermore, glutathione reductase, r-reducing back one molecule of GSSG to two molecules of GSH at the expense of reducing equivalents such as NADPH, possesses two catalytically important thiols that can undergo *S*-nitrosation, thereby leading to enzyme inhibition, causing in turn a decrease in cellular levels or reduced glutathione. On the other hand, NO has been shown to upregulate glutathione synthesis, which could be beneficial in the presence of oxidative stress (Xiong et al. 2011; Arora et al. 2016). Formation of GSNO will scavenge NO, preventing it from further participation in signaling events. *S*-nitrosoglutathione, on the other hand, can act not only as a molecular sink of NO, but also as a source of it, and also act as a mediator of certain effects of NO (Yun et al. 2016), showing distinct and overlapping molecular targets in comparison to NO itself.

S-nitrosoglutathione has also been considered as a physiologically significant mechanism for the long-distance transport of the NO signal, possibly via the plant vascular system. To stop *S*-nitrosoglutathione-mediated signaling, its removal by means of GSNOR is needed (Hancock and Neill 2019), which would lower the bioavailability of NO. A number of different stresses, either biotic or abiotic, also have been shown to influence GSNOR activity and modulate its expression (Chaki et al. 2011; Corpas 2011; Gong et al. 2015; Arora et al. 2016). It has been proposed (Lindermayr 2018) that ROS can fine-tune GSNOR activity, thereby affecting the cell concentrations of NO/SNO, which activate protection against oxidative stress.

NO-induced PTM appears to increase the activities of the antioxidant-regenerating enzymes involved in the ascorbate–GSH cycle, but not in response to NO or drying, at least in seeds, where only slight changes in the activity of these enzymes are seen (Bai et al. 2011). By enhancing the activities of mono DHA- and DHA reductase, NO has also been claimed to be involved in the regeneration of ascorbate in response to salinity (Hasanuzzaman et al. 2011; Arora et al. 2016).

In summary, the available evidence points to a role for NO-induced PTMs, by targeting the enzymes involved, in modulating both ROS formation and removal, which implies that NO can strictly control ROS homeostasis. Aside from acting as a direct agent of PTM, NO can also be involved in a competition for target proteins with molecules with similar properties, including ROS and hydrogen sulfide, suggesting the existence of multiple regulatory levels.

Nitro-fatty acids behave on their own as signaling molecules, with their formation following the reaction between unsaturated fatty acids, often oleic or linolenic acids, and NO and related reactive species (Corpas et al. 2013; Gupta et al. 2020b). Nitro-oleic acid has been recently reported to increase the activity of NADPH oxidase, thus affecting ROS formation (Di Palma et al. 2020), which points to a previously uncovered level of crosstalk among the signals generated by NO and ROS.

An additional source of interference and crosstalk with NO-generated signals could arise from H_2S, which has been recently claimed to be an important signal-generating molecule on its own in both plant and animal systems (Lisjak et al. 2013; Olas 2015; Kimura 2017; Zhou et al. 2021b). Hydrogen sulfide is formed following plant exposure to a variety of stresses, including cadmium (Mostofa et al. 2015). Hydrogen sulfide can participate in a direct reaction with NO, from which nitrosothiols are formed. As previously seen in the case of the reaction between NO and GSH, the direct chemical interaction between hydrogen sulfide and nitric oxide could decrease the availability of both in terms of direct signaling capacity, but on the other hand it would form a new signal, with its own signaling potential (Hancock and Neill 2019). Hydrogen sulfide may also enhance the cellular contents of glutathione (Hancock and Neill 2019), thereby potentially altering the cellular levels of *S*-nitrosoglulathione. Hydrogen sulfide and H_2O_2 can also cause PTM of proteins by interacting with catalytically important thiols on cysteinyl residues. Hydrogen peroxide is able to oxidize the -SH group of cysteinyl residues in proteins to form sulfenic acid, a reaction denoted as *S*-sulfenylation (Huang et al. 2019), while hydrogen sulfide can bring about persulfidation (Hancock 2019; Corpas et al. 2019; Zhou et al. 2021b). Interestingly, several of the enzymes targeted by the above reaction actively participate in ROS formation or removal.

Finally, hydrogen gas has recently been claimed to act as a signal in both plants and animals (Wilson et al. 2017; Hancock and Neill 2019). Molecular hydrogen can influence the cellular levels of antioxidants (Ohta 2015), thereby affecting the metabolism of NO in an indirect manner, although it has also been proposed to directly interact with certain compounds of nitrogen (Hancock and Hancock 2018). Certainly, hydrogen gas can affect plant metabolism (Lin et al. 2014; Wu et al. 2015; Su et al. 2018), to the point where it has been envisaged as a future protective/biostimulating agent for treating plants (Zeng et al. 2014). NO has been shown to be required for the occurrence of certain effects attributable to hydrogen gas (Zhu et al. 2016a, 2016b) and therefore it is highly probable that further studies will reveal more about interactions and crosstalk in signaling between NO and H_2.

7.5 Crosstalk in Abiotic Stress

7.5.1 Heavy Metals

It has been proposed that the interaction between NO and ROS is an important determinant of tolerance against heavy metals (Wang et al. 2014; Feigl et al. 2015; Silveira et al. 2015; Thao et al. 2015; Farnese et al. 2016; Rizwan et al. 2018; Okant and Kaya 2019; Singh and Prasad 2019), although the relevant mechanisms according to which such interactions take place have not been elucidated yet. An increases in ROS production is believed to be among the first cellular responses to excess heavy metals, which have been shown to be able to stimulate apoplastic NADPH oxidases, but also to promote oxidative stress in mitochondria and chloroplasts (Chmielowska-Bak et al. 2014). As far as the synthesis of NO is concerned, heavy metals are known to enhance it, although the mechanisms are thought to be metal-specific. Lead, for example, has been shown to increase NR activity in the cytosol (Yu et al. 2012b), whereas cadmium-induced enhancement of cellular NO levels appears to be associated with iron deficiency (Farnese et al. 2016). Instead, in response to certain metalloids, the operating level of the signal will be set by the balance between NO synthesis and GSNOR activity, both of which were found to be stimulated (Leterrier et al. 2012). Furthermore, an increased synthesis of NO might be fueled by ROS-mediated signaling following the ROS overproduction unleashed by heavy metal stress (Yun et al. 2011).

As discussed earlier, ROS and NO can behave as stressing agents or signals, depending on their concentration in the cellular milieu. At low concentrations, NO can promote tolerance to heavy metals by favoring the metal-binding capacity of the cell wall, so reducing their net influx into the cell (Singh et al. 2011) or increasing heavy metal sequestration into the vacuole, either via stimulation of phytochelatin biosynthesis or by influencing the electrochemical gradient of vacuolar proton pumps in order to favor the vacuolar uptake of heavy metals (see Farnese et al. 2016 and references therein). Furthermore, NO can ameliorate plant antioxidant capacity, thereby limiting the heavy metal-induced oxidative stress (Cheng et al. 2015; Andrade et al. 2016). Last but not least, NO-dependent PTM can downregulate the activity/level of ROS-generating enzymes, such as NADPH oxidase and glycolate oxidase, thus supporting the plant cell in maintaining or recovering its homeostasis in redox chemistry (Yun et al. 2011; Quiang et al. 2012).

7.5.2 Water Shortage

To tolerate drought, plants must have a precise and timely control of stomatal behavior and a finely tuned balance in cellular metabolism, in both of which NO and ROS are deeply involved (Osakabe et al. 2014). That NO is strictly required for generating the signal following exposure to water shortage can be argued by the fact that drought-dependent NO stimulation has been seen in a wide variety of both higher and lower plant species (Santisree et al. 2015; Nabi et al. 2019; Shams et al. 2019). The metabolic pathways participating in drought-dependent NO stimulation, however, are still mostly unknown, even if roles for NR (Farnese et al. 2016) and for xanthine oxidoreductase (Yu et al. 2014) have been suggested. On the other hand, it is well known that drought stimulates ROS generation and this can depend on the overreduction of the photosynthetic electron transport chain, resulting as a backward effect from the stress-dependent impairment of Rubisco activity and the regeneration of $NADP^+$ operated by the Calvin cycle. Apart from the ROS overproduction originating from the photosynthetic machinery, water stress has been shown to activate an oxidative burst via NADPH oxidases, which is key for the deployment of defense/signaling in response to water shortage (Wang X et al. 2015; Farnese et al. 2016).

A reduction in stomatal opening is the most important and easily predictable effect of water shortage; stomatal behavior appears to respond to the complex interplay between ROS, NO, and abscisic acid (ABA). By acting as an the first element of this regulatory chain, ABA induces the synthesis of NO, which, in synergy with ROS, acts in turn upon stomatal closure by producing 8-nitro-cyclic guanosine monophosphate (cGMP; Pasqualini et al. 2009). ROS overproduction and the resulting alteration of the cellular redox state of the cell are thought to be essential for transducing the *ab initio* ABA signal (Joudoi et al. 2013).

Alone, and alongside ABA and ROS, under conditions of water shortage NO is thought to influence transcriptional activation, with beneficial effects in terms of biomass production and survival, which suggests a protective role for this molecule (Shi et al. 2014). DNA methylation, and its consequences in terms of epigenetic changes, is another of the effects attributed to NO in plants exposed to water stress: as shown by Fan and Co-workers (2012), a NO donor such as sodium nitroprusside supplied exogenously caused an overall decrease in the levels of DNA methylation in *Dendrobium huoshanense*, which resulted in an increased transcription of genes coding for antioxidant enzymes.

7.5.3 Flooding

In the mechanisms underlying the perception of reduced oxygen availability in plants, changes in the balances between ROS/RNS formation and removal may participate, via the inhibitory effect of hypoxia on the mitochondrial electron transport chain (complex III) and mediated by a transient mitogen-activated protein kinase (MAPK) signaling cascade (Chang et al. 2012). Furthermore, the ROS-dependent response regulation under conditions of micro-aerobiosis or anaerobiosis might imply, at least partially, the participation of ethylene, since both have been found to stimulate the formation of adventitious roots in rice plants exposed to oxygen deprivation (Steffens et al. 2012). On the other hand, exogenous NO application can alleviate flooding stress in maize plants by upregulating the activities of antioxidant enzymes and of alcohol dehydrogenase, and the levels of phenols and other nonenzymatic antioxidants (ascorbate) as well (Jaiswal and Srivastava 2018).

The mechanisms of NO synthesis under oxygen deprivation are still unclear, even if there is evidence of involvement of nitrite and ascorbate, within the context of reactions occurring in the mitochondria (Wang and Hargrove 2013). Nitrite could accept electrons from mitochondrial complex IV and complex III, and from alternative oxidases as well, with the purpose of limiting ATP production under hypoxic conditions and so keep in balance the energy status of the cell (Gupta et al. 2011). The NO produced in the mitochondria could diffuse into the cytosol, being converted to nitrate by phytoglobins. Nitrate could then be reduced to nitrite by the action of NR (Dordas 2015), and nitrite, in turn, could undergo mitochondrial conversion again to NO, so dissipating a potentially dangerous overflow of mitochondrial electrons under conditions of scarcity of the terminal electron acceptor (Gupta et al. 2011).

7.5.4 Salinity Stress

Arabidopsis mutants deficient in NO synthesis show a salt-hypersensitive phenotype, and this points to an essential role for NO in the tolerance of plants to salinity stress. NO could ameliorate tolerance to salinity by increasing the activities of ATPase and of H^+-pyrophosphatase at the plasma membrane and the tonoplast levels, which could contribute to keeping in balance the Na^+/K^+ ratio (see Farnese et al. 2016 and references therein). Furthermore, NO appears to be able to activate the transcription of stress-related genes and to ostensibly increase the activity of antioxidant enzymes, both the ROS-scavenging and the antioxidant-recycling ones (Hasanuzzaman et al. 2011; Arora et al. 2016; Khoshbakht et al. 2018). Likewise, hydrogen peroxide also modulates the transcription of stress-related genes in plants exposed to salinity, especially in the root cells, and stimulates the activity of specific enzymes. It is noteworthy that the compensatory changes promoted by ROS and NO do not remain confined within the tissues where these molecules are primarily produced, and this is reminiscent of the ROS/RNS "wave" propagating mechanisms discussed before: in this respect, as a matter of fact, both NO and GSNO, whose role in NO transport have been described earlier, have been shown to be present in the vasculature of plants undergoing salt stress (see Farnese et al. 2016 and references therein).

NO and H_2O_2 could also participate in the development of acquisition of plant immunity against salt stress. In such process, denoted as priming or hardening, the previous exposure of a plant to sublethal doses of a given abiotic stress accelerates and potentiates its defense capacity upon re-exposure to that same stressor (see Farnese et al. 2016 and references therein). As a matter of fact, plants given low levels of NO and hydrogen peroxide showed an upregulation of their physiological and metabolic responses upon exposure to salinity (Molassiotis et al. 2016). Consistently, in primed wheat roots a stimulation in the production of H_2O_2 and NO was observed, which was associated with a priming-dependent induction of tolerance to drought (Wang et al. 2020). This study also pointed to a direct interaction between H_2O_2 biosynthesis deriving from the stress-induced activation of NADPH oxidase and the downstream signaling operated by NO during the H_2O_2-mediated accumulation of proline and glycine betaine. In the same vein, halophytes and glycophytes showed an increase in their tolerance to salt stress when previously treated with NO donors. This evidence suggests that NO and hydrogen peroxide are able to prime for increased tolerance to salt stress, protecting the plant from undesirable effects upon subsequent exposure to this stressing agent (see Farnese et al. 2016 and references therein).

7.5.5 Dystermia

Dystermia, often in combination with drought, mostly determines the plant biodiversity worldwide, and even more so under the present era of climate change. Even if exposure to high or low temperatures evokes quite distinct metabolic patterns, a commonality exists among the two opposite circumstances, i.e. ROS overproduction and the derived challenge to the plant cell structure and function (Zhou et al. 2012; Bita and Gerats 2013; Farnese et al. 2016). As in the cases of other abiotic stresses in plants, ROS overproduction arising from dystermia appears to be a deleterious consequence of the loss of capacity to keep pace between energy-capturing and energy-processing reactions (Hasanuzzaman et al. 2013). In contrast, experimental evidence has been increasingly demonstrating that ROS, RNS, and their crosstalk, as well as their capacity to induce posttranscriptional modification of proteins (see earlier), are essential components of acclimatization (Habibi 2014; Yu et al. 2014; Hossain et al. 2015; Zhou et al. 2021a). As a matter of fact, treating plants with low levels of NO and hydrogen peroxide allows them to perform better when challenged by thermal stress (Cantrel et al. 2011; Dong et al. 2018).

High temperatures usually cause an increase in the production of endogenous NO, a response thought to be important for plant acclimation (Yu et al. 2014). Treating plants with an exogenous source of NO during their exposure to high temperature stress stimulates the level/activity of both nonenzymatic and enzymatic antioxidants, which reduces the extent of stress-induced cellular damage (Hasanuzzaman et al. 2013; Rai et al. 2018). Symmetrically, supplying plants with a NO scavenger, such as 2-4-carboxyphenyl-4,4,5,5-tetramethylimidazoline-1-oxyl-3-oxide (cPTIO), suppresses the aforementioned protective effects of NO against high temperature stress, which further supports the idea that this molecule could be an important determinant for the onset of tolerance (see Farnese et al. 2016 and references therein). Likewise, accumulation of ROS, and in particular of hydrogen peroxide, is reputed to bring about the signal transduction leading to the transcriptional activation of heat shock genes, encoding proteins reputed to have key roles in the preservation of homeostasis at the cellular level under stressing circumstances (see Farnese et al. 2016 and references therein). Rai et al. (2020) proposed that in plants the heat stress-induced shock response is brought about via synthesis and distribution of NO and salicylate (SA), through calcium signaling.

As in the case of elevated temperatures, cold stress can also lead to a quick increase in the cellular levels of NO and ROS (Cantrel et al. 2011). Working with plant mutants defective in NR expression suggested that NO most likely participates in the onset of tolerance to cold, since mutant lines showed reduced NO levels if compared with the wild type and a concomitantly higher incidence of damage from cold stress (Cantrel et al. 2011). The promotive action of NO toward tolerance to low temperatures may imply effects at the gene expression level, seen in the transcriptional regulation of both biosynthetic (*P5CS1*) and catabolic (*ProDH*) proline genes, resulting in the accumulation of this compatible solute and osmoprotectant.

Aside from the effects attributable to NO, the role of hydrogen peroxide has been the subject of many studies, showing that exogenous application of H_2O_2 attenuates the damaging effects of cold stress and increases the poststress survival rates (Farnese et al. 2016; Dong et al. 2018). This beneficial impact of H_2O_2 appears to rely again on the promotion of antioxidant defense, both enzymatic, through the activation of CAT, GPX, and APX, and nonenzymatic, through the increase of glutathione levels (Iseri et al. 2013; Wang et al. 2013). In this case, the signaling pathway leading to the mitigation of cold stress effects might involve, at least in part, transitory "waves" of Ca^{2+} ions in the cytosol deriving from the hydrogen peroxide-mediated opening of Ca^{2+} channels at the plasma membrane level (Farnese et al. 2016). Although the studies suggesting that ROS and NO are central players in raising plant tolerance to dystermia have been many, a comparatively scanty amount of work has been dedicated to unraveling the molecular mechanisms underpinning the signaling cascades involved. In this respect, it has been reported that in some plants cold stress could evoke NO formation by activating ion channels causing the downstream upregulation of TFs such as the inducer of CBF (ICE) proteins and the C-repeat binding factor/Dehydration-responsive element binding-1 (CBF /DREB1). These TFs are known to be under the regulation brought about by several miRNA candidates, including miR172, miR166, and miR397, and by certain regulators of ABA signaling, such as miR168 (Kumar 2014). The CBF/DREB TFs are known to work by binding to the C-repeat elements/low-temperature response *cis*-acting elements (CRT/LTRE) in the promoter regions of cold-responsive genes assumed to confer tolerance to cold stress in plants (Kumar 2014). In several plant species, the activities of the aforementioned miRNAs were found to be affected by NO and are thought to be involved in the development of cold tolerance (Kumar 2014; Singh et al. 2017; Prakash et al. 2019).

7.6 Crosstalk in Biotic Stress

A landmark paper about the crosstalk between ROS and NO in plant–host interactions was published at the very end of the previous century, in which NO was found to participate in the ROS-induced PCD resulting from the HR raised by *Arabidopsis thaliana* plants infected by virulent *Pseudomonas syringae* (Delledonne et al. 1998). Already at that time, it was apparent that GSNO behaves as a long-distance signal in the development of systemic acquired resistance in plants. It then became apparent that, unlike in animal cells, HR-induced PCD in plants includes the interaction of NO with H_2O_2 generated by SOD, and not a $NO\text{-}\bullet O_2^-$ interaction. However, a ROS-independent PCD pathway during compatible plant–host interaction was also proposed, in which a cGMP-dependent pathway involving MAPK was envisaged (see Kolbert et al. 2019 and references therein).

A new pathway concerning the crosstalk between NO and ROS during plant–pathogen interaction was also proposed, based on the inhibition of the two major H_2O_2-scavenging enzymes, namely CAT and APX, exerted by NO and $ONOO^-$. In tobacco BY-2 cells, it was also shown that the ascorbate and glutathione redox couples fuel the NO- and

ROS-dependent transduction signaling leading to PCD and plant defense responses. In some cases, however, NO and ROS were found not to participate as signal molecules in the early stage of HR initiation, being involved instead in the subsequent stage of HR spreading to the cells surrounding the infection entry point (see Kolbert et al. 2019 and references therein).

Mitochondrial nitrite reduction (see earlier), together with NOS (see earlier) and NR activities, were found to cooperate to generate NO following infection of *A. thaliana* leaves brought about by *Pseudomonas syringae*. Likewise, NR was shown to originate the NO produced upon induction of infestin, the main eliciting molecule produced by *Phytophthora infestans* when attacking *Nicotiana benthamiana*. In this same pathosystem, MAPKs were found to be involved in the signaling cascade initiated by NO- and RBOH-dependent oxidative burst. Furthermore, a prosthetic group in common among NOG, NR, and NADPH oxidase, namely flavin, was found to be required for both NO and ROS waves leading to HR-induced PCD (see Kolbert et al. 2019 and references therein).

Crosstalk between NO and GSH was proposed concerning the *S*-nitrosation of regulatory elements, such as NPR1 (non-expressor of pathogenesis-related gene 1), participating in the salicylic acid (SA)-induced activation of pathogenesis-related genes, and nitrosative PTMs were also shown to target ROS-producing enzymes and components of the SA-dependent signaling (Sun et al. 2012; Kovacs et al. 2015). Oligogalacturonides eliciting defense responses in *Arabidopsis* may promote NR-dependent NO production. This can negatively affect the capacity of RBOH to initiate the HR-dependent oxidative burst, because, under high *S*-nitrosothiol levels, NO can bring about *S*-nitrosation of the enzyme at a specific and conserved cysteinyl residue, resulting in RBOH inhibition (Yun et al. 2011; Rasul et al. 2012).

In summary, a considerable amount of literature testifies that both NO and ROS, both separately and through crosstalk, participate in the molecular mechanisms following plant–host interaction, from the early pathogen recognition leading to HR, to the later signaling cascades resulting in the reprogramming of gene expression aimed at local as well as systemic defense and immunity. Mechanisms and pathways, however, may strongly depend on the phytopathogen considered, its plant host(s), including its genotype, and the resistance strategies deployed upon plant–pathogen interaction. As the majority of available evidence has been derived from herbaceous plant models, such as *Arabidopsis* and tobacco and their available mutant lines, a considerable amount of further work is needed on the one hand for transferring this into practical applications focused on agriculturally important crops, and on the other for understanding NO/ROS crosstalk during plant–pathogen interactions in nonmodel species, including those belonging to natural or semi-natural plant communities, e.g. forest trees.

7.7 Conclusion and Perspectives

The recent literature clearly supports the notion that redox signaling plays key roles in almost all the aspects concerning normal plant growth and development, as well as those associated with beneficial, such as symbiosis, or potentially detrimental plant–environment interactions, as under biotic or abiotic stress. The most relevant redox molecules are

ROS – such as the superoxide anion, singlet oxygen, hydrogen peroxide, and the hydroxyl radical – and NO. In plants, they play key signaling functions and participate in the regulation of gaseous exchange, defense, cell death, seed germination, and the transition from the vegetative to the reproductive stage, among others.

The diversity, as well as the outreach, of these functions becomes comprehensible if one considers that ROS and NO interact rapidly to form a number of RNS. Besides the direct interactions of these redox molecules, both ROS and NO can modify the activity or function of proteins involved in their own generation and signaling, as well as in their metabolism and homeostasis.

The studies reported earlier suggest the importance of the interplay and crosstalk between NO, ROS, and H_2S. They recall that the signaling brought about by NO and ROS can be influenced by either synergistic or antagonistic interactions, or instead can function in parallel. In addition, the available evidence supports the notion that the NO-dependent influence on the activity of ROS-scavenging and antioxidant-recycling enzymes plays a key role in the regulation of oxidative stress and ROS homeostasis.

NO is produced in the plant cell in response to the same stimuli able to promote ROS and H_2S overproduction, so that NO will never work alone. The reaction of NO with ROS and/or with hydrogen sulfide will decrease the cellular NO level, but also give rise to new signaling molecules, such as $ONOO^-$ and SNOs, able to act themselves as signal molecules, with possibly distinct signaling outcomes. The impact of NO on the cellular redox balance, especially within the plant cell compartments, needs to be considered, as well as, and reciprocally, the effects of the inherent redox environment on the NO metabolism that may result. The ways in which NO is interconnected with the signaling behavior of other key reactive chemicals need to be deciphered.

Much more information is also needed to understand how and to what extent these signaling molecules regulate and modulate gene transcription, by unraveling, for example, their influence on epigenetic processes. Along with present-day and future progress in the field of redox regulation in plants, a better understanding as well as new practical applications will become attainable in relation to plant–environment interactions, protection of biodiversity, agriculture, and plant biotechnology.

References

Ageeva-Kieferle, A., Georgii, E., Winkler, B. et al. (2021). Nitric oxide coordinates growth, development, and stress response via histone modification and gene expression. *Plant Physiology*, kiab222. doi:10.1093/plphys/kiab222.

Akter, S., Huang, J., Waszczak, C. et al. (2015). Cysteines under ROS attack in plants: a proteomics view. *Journal of Experimental Botany* 66: 2935–2944. doi:10.1093/jxb/erv044.

Andrade, H.M., Oliveira, J.A., Farnese, F.S. et al. (2016). Arsenic toxicity: cell signaling and the attenuating effect of nitric oxide in *Eichhornia crassipes*. *Biologia Plantarum* 60: 173–180. doi:10.1007/s10535-015-0572-4.

Arasimowicz-Jelonek, M., Floryszak-Wieczorek, J., Drzewiecka, K. et al. (2014). Aluminum induces cross-resistance of potato to *Phytophthora infestans*. *Planta* 239: 679–694. doi:10.1007/s00425-013-2008-8.

Aroca, A., Schneider, M., Scheibe, R. et al. (2017). Hydrogen sulfide regulates the cytosolic/nuclear partitioning of glyceraldehyde-3-phosphate dehydrogenase by enhancing its nuclear localization. *Plant and Cell Physiology* 58: 983–992.

Arora, D. and Bhatla, S.C. (2015). Nitric oxide triggers a concentration dependent differential modulation of superoxide dismutase (FeSOD and Cu/ZnSOD) activity in sunflower seedling roots and cotyledons as an early and long distance signaling response to NaCl stress. *Plant Signaling and Behavior* 10: e1071753. doi:10.1080/15592324.2015.1071753.

Arora, D., Jain, P., Singh, N. et al. (2016). Mechanisms of nitric oxide crosstalk with reactive oxygen species scavenging enzymes during abiotic stress tolerance in plants. *Free Radical Research* 50 (3): 291–303. doi:10.3109/10715762.2015.1118473.

Astier, J., Gross, I., and Durner, J. (2018). Nitric oxide production in plants: an update. *Journal of Experimental Botany* 69: 3401–3411.

Astier, J., Loake, G., Velikova, V. et al. (2016). Editorial: interplay between NO signaling, ROS, and the antioxidant system in plants. *Frontiers in Plant Science* 7: 1731. doi:10.3389/fpls.2016.01731.

Bai, X., Yang, L., Tian, M. et al. (2011). Nitric oxide enhances desiccation tolerance of recalcitrant *Antiaris toxicaria* seeds via protein S-nitrosylation and carbonylation. *PLoS ONE* 6: e20714.

Bauer, G. (2015). Increasing the endogenous NO level causes catalase inactivation and reactivation of intercellular apoptosis signaling specifically in tumor cells. *Redox Biology* 6: 353–371.

Begara-Morales, J.C., Chaki, M., Valderrama, R. et al. (2018). Nitric oxide buffering and conditional nitric oxide release in stress response. *Journal of Experimental Botany* 69: 3425–3438.

Bhuyan, M.H.M.B., Hasanuzzaman, M., Parvin, K. et al. (2020). Nitric oxide and hydrogen sulfide: two intimate collaborators regulating plant defense against abiotic stress. *Plant Growth Regulation* 90: 409–424. doi:10.1007/s10725-020-00594-4.

Bita, C.E. and Gerats, T. (2013). Plant tolerance to high temperature in a changing environment: scientific fundamentals and production of heat stress-tolerant crops. *Frontiers in Plant Science* 4: 273. doi:10.3389/fpls.2013.00273.

Camejo, D., Romero-Puertas, M.C., Rodriguez-Serrano, M. et al. (2013). Salinity-induced changes in S-nitrosylation of pea mitochondrial proteins. *Journal of Proteomics* 79: 87–99. doi:10.1016/j.jprot.2012.12.003.

Cantrel, C., Vazquez, T., Puyaubert, J. et al. (2011). Nitric oxide participates in cold-responsive phosphophingolipid formation and gene expression in *Arabidopsis thaliana*. *The New Phytologist* 189: 415–427. doi:10.1111/j.1469-8137.2010.03500.x.

Chaki, M., Shekariesfahlan, A., Ageeva, A. et al. (2015). Identification of nuclear target proteins for S-nitrosylation in pathogen-treated *Arabidopsis thaliana* cell cultures. *Plant Science* 238: 115–126. doi:10.1016/j.plantsci.2015.06.011.

Chaki, M., Valderrama, R., Fernández-Ocanã, A.M. et al. (2011). Mechanical wounding induces a nitrosative stress by downregulation of GSNO reductase and a rise of S-nitrosothiols in sunflower (*Helianthus annuus*) seedlings. *Journal of Experimental Botany* 62: 1803–1813.

Chamizo-Ampudia, A., Sanz-Luque, E., Llamas, A., Galvan, A., and Fernandez, E. (2017). Nitrate reductase regulates plant nitric oxide homeostasis. *Trends in Plant Science* 22: 163–174. doi:10.1016/j.tplants.2016.12.001.

Chang, R., Jang, C.J., Branco-Price, C. et al. (2012). Transient MPK6 activation in response to oxygen deprivation and reoxygenation is mediated by mitochondria and aids seedling survival in *Arabidopsis*. *Plant Molecular Biology* 78: 109–122. doi:10.1007/s11103-011-9850-5.

Chen, J., Wu, F.H., Shang, Y.T. et al. (2015a). Hydrogen sulphide improves adaptation of *Zea mays* seedlings to iron deficiency. *Journal of Experimental Botany* 66(21): 6605–22. doi:10.1093/jxb/erv368.

Chen, L., Luan, Y., and Zhai, J. (2015b). Sp-miR396a-5p acts as a stress-responsive genes regulator by conferring tolerance to abiotic stresses and susceptibility to *Phytophthora nicotianae* infection in transgenic tobacco. *Plant Cell Reports* 34: 2013–2025. doi:10.1007/s00299-015-1847-0.

Cheng, T.L., Chen, J.H., Abd Allah, E.F. et al. (2015). Quantitative proteomics analysis reveals that S-nitrosoglutathione reductase (GSNOR) and nitric oxide signaling enhance poplar defense against chilling stress. *Planta* 242: 1361–1390. doi:10.1007/s00425-015-2374-5.

Chmielowska-Bak, J., Gzyl, J., Rucinska-Sobkowiak, R. et al. (2014). The new insights into cadmium sensing. *Frontiers in Plant Science* 5: 245. doi:10.3389/fpls.2014.00245.

Corpas, F.J. (2011). Function of S-nitrosoglutathione reductase (GSNOR) in plant development and under biotic/abiotic stress. *Plant Signaling and Behavior* 6: 789–793.

Corpas, F.J., Freschi, L., Rodríguez-Ruiz, M. et al. (2018). Nitro-oxidative metabolism during fruit ripening. *Journal of Experimental Botany* 69: 3449–3463.

Corpas, F.J., Gonzalez-Gordo, S., Cañas, A. et al. (2019). Nitric oxide and hydrogen sulfide in plants: which comes first? *Journal of Experimental Botany* 70: 4391–4404. doi:https://doi.org/10.1093/jxb/erz031.

Corpas, F.J., Palma, J.M., del Río, L.A. et al. (2013). Protein tyrosine nitration in higher plants grown under natural and stress conditions. *Frontiers in Plant Science* 4: 29.

Correa-Aragunde, N., Foresi, N., and Lamattina, L. (2015). Nitric oxide is an ubiquitous signal for maintaining redox balance in plant cells: regulation of ascorbate peroxidase as a case study. *Journal of Experimental Botany* 66: 2913–2921. doi:10.1093/jxb/erv073.

Couturier, J., Chibani, K., Jacquot, J.P. et al. (2013). Cysteine-based redox regulation and signaling in plants. *Frontiers in Plant Science* 4: 105. doi:10.3389/fpls.2013.00105.

da Silva, C.J., Mollica, D.C., Vicente, M.H. et al. (2018). NO, hydrogen sulfide does not come first during tomato response to high salinity. *Nitric Oxide* 76: 164–173.

Das, K. and Roychoudhury, A. (2014). Reactive oxygen species (ROS) and response of antioxidants as ROS-scavengers during environmental stress in plants. *Frontiers in Environmental Science* 2: 53. doi:10.3389/fenvs.2014.00053.

de Dios Barajas-Lopez, J., Blanco, N.E., and Strand, Å. (2013). Plastid-to-nucleus communication: signals controlling the running of the plant cell. *Biochimica et Biophysica Acta* 1833: 425–437. doi:10.1016/j.bbamcr.2012.06.020.

Del Castello, F., Nejamkin, A., Cassia, R. et al. (2019). The era of nitric oxide in plant biology: twenty years tying up loose ends. *Nitric Oxide*85: 17–27.

del Río, L.A. (2015). ROS and RNS in plant physiology: an overview. *Journal of Experimental Botany* 66: 2827–2837. doi:10.1093/jxb/erv099.

Delledonne, M., Xia, Y., Dixon, R.A. et al. (1998). Nitric oxide functions as a signal in plant disease resistance. *Nature* 394: 585–588.

Demidchik, V. (2015). Mechanisms of oxidative stress in plants: from classical chemistry to cell biology. *Environmental and Experimental Botany* 109: 212–228. doi:10.1016/j.envexpbot.2014.06.021.

Dietz, K.-J. (2014). Redox regulation of transcription factors in plant stress acclimation and development. *Antioxidants & Redox Signaling* 21: 1356–1372. doi:10.1089/ars.2013.5672.

Di Palma, A.A., Di Fino, L.M., Salvatore, S.R. et al. (2020). Nitro-oleic acid triggers ROS production via NADPH oxidase activation in plants: a pharmacological approach. *Journal of Plant Physiology* 246–247: 153128.

Dong, N., Li, Y., Qi, J. et al. (2018). Nitric oxide synthase dependent nitric oxide production enhances chilling tolerance of walnut shoots in vitro via involvement chlorophyll fluorescence and other physiological parameter levels. *Scientia Horticulturae* 230: 68–77.

Dordas, C. (2015). Nitric oxide and plant hemoglobins improve the tolerance of plant to hypoxia. In: *Nitric Oxide Action in Abiotic Stress Response to Plants* (eds. M.N. Khan, M. Mobin, F. Mohammad, and F.J. Corpas). Cham, Switzerland: Springer International Publishing. 115–128. doi:10.1007/978-3-319-1784-2_1.

Fan, H., Li, T., Guan, L. et al. (2012). Effects of exogenous nitric oxide on antioxidation and DNA methylation of *Dendrobium huoshanense* grown under drought stress. *Plant Cell, Tissue and Organ Culture* 109: 307–314. doi:10.1007/s11240-011-0096-3.

Fancy, N.N., Bahlmann, A.K., and Loake, G.J. (2017). Nitric oxide function in plant abiotic stress. *Plant, Cell & Environment* doi:10.1111/pce.12707.

Farnese, F.S., Menezes-Silva, P.E., Gusman, G.S. et al. (2016). When bad guys become good ones: the key role of reactive oxygen species and nitric oxide in the plant responses to abiotic stress. *Frontiers in Plant Science* 7: 471. doi:10.3389/fpls.2016.00471.

Feigl, G., Lehotai, N., Molnár, A., Ördög, A., Rodríguez-Ruiz, M., Palma, JM., Corpas, F.J., Erdei, L., and Kolbert, Z. (2015). Zinc induces distinct changes in the metabolism of reactive oxygen and nitrogen species (ROS and RNS) in the roots of two Brassica species with different sensitivity to zinc stress. Ann Bot. 116(4): 613–25. doi: 10.1093/aob/mcu246.

Floryszak-Wieczorek, J., Arasimowicz-Jelonek, M., Milczarek, G. et al. (2012). Nitric oxide-mediated stress imprint in potato as an effect of exposure to a priming agent. *Molecular Plant–Microbe Interactions* 25: 1469–1477. doi:10.1094/MPMI-02-12-0044-R.

Foresi, N., Correa-Aragunde, N., and Lamattina, L. (2017). Synthesis, actions, and perspectives of nitric oxide in photosynthetic organisms. In: *Nitric Oxide*, 3e (eds. L.J. Ignarro and B.A. Freeman), 125–136. Academic Press. ISBN 9780128042731.

Freschi, L. (2013). Nitric oxide and phytohormone interactions: current status and perspectives. *Frontiers in Plant Science* 4: 398. doi:10.3389/fpls.2013.00398.

Fukudome, M., Calvo-Begueria, L., Kado, T. et al. (2016). Hemoglobin LjGlb1-1 is involved in nodulation and regulates the level of nitric oxide in the *Lotus japonicus–Mesorhizobium loti* symbiosis. *Journal of Experimental Botany* 67: 5275–5283. doi:10.1093/jxb/erw290.

Fungillo, L., Skelly, M.J., Loake, G.J. et al. (2014). S-nitrosothiols regulate nitric oxide production and storage in plants through the nitrogen assimilation pathway. *Nature Communications* 5: 54001. doi:10.1038/ncomms6401.

Gilroy, S., Suzuki, N., Miller, G. et al. (2014). A tidal wave of signals: calcium and ROS at the forefront of rapid systemic signaling. *Trends in Plant Science* 19: 623–630. doi:10.1016/j.tplants.2014.06.013.

Gong, B., Li, X., Bloszies, S. et al. (2014). Sodic alkaline stress mitigation by interaction of nitric oxide and polyamines involves antioxidants and physiological strategies in *Solanum lycopersicum*. *Free Radical Biology* 71: 36–48.

Gong, B., Wen, D., Wang, X. et al. (2015). S-nitrosoglutathione reductase-modulated redox signaling controls sodic alkaline stress responses in *Solanum lycopersicum* L. *Plant and Cell Physiology* 56: 790–802.

Green, J., Rolfe, M.D., and Smith, L.J. (2014). Transcriptional regulation of bacterial virulence gene expression by molecular oxygen and nitric oxide. *Virulence* 5: 794–809. doi:10.4161/viru.27794.

Groß, F., Dumer, J., and Gaupels, F. (2013). Nitric oxide, antioxidants and prooxidants in plant defence responses. *Frontiers in Plant Science* 29: 419. doi:10.3389/fpls.2013.00419.

Gupta, K.J. and Igamberdiev, A.U. (2011). The anoxic plant mitochondrion as a nitrite: NO reductase. *Mitochondrion* 11: 537–543. doi:10.1016/j.mito.2011.03.005.

Gupta, K.L. and Igamberdiev, A.U. (2015). Compartmentalization of reactive oxygen species and nitric oxide production in plant cells: an overview. In: *Reactive Oxygen and Nitrogen Species Signaling and Communication in Plants* (eds. K.J. Gupta and A.U. Igamberdiev), 215–237. Cham, Switzerland: Springer International Publishing. doi:10.1007/978-3-319-10079-1_11.

Gupta, K.J., Igamberdiev, A.U., Manjunatha, G. et al. (2011). The emerging roles of nitric oxide (NO) in plant mitochondria. *Plant Science* 181: 520–526. doi:10.1016/j.plantsci.2011.03.018.

Gupta, K.J., Kolbert, Z., Durner, J. et al. (2020a). Regulating the regulator: nitric oxide control of post-translational modifications. *The New Phytologist* 227: 1319–1325. doi:10.1111/nph.16622.

Gupta, K.J., Mur, L., Wany, A. et al. (2020b). The role of nitrite and nitric oxide under low oxygen conditions in plants. *The New Phytologist* 225: 1143–1151. doi:10.1111/nph.15969.

Habibi, G. (2014). Hydrogen peroxide (H_2O_2) generation, scavenging and signaling in plants. In: *Oxidative Damage to Plants: Antioxidant Networks and Signalin* (ed. P. Ahmad), 557–584. San Diego, CA: Elsevier. doi:10.1016/B978-0-12-799963-0.00019-8.

Hancock, J.T. (2012). NO synthase? Generation of nitric oxide in plants. *Periodicum Biologorum* 114: 19–24.

Hancock, J.T. (2019). Considerations of the importance of redox state for reactive nitrogen species action. *Journal of Experimental Botany* 70: 4323–4331.

Hancock, J.T. (2020). Nitric oxide signaling in plants. *Plants* 9: 1550. doi:10.3390/plants9111550.

Hancock, J.T., Craig, T., and Whiteman, M. (2017). Competition of reactive signals and thiol modifications of proteins. *Journal of Cellular Signalling* 2: 170.

Hancock, J.T. and Hancock, T.H. (2018). Hydrogen gas, ROS metabolism and cell signaling: are hydrogen spin states important? *Reactive Oxygen Species* 6: 389–395.

Hancock, J.T. and Neill, S.J. (2014). NO synthase in plants? *CAB Reviews* 9: 1–9.

Hancock, J.T. and Neill, S.J. (2019). Nitric oxide: its generation and interactions with other reactive signaling compounds. *Plants* 8: 41. doi:10.3390/plants8020041.

Hasanuzzaman, M., Anwar, M., and Fujita, M. (2011). Nitric oxide modulates antioxidant defense and the methylglyoxal detoxification system and reduces salinity-induced damage of wheat seedlings. *Plant Biotechnology Reports* 5: 353. doi:https://doi.org/10.1007/s11816-011-0189-9.

Hasanuzzaman, M., Nahar, K., Alam, M.M. et al. (2013). Physiological, biochemical, and molecular mechanisms of heat stress tolerance in plants. *International Journal of Molecular Sciences* 14: 9643–9684. doi:10.3390/ijms14059643.

Hasanuzzaman, M., Oku, H., Nahar, K. et al. (2018). Nitric oxide-induced salt stress tolerance in plants: ROS metabolism, signaling, and molecular interactions. *Plant Biotechnology Reports* 12: 77–92. doi:10.1007/s11816-018-0480-0.

Hichri, I., Boscari, A., Meilhoc, E. et al. (2016). Nitric oxide: a multitask player in plant-microorganism symbioses. In: *Gasotransmitters in Plants: The Rise of a New Paradigm in Cell Signaling* (eds. L. Lamattina and C. Garcia-Mata) Springer International Publishing Switzerland 2016, 239–268. ISBN: 9783319407111

Holzmeister, C., Gaupels, F., Geerlof, A. et al. (2015). Differential inhibition of Arabidopsis superoxide dismutases by peroxynitrite-mediated tyrosine nitration. *Journal of Experimental Botany* 66: 989–999.

Horchani, F., Prevot, M., Boscari, A. et al. (2011). Both plant and bacterial nitrate reductases contribute to nitric oxide production in *Medicago truncatula* nitrogen-fixing nodules. *Plant Physiology* 155: 1023–1036. doi:10.1104/pp.110.166140.

Hossain, M.A., Bhattacharjee, S., Armin, S.M. et al. (2015). Hydrogen peroxide priming modulates abiotic oxidative stress tolerance: insights from ROS detoxification and scavenging. *Frontiers in Plant Science* 6: 420. doi:10.3389/fpls.2015.00420.

Huang, J., Willems, P., Wei, B. et al. (2019). Mining for protein S-sulfenylation in Arabidopsis uncovers redox-sensitive sites. *Proceedings of the National Academy of Sciences of the United States of America* 116: 21256–21261.

Iseri, O.D., Korpe, D.A., Sahin, F.I. et al.(2013). Hydrogen peroxide pretreatment of roots enhanced oxidative stress response of tomato under cold stress. *Acta Physiologiae Plantarum* 35: 1905–1913. doi:10.1007/s11738-013-1228-7.

Jafari, M. and Daneshvar, M.H. (2020). Effects of sodium nitroprusside on indirect shoot organogenesis and in vitro root formation of *Tagetes erecta*: an important medicinal plant. *Polish Journal of Applied Sciences* 5: 14–19. doi:10.34668/PJAS.2019.5.3.03.

Jahnová, J., Luhová, L., and Petřivalský, M. (2019). S-nitrosoglutathione reductase – the master regulator of protein S-nitrosation in plant NO signaling. *Plants* 8: 48. doi:10.3390/plants8020048.

Jaiswal, A. and Srivastava, J.P. (2018). Changes in reactive oxygen scavenging systems and protein profiles in maize roots in response to nitric oxide under waterlogging stress. *Indian Journal of Biochemistry & Biophysics* 55: 26–33.

Jajic, I., Sarna, T., and Strzalka, K. (2015). Senescence, stress, and reactive oxygen species. *Planta* 4: 393–411. doi:10.3390/plants4030393.

Jeandroz, S., Wipf, D., Stuehr, D.J. et al. (2016). Occurrence, structure, and evolution of nitric oxide synthase-like proteins in the plant kingdom. *Science Signaling* 9: re2.

Jedelska, T., Luhová, L., and Petřivalský, M. (2020). Nitric oxide signalling in plant interactions with pathogenic fungi and oomycetes. *Journal of Experimental Botany* 72: 848–863. doi:10.1093/jxb/eraa596.

Jiménez, A., Martí, M.C., Camejo, D. et al. (2019). Hydrogen peroxide and nitric oxide metabolism in chloroplasts. In: *Nitric Oxide and Hydrogen Peroxide Signaling in Higher Plants* (eds. D. Gupta, J. Palma, and F. Corpas). Cham, Switzerland: Springer. doi:https://doi.org/10.1007/978-3-030-11129-8_6.

Joudoi, T., Shichiri, Y., Kamizono, N. et al. (2013). Nitrated cyclic GMP modulates guard cell signaling in *Arabidopsis*. *The Plant Cell* 25: 558–571. doi:10.1105/tpc.112.105049.

Kalinina, E. and Novichkova, M. (2021). Glutathione in protein redox modulation through S-glutathionylation and S-nitrosylation. *Molecules* 26: 435. doi:10.3390/molecules26020435.

Kaya, C., Ashraf, M., Alyemeni, M.N. et al.(2020). Responses of nitric oxide and hydrogen sulfide in regulating oxidative defence system in wheat plants grown under cadmium stress. *Physiologia Plantarum*. 168(2): 345–360. doi:10.1111/ppl.13012.

Khan, M.N., Alamri, S., Al-Amri, A.A. et al. (2020). Effect of nitric oxide on seed germination and seedling development of tomato under chromium toxicity. *J Plant Growth Regul 40*, 2358–2370. https://doi.org/10.1007/s00344-020-1021.

Khoshbakht, D., Asghari, M.R., and Haghighi, M. (2018). Effects of foliar applications of nitric oxide and spermidine on chlorophyll fluorescence, photosynthesis and antioxidant enzyme activities of citrus seedlings under salinity stress. *Photosynthetica* 56: 1313–1325. doi:10.1007/s11099-018-0839-z.

Kimura, H. (2017). Hydrogen sulfide and polysulfide signaling. *Antioxidants and Redox Signaling* 27: 619–621.

Klein, A., Hüsselmann, L., Keyster, M. et al. (2018). Exogenous nitric oxide limits salt-induced oxidative damage in maize by altering superoxide dismutase activity. *South African Journal of Botany* 115: 44–49.

Kneeshaw, S., Gelineau, S., Tada, Y. et al. (2014). Selective protein denitrosylation activity of thioredoxin-h5 modulates plant immunity. *Molecular Cell* 56: 153–162.

Kolbert, Z. and Feigl, G. (2017). Cross-talk of reactive oxygen species and nitric oxide in various processes of plant development. In: *Reactive Oxygen Species in Plants* (eds. V.P. Singh, S. Singh, D.K. Tripathi, S.M. Prasad, and D. Chauhan), 261–289. Hoboken, NJ: John Wiley & Sons.

Kolbert, Z., Feigl, G., Bordé, Á. et al. (2017). Protein tyrosine nitration in plants: present knowledge, computational prediction and future perspectives. *Plant Physiology and Biochemistry* 113: 56–63.

Kolbert, Z., Barroso, J.B., Brouquisse, R. et al. (2019). A forty year journey: the generation and roles of NO in plants. *Nitric Oxide* 93: 53–70. doi:10.1016/j.niox.2019.09.006.

Kovacs, I., Durner, J., and Lindermayr, C. (2015). Crosstalk between nitric oxide and glutathione is required for NONEXPRESSOR OF PATHOGENESIS-RELATED GENES 1 (NPR1)-dependent defense signaling in *Arabidopsis thaliana*. *The New Phytologist* 208: 860–872.

Krasensky-Wrzaczek, J. and Kangasjärvi, J. (2018). The role of reactive oxygen species in the integration of temperature and light signals. *Journal of Experimental Botany* 69: 3347–3358.

Kumar, R. (2014). Role of MicroRNAs in biotic and abiotic stress responses in crop plants. *Applied Biochemistry and Biotechnology* 174: 93–115. doi:10.1007/s12010-014-0914-2.

Kumar, S. and Trivedi, P.K. (2016). Heavy metal stress signaling in plants. In: *Plant Metal Interaction – Emerging Remediation Techniques* (ed. P. Ahmad), 585–603. Amsterdam: Elsevier. doi:10.1016/B978-0-12-803158-2.00025-4.

Kushwaha, B.K., Singh, S., Tripathi, D.K. et al. (2019). New adventitious root formation and primary root biomass accumulation are regulated by nitric oxide and reactive oxygen species in rice seedlings under arsenate stress. *Journal of Hazardous Materials* 361: 134–140.

Lamotte, O., Bertoldo, J.B., Besson-Bard, A. et al. (2015). Protein S-nitrosylation: specificity and identification strategies in plants. *Frontiers in Chemistry* 2: 114. doi:10.3389/fchem.2014.00114.

Lazaro, J.J., Jimenzez, A., Camejo, D. et al. (2013). Dissecting the integrative antioxidant and redox systems in plant mitochondria. Effect of stress and S-nitrosylation. *Frontiers in Plant Science* 4: 460. doi:10.3389/fpls.2013.00460.

Leon, J. and Costa-Broseta, A. (2020). Present knowledge and controversies, deficiencies, and misconceptions on nitric oxide synthesis, sensing, and signaling in plants. *Plant, Cell and Environment* 43: 1–15. doi:10.1111/pce.13617.

Leterrier, M., Airaki, M., Palma, J.M. et al. (2012). Arsenic triggers the nitric oxide (NO) and S-nitrosoglutathione (GSNO) metabolism in *Arabidopsis*. *Environmental Pollution* 166: 136–143. doi:10.1016/j.envpol.2012.03.012.

Lin, Y., Zhang, W., Qi, F. et al. (2014). Hydrogen-rich water regulates cucumber adventitious root development in a heme oxygenase-1/carbon monoxide-dependent manner. *Journal of Plant Physiology* 171: 1–8.

Lindermayr, C. (2018). Crosstalk between reactive oxygen species and nitric oxide in plants: key role of S-nitrosoglutathione reductase. *Free Radical Biology and Medicine* 122: 110–115. doi:10.1016/j.freeradbiomed.2017.11.027.

Lisjak, M., Teklic, T., Wilson, I.D. et al. (2013). Hydrogen sulfide: environmental factor or signaling molecule? *Plant, Cell and Environment* 36: 1607–1616.

Liu, T., Xu, J., Li, J. et al. (2019). NO is involved in JA-and H_2O_2-mediated ALA-induced oxidative stress tolerance at low temperatures in tomato. *Environmental and Experimental Botany* 161: 334–343.

Locato, V., Cimini, S., and De Gara, L. (2018). ROS and redox balance as multifaceted players of cross-tolerance: epigenetic and retrograde control of gene expression. *Journal of Experimental Botany* 69: 3373–3391.

Lushchak, V.I. (2015). Free radicals, reactive oxygen species, oxidative stress and its classification. *Chemico-Biological Interactions* 224: 164–165. doi:10.1016/j.cbi.2014.10.016.

Ma, M., Wendehenne, D., Philippot, L. et al. (2020). Physiological significance of pedospheric nitric oxide for root growth, development and organismic interactions. *Plant, Cell & Environment* 43: 2336–2354. doi:10.1111/pce.13850.

Malik, S.I., Hussain, A., Yun, B.-W. et al. (2011). GSNOR-mediated de-nitrosylation in the plant defence response. *Plant Science* 181: 540–544. doi:10.1016/j.plantsci.2011.04.004.

Martinez-Medina, A., Pescador, L., Fernande, I. et al. (2019). Nitric oxide and phytoglobin PHYTOGB1 are regulatory elements in the *Solanum lycopersicum–Rhizophagus irregularis* mycorrhizal symbiosis. 223: 1560–1574. doi:10.1111/nph.15898.

Mata-Perez, C. and Spoel, S.H. (2019). Thioredoxin-mediated redox signaling in plant immunity. *Plant Science* 279: 27–33. doi:10.1016/j.plantsci.2018.05.001.

Mengel, A., Chaki, M., Shekariesfahlan, A. et al. (2013). Effect of nitric oxide on gene transcription – S-nitrosylation of nuclear proteins. *Frontiers in Plant Science* 4: 293. doi:10.3389/fpls.2013.00293.

Minguez, P., Parca, L., Diella, F. et al. (2012). Deciphering a global network of functionally associated post-translational modifications. *Molecular Systems Biology* 8: 599.

Mittler, R. and Blumwald, E. (2015). The roles of ROS and ABA in systemic acquired acclimation. *The Plant Cell* 27: 64–70. doi:10.1105/tpc.114.133090.

Molassiotis, A., Job, D., Ziogas, V. et al. (2016). Citrus plants: a model system for unlocking the secrets of NO and ROS-inspired priming against salinity and drought. *Frontiers in Plant Science* 7: 229. doi:10.3389/fpls.2016.00229.

Moni, A., Islam, M., and Uddin, M. (2018). Role of auxin and nitric oxide on growth and development of lateral root of plants: possible involvement of exogenously induced Phot1. *Journal of Advanced Biotechnology and Experimental Therapeutics* 1: 61. doi:10.5455/jabet.2018.d11.

Montilla-Bascon, G., Rubiales, D., Hebelstrup, K.H. et al. (2017). Reduced nitric oxide levels during drought stress promote drought tolerance in barley and is associated with elevated polyamine biosynthesis. *Scientific Reports* 7: 1–15. doi:10.1038/s41598-017-13458-1.

Morales, R.C., Bahnson, E.S.M., Havelka, G.E. et al. (2015). Sex-based differential regulation of oxidative stress in the vasculature by nitric oxide. *Redox Biology* 4: 226–233. doi:10.1016/j.redox.2015.01.007.

Mostofa, M.G., Rahman, A., Ansary, M.M. et al. (2015). Hydrogen sulfide modulates cadmium-induced physiological and biochemical responses to alleviate cadmium toxicity in rice. *Scientific Reports* 5: 14078.

Mukherjee, S. and Corpas, F.J. (2020). Crosstalk among hydrogen sulfide (H_2S), nitric oxide (NO) and carbon monoxide (CO) in root system development and its rhizosphere interactions: a gaseous interactome. *Plant Physiology and Biochemistry*. 155: 800–814. doi:10.1016/j.plaphy.2020.08.020.

Munawar, A., Akram, N.A., Ahmad, A. et al.(2019). Nitric oxide regulates oxidative defense system, key metabolites and growth of broccoli (*Brassica oleracea* L.) plants under water limited conditions. *Scientia Horticulturae* 254: 7–13.

Mur, L.A.J., Mandon, J., Persijn, S. et al. (2013a). Nitric oxide in plants: an assessment of the current state of knowledge. *AoB Plants* 5: pls052. doi:10.1093/aobpla/pls052.

Mur, L.A.J., Prats, E., Pierre, S. et al. (2013b). Integrating nitric oxide into salicylic acid and jasmonic acid/ethylene plant defense pathways. *Frontiers in Plant Science* 4: 215. doi:10.3389/fpls.2013.00215.

Nabi, R.B.S., Tayade, R., Hussain, A. et al. (2019). Nitric oxide regulates plant responses to drought, salinity, and heavy metal stress. *Environmental and Experimental Botany* 161: 120–133. doi:10.1016/j.envexpbot.2019.02.003.

Ohta, S. (2015). Molecular hydrogen as a novel antioxidant: overview of the advantages of hydrogen for medical applications. *Methods in Enzymology* 555: 289–317.

Okant, M. and Kaya, C. (2019). The role of endogenous nitric oxide in melatonin-improved tolerance to lead toxicity in maize plants. *Environmental Science and Pollution Research* 26 (12): 11864–11874.

Olas, B. (2015). Hydrogen sulfide and signaling pathways. *Clinica Chimica Acta* 439: 212–218.

Osakabe, Y., Osakabe, K., Shinozaki, K. et al. (2014). Response of plants to water stress. *Frontiers in Plant Science* 5: 86. doi:10.3389/fpls.2014.00086.

Ozgur, R., Uzilday, B., Iwata, Y. et al. (2018). Interplay between unfolded protein response and reactive oxygen species: a dynamic duo. *Journal of Experimental Botany* 69: 3333–3345.

Parí, R., Iglesias, M.J., Terrile, M.C. et al. (2013). Functions of S-nitrosylation in plant hormone networks. *Frontiers in Plant Science* 4: 294.

Pasqualini, S., Meier, S., Gehring, C. et al. (2009). Ozone and nitric oxide induce cGMP-dependent and -independent transcription of defence genes in tobacco. *The New Phytologist* 181: 860–870. doi:10.1111/j.1469-8137.2008.02711.x.

Peng, R., Bian, Z., Zhou, L. et al. (2016). Hydrogen sulfide enhances nitric oxide-induced tolerance of hypoxia in maize (*Zea mays* L.). *Plant Cell Reports* 35: 2325–2340.

Prakash, V., Singh, V.P., Tripathi, D.K. et al. (2019). Crosstalk between nitric oxide (NO) and abscisic acid (ABA) signaling molecules in higher plants. *Environmental and Experimental Botany* 161: 41–49. doi:10.1016/j.envexpbot.2018.10.033.

Puyaubert, J., Fares, A., Reze, N. et al. (2014). Identification of endogenously S-nitrosylated proteins in *Arabidopsis* plantlets: effect of cold stress on cysteine nitrosylation level. *Plant Science* 21: 150–156. doi:10.1016/j.plantsci.2013.10.014.

Quiang, J., Chen, F., Kovalenkov, Y. et al. (2012). Nitric oxide reduces NADPH oxidase 5 (Nox5) activity by reversible S-nitrosylation. *Free Radical Biology & Medicine* 52: 1806–1819. doi:10.1016/j.freeradbiomed.2012.02.029.

Rai, K.K., Pandey, N., and Rai, S.P. (2020). Salicylic acid and nitric oxide signaling in plant heat stress. *Physiologia Plantarum* 168: 241–255. doi:10.1111/ppl.12958.

Rai, K.K., Rai, N., and Rai, S.P. (2018). Salicylic acid and nitric oxide alleviate high temperature induced oxidative damage in *Lablab purpureus* L. plants by regulating bio-physical processes and DNA methylation. *Plant Physiology and Biochemistry* 128: 72–88.

Rasul, S., Dubreuil-Maurizi, C., Lamotte, O. et al. (2012). Nitric oxide production mediates oligogalacturonide-triggered immunity and resistance to *Botrytis cinerea* in *Arabidopsis thaliana*. *Plant, Cell and Environment* 35: 1483–1499.

Raya-Gonzalez, J., Lopez-Bucio, J.S., and Lopez-Bucio, J. (2019). Nitric oxide and hydrogen peroxide in root organogenesis. In: *Nitric Oxide and Hydrogen Peroxide Signaling in Higher Plants* (eds. D. Gupta, J. Palma, and F. Corpas), 157–173. Cham, Switzerland: *Springer International Publishing*. doi:10.1007/978-3-030-11129-8_8.

Ren, Y., Wang, W., He, J. et al. (2020). Nitric oxide alleviates salt stress in seed germination and early seedling growth of pak choi (*Brassica chinensis* L.) by enhancing physiological and biochemical parameters. *Ecotoxicology and Environmental Safety* 187: 109785. doi:10.1016/j.ecoenv.2019.109785.

Rizwan, M., Mostofa, M.G., Ahmad, M.Z. et al. (2018). Nitric oxide induces rice tolerance to excessive nickel by regulating nickel uptake, reactive oxygen species detoxification and defense related gene expression. *Chemosphere* 191: 23–35.

Romero-Puertas, M.C. and Sandalio, L.M. (2016). Nitric oxide level is self-regulating and also regulates its ROS partners. *Frontiers in Plant Science* 7: 316. doi:10.3389/fpls.2016.00316.

Rubbo, H. (2013). Nitro-fatty acids: novel anti-inflammatory lipid mediators. *Brazilian Journal of Medical and Biological Research* 46: 728–734. doi:10.1590/1414-431X20133202.

Saddhe, A.A., Malvankar, M.R., Karle, S.B. et al. (2019). Reactive nitrogen species: paradigms of cellular signaling and regulation of salt stress in plants. *Environmental and Experimental Botany* 161: 86–97.

Sahay, S. and Gupta, M. (2017). An update on nitric oxide and its benign role in plant responses under metal stress. *Nitric Oxide* 67: 39–52. doi:10.1016/j.niox.2017.04.011.

Sakamoto, A., Ueda, M., and Morikawa, H. (2002). Arabidopsis glutathione-dependent formaldehyde dehydrogenase is an S-nitrosoglutathione reductase. *FEBS Letters* 515: 20–24.

Sami, F., Faizan, M., Faraz, A. et al. (2018). Nitric oxide-mediated integrative alterations in plant metabolism to confer abiotic stress tolerance, NO crosstalk with phytohormones and NO-mediated post translational modifications in modulating diverse plant stress. *Nitric Oxide* 73: 22–38.

Sandalio, L.M. and Romero-Puertas, M.C. (2015). Peroxisomes sense and respond to environmental cues by regulating ROS and RNS signaling networks. *Annals of Botany* 116: 475–485. doi:10.1093/aob/mcv074.

Santisree, P., Bhatnagar-Mathur, P., and Sharma, K.K. (2015). NO to drought – multifunctional role of nitric oxide in plant drought: do we have all the answers? *Plant Science* 239: 44–55. doi:10.1016/j.plantsci.2015.07.012.

Santolini, J., André, F., Jeandroz, S. et al. (2017). Nitric oxide synthase in plants: where do we stand? *Nitric Oxide* 63: 30–38.

Sanz, L., Albertos, P., Mateos, I. et al. (2015). Nitric oxide (NO) and phytohormones crosstalk during early plant development. *Journal of Experimental Botany* 66: 2857–2868.

Sevilla, F., Camejo, D., Ortiz-Espin, A. et al. (2015). The thioredoxin/peroxiredoxin/sulfiredoxin system: current overview on its redox function in plants and regulation by reactive oxygen and nitrogen species. *Journal of Experimental Botany* 66: 2945–2955. doi:10.1093/jxb/erv146.

Shams, M., Ekinci, M., Ors, S. et al. (2019). Nitric oxide mitigates salt stress effects of pepper seedlings by altering nutrient uptake, enzyme activity and osmolyte accumulation. *Physiology and Molecular Biology of Plants* 25 (5): 1149–1161.

Shapiguzov, A., Vainonen, J.P., Wrzaczek, M. et al. (2012). ROS-talk – how the apoplast, the chloroplast, and the nucleus get the message through. *Frontiers in Plant Science* 3: 292. doi:10.3389/fpls.2012.00292.

Sharma, A., Soares, C., Sousa, B. et al. (2019). Nitric oxide-mediated regulation of oxidative stress in plants under metal stress: a review on molecular and biochemical aspects. *Physiologia Plantarum* 168: ppl13004. doi:10.1111/ppl.13004.

Shi, H., Ye, T., Zhu, J.-K. et al. (2014). Constitutive production of nitric oxide leads to enhanced drought stress resistance and extensive transcriptional reprogramming in *Arabidopsis*. *Journal of Experimental Botany* 65: 4119–4131. doi:10.1093/jxb/eru184.

Silveira, N.M., Oliveira, J.A., Ribeiro, C. et al. (2015). Nitric oxide attenuates oxidative stress induced by arsenic in lettuce (*Lactuca sativa*) leaves. *Water, Air, and Soil Pollution* 226: 379. doi:10.1007/s11270-015-2630-0.

Singh, R., Gautam, N., Mishra, A. et al. (2011). Heavy metals and living systems: an overview. *Indian Journal of Pharmacology* 43: 246–253. doi:10.4103/0253-7613.81505.

Singh, R., Parihar, P., Singh, S. et al. (2017). Micro RNAs and nitric oxide cross talk in stress tolerance in plants. *Plant Growth Regulation* 83: 199–205. doi:10.1007/s10725-016-0190-y.

Singh, S. and Prasad, S.M. (2019). Management of chromium (VI) toxicity by calcium and sulfur in tomato and brinjal: implication of nitric oxide. *Journal of Hazardous Materials* 373: 212–223.

Speckmann, B., Steinbrenner, H., Grune, T. et al. (2016). Peroxynitrite: from interception to signaling. *Archives of Biochemistry and Biophysics* 595: 153–160.

Steelheart, C., Galatro, A., Bartoli, C.G. et al. (2019). Nitric oxide and hydrogen peroxide: signals in fruit ripening. In: *Nitric Oxide and Hydrogen Peroxide signaling in Higher Plants* (eds. D. Gupta, J. Palma, and F. Corpas). Cham, Switzerland: Springer. 175–199. doi:10.1007/978-3-030-11129-8_9.

Steffens, B., Kovalev, A., Gorb, S.N. et al.(2012). Emerging roots alter epidermal cell fate through mechanical and reactive oxygen species signaling. *The Plant Cell* 24: 3296–3306. doi:10.1105/tpc.112.101790.

Su, J., Zhang, Y., Nie, Y. et al. (2018). Hydrogen-induced osmotic tolerance is associated with nitric oxide-mediated proline accumulation and reestablishment of redox balance in alfalfa seedlings. *Environmental and Experimental Botany* 147: 249–260.

Sun, A., Nie, S., and Xing, D. (2012). Nitric oxide-mediated maintenance of redox homeostasis contributes to NPR1-dependent plant innate immunity triggered by lipopolysaccharides. *Plant Physiology* 160: 1081–1096.

Sun, C.L., Liu, L.J., Lu, L.L. et al. (2018). Nitric oxide acts downstream of hydrogen peroxide in regulating aluminum-induced antioxidant defense that enhances aluminum resistance in wheat seedlings. *Environmental and Experimental Botany* 145: 95–103.

Tavares, C.P., Vernal, J., Delena, R.A. et al. (2014). S-nitrosylation influences the structure and DNA binding activity of AtMYB30 transcription factor from *Arabidopsis thaliana*. *Biochimica et Biophysica Acta* 1844: 810–817. doi:10.1016/j.bbapap.2014.02.015.

Thao, N.P., Khan, M.I.R., Thu, N.B.A. et al. (2015). Role of ethylene and its cross talk with other signaling molecules in plant responses to heavy metal stress. *Plant Physiology* 169: 73–84. doi:10.1104/pp.15.00663.

Turkan, I., Uzilday, B., Dietz, K.J. et al. (2018). Reactive oxygen species and redox regulation in mesophyll and bundle sheath cells of C4 plants. *Journal of Experimental Botany* 69: 3321–3331.

Umbreen, S., Lubega, J., Cui, B. et al. (2018). Specificity in nitric oxide signalling. *Journal of Experimental Botany* 69 (14): 3439–3448. doi:10.1093/jxb/ery184.

Viehweger, K. (2014). How plants cope with heavy metals. *Botanical Studies* 55: 1–12. doi:10.1186/1999-3110-55-35.

Wang, D., Liu, Y., Tan, X. et al. (2015). Effect of exogenous nitric oxide on antioxidative system and S-nitrosylation in leaves of *Boehmeria nivea* (L.) Gaud under cadmium stress. *Environmental Science and Pollution Research* 22: 3489–3497. doi:10.1007/s11356-014-3581-5.

Wang, L., Su, H., Han, L. et al. (2014). Differential expression profiles of poplar MAP kinase kinases in response to abiotic stresses and plant hormones, and overexpression of PtMKK4 improves the drought tolerance of poplar. *Gene* 545: 141–148. doi:10.1016/j.gene.2014.04.058.

Wang, X. and Hargrove, M.S. (2013). Nitric oxide in plants: the roles of ascorbate and hemoglobin. *PLoS ONE* 8: e82611. doi:10.1371/journal.pone.0082611.

Wang, X., Hou, C., Liu, J. et al. (2013). Hydrogen peroxide is involved in the regulation of rice (*Oryza sativa* L.) tolerance to salt stress. *Acta Physiologiae Plantarum* 35: 891–900. doi:10.1007/s11738-012-1132-6.

Wang, X., Li, Q., Yang, M. et al. (2020). Crosstalk between hydrogen peroxide and nitric oxide mediates priming-induced drought tolerance in wheat. *Journal of Agronomy and Crop Science* doi:10.1111/jac.12458.

Wang, X., Zhang, M.M., Wang, Y.J. et al. (2015). The plasma membrane NADPH oxidase OsRbohA plays a crucial role in developmental regulation and drought-stress response in rice. *Physiologia Plantarum* 24: 1–34. doi:10.1111/ppl.12389.

Waszczak, C., Akter, S., Eeckhout, D. et al. (2014). Sulfenome mining in *Arabidopsis thaliana*. *Proceedings of the National Academy of Sciences of the United States of America* 111: 11545–11550. doi:10.1073/pnas.1411607111.

Waszczak, C., Akter, S., Jacques, S. et al. (2015). Oxidative post-translational modifications of cysteine residues in plant signal transduction. *Journal of Experimental Botany* 66: 2923–2934. doi:10.1093/jxb/erv084.

Williams, E., Pead, S., Whiteman, M. et al. (2015). Detection of thiol modifications by hydrogen sulfide. *Methods in Enzymology* 555: 233–251.

Williams, E., Whiteman, M., Wood, M.E. et al. (2019). Investigating ROS, RNS and H_2S sensitive signalling proteins. In: *Redox Signal Transduction: Methods and Protocols* (eds. J.T. Hancock and M. Conway). Berlin: Springer. 27–42.

Wilson, H.R., Veal, D., Whiteman, M. et al. (2017). Hydrogen gas and its role in cell signaling. *CAB Reviews* 12: 1–3.

Wu, Q., Su, N., Cai, J. et al. (2015). Hydrogen-rich water enhances cadmium tolerance in Chinese cabbage by reducing cadmium uptake and increasing antioxidant capacities. *Journal of Plant Physiology* 175: 174–182.

Xiong, Y., Uys, J.D., Tew, K.D. et al. (2011). S-glutathionylation: from molecular mechanisms to health outcomes. *Antioxidants and Redox Signaling* 15: 233–270.

Yang, H., Mu, J., Chen, L. et al. (2015). S-nitrosylation positively regulates ascorbate peroxidase activity during plant stress responses. *Plant Physiology* 167: 1604–1615.

Yu, M., Lamattina, L., Spoel, S.H. et al. (2014). Nitric oxide function in plant biology: a redox cue in deconvolution. *The New Phytologist* 202: 1142–1156. doi:10.1111/nph.12739.

Yu, M., Yun, B.W., Spoel, S.H. et al. (2012a). A sleigh ride through the SNO: regulation of plant immune function by protein S-nitrosylation. *Current Opinion in Plant Biology* 15: 424–430.

Yu, Q., Sun, L., Jin, H. et al. (2012b). Lead-induced nitric oxide generation plays a critical role in lead uptake by *Pogonatherum crinitum* root cells. *Plant and Cell Physiology* 53: 1728–1736. doi:10.1093/pcp/pcs116.

Yun, B.W., Feechan, A., Yin, M. et al. (2011). S-nitrosylation of NADPH oxidase regulates cell death in plant immunity. *Nature* 478: 264–268. doi:10.1038/nature10427.

Yun, B.W., Skelly, M.J., Yin, M. et al. (2016). Nitric oxide and S-nitrosoglutathione function additively during plant immunity. *The New Phytologist* 211: 516–526.

Zagorchev, L., Seal, C.E., Kranner, I. et al. (2013). A central role for thiols in plant tolerance to abiotic stress. *International Journal of Molecular Sciences* 14: 7405–7432. doi:10.3390/ijms14047405.

Zeng, J., Ye, Z., and Sun, X. (2014). Progress in the study of biological effects of hydrogen on higher plants and its promising application in agriculture. *Medical Gas Research* 4: 1–7.

Zhou, J., Wang, J., Shi, K. et al. (2012). Hydrogen peroxide is involved in the cold acclimation-induced chilling tolerance of tomato plants. *Plant Physiology and Biochemistry* 60: 141–149. doi:10.1016/j.plaphy.2012.07.010.

Zhou, K. and Zhang, J. (2014). Nitric oxide in plants and its role in regulating flower development. *Yi Chuan* 36: 661–668.

Zhou, X., Joshi, S., Khare, T. et al. (2021). Nitric oxide, crosstalk with stress regulators and plant abiotic stress tolerance. *Plant Cell Reports*. 40(8): 1395–1414. doi:https://doi.org/10.1007/s00299-021-02705-5.

Zhou, X., Joshi, S., Patil, S., Khare, T., and Kumar, V. (2022). Reactive Oxygen, Nitrogen, Carbonyl and Sulfur Species and Their Roles in Plant Abiotic Stress Responses and Tolerance. *Journal of plant growth regulation*. 41(1), 119–142.

Zhu, Y., Liao, W., Niu, L. et al. (2016a). Nitric oxide is involved in hydrogen gas-induced cell cycle activation during adventitious root formation in cucumber. *BMC Plant Biology* 16: 146.

Zhu, Y., Liao, W., Wang, M. et al. (2016b). Nitric oxide is required for hydrogen gas-induced adventitious root formation in cucumber. *Journal of Plant Physiology* 195: 50–58.

8

Ascorbate–Glutathione Cycle

Nitric Oxide and Phytohormone Interactions for Plant Stress Tolerance

Priti Chauhan[1], Rayees Ahmad Mir[2], and Mushtaq Ahmad Khah[3]

[1]*Department of Bioscience and Biotechnology, Banasthali Vidyapith, Rajasthan, India*
[2]*School of Studies in Botany, Jiwaji University, Gwalior, India*
[3]*Department of Botany, Shri JJT University, Jhunjhunu, Rajasthan, India*

8.1 Introduction

Global warming is a potential threat to agriculture the world over and leads to the occurrence of various stresses (biotic and abiotic) that alter plant growth, ultimately affecting and diminishing agricultural productivity (Shafi and Zahoor 2020). These conditions induce an overproduction of reactive oxygen species (ROS), which are extremely toxic and can disrupt proteins, lipids, carbohydrates, and DNA in plants, affecting various cellular processes and ultimately resulting in oxidative damage (Gill and Tuteja 2010; Ahanger et al. 2017, 2019). To fight these hostile conditions, plants are equipped with well0developed defense systems like antioxidants, osmolytes, etc., which protect growth and development by scavenging excess ROS. Among the antioxidants, ascorbate (AsA), glutathione (GSH), tocopherol, flavonoids, alkaloids, and carotenoids are part of the nonenzymatic defense system (Ahanger et al. 2017; Hasanuzzaman et al. 2019), with enzymatic components including superoxide dismutase (SOD), catalase (CAT), ascorbate peroxidase (APX), dehydroascorbate reductase (DHAR), monodehydroascorbate reductase (MDHAR), glutathione reductase (GR), glutathione peroxidase (GPX), and peroxiredoxin (PRX) (Pandey et al. 2015; Hasanuzzaman et al. 2017; Ahanger et al. 2017, 2021; Iqbal et al. 2021).

Among the nonenzymatic antioxidant components, AsA and GSH are essential molecules regulating various fundamental plant processes including the AsA-GSH cycle and maintenance of cellular redox homeostasis (Noctor and Foyer 1998, 2012). This cycle is operational in different cellular compartments in plants (Mittler et al. 2004) and the enzymes of this pathway exist in multiple isoforms, whose expression is regulated both developmentally and spatially; however, the expression of these enzyme isoforms is affected under different environmental conditions, which helps to maintain the redox balance under extreme/stressful environments (Pandey et al. 2015). There are various reports available which have shown that upregulation of different components of the AsA-GSH cycle (enzymatic and nonenzymatic) confers better stress tolerance in plants by efficiently reducing ROS (see Hasanuzzaman et al. 2019).

In plants, nitric oxide (NO) is another vital endogenous free radical molecule with multifaceted physiological roles. NO controls a variety of physiological functions ranging from seed germination to senescence. NO is also a primary molecule that regulates signaling and metabolic processes of plants and enables them to combat stressful conditions (Asgher et al. 2017; Singh et al. 2017; Corpas and Palma 2018; Kushwaha et al. 2019). In addition, it has been speculated for various plant species that NO is involved in regulating the AsA-GSH cycle (Shan et al. 2012; Wu et al. 2019). Endogenous NO molecules also induce enzymatic activities of the AsA-GSH cycle, which imparts stress tolerance to plants (Ahmad et al. 2018; Sehar et al. 2019). NO is a ubiquitous key molecule that plays a significant role in a variety of environmental reactions and in endogenous cues as well. NO interacts with phytohormones and other endogenous compounds regularly during the signaling events and recent studies have shown that all major plant hormones may influence the endogenous levels of NO to some extent (Freschi 2013).

Phytohormonal signals are triggered by plants during growth, development, and environmental responses. Events that are regulated by their action are confirmed by mutation and overexpression studies of phytohormone metabolism (He et al. 2012; Domingos et al. 2015). NO been suggested as a rapidly induced and efficient regulator for plant growth over recent years. In hormone signaling, NO serves as a crossroads to activate metabolic and physiological responses. Since plant hormones may induce root elongation, research into the relationship of NO with plant hormones will make it easier to understand the function of NO (Campos et al. 2019).

Plant hormones are small organic molecules or compounds, naturally present in very low concentrations that affect various physiological processes (Davies 2004). These are chemical messengers that control cellular processes in plants (Fleet and Williams 2011). There are about nine groups of phytohormones reported so far, namely auxins (AUXs), cytokinins (CKs), ethylene (ET), gibberellins (GAs), abscisic acid (ABA), salicylates (SAs), jasmonates (JAs), brassinosteroids (BRs), and strigolactones (SLs) (Su et al. 2017). Recent studies have confirmed that NO interacts with signaling pathways of plant hormones and regulates plant growth, development, and metabolism. NO-regulated plant genes code for glutathione synthesis enzymes, jasmonic acid and ET synthesis enzymes, phenylpropanoid synthesis enzymes, and AUX-responsive proteins (Wendehenne et al. 2004; Innocenti et al. 2007). Liu et al. (2005) confirmed the involvement of JA in signal transduction. Under drought conditions, exogenous application of JA generates NO, which upregulates the AsA-GSH cycle activity in wheat leaves (Shan et al. 2015). This chapter discusses the recent developments in plant science research regarding the regulation of the AsA-GSH pathway in plants under extreme environmental conditions and the significant role played by NO in phytohormonal signaling, to regulate growth and development in plants under such conditions.

8.2 AsA-GSH Cycle

The AsA-GSH cycle, also known as the Halliwell–Asada cycle, is an effective mechanism for cells to detoxify H_2O_2 in cellular compartments when no catalysts are present (Hasanuzzaman et al. 2012, 2019). It consists of four enzymatic components in a sequence of reactions, APX, MDHAR, DHAR, and GR; and two nonenzymatic components, ascorbic acid and glutathione (Pandey et al. 2015; Hu et al. 2019). This pathway exists in chloroplasts, mitochondria, cytoplasm, and peroxisomes (Hasanuzzaman et al. 2017). APX is the primary enzyme of the AsA-GSH pathway that detoxifies the H_2O_2 via AsA peroxidation and MDHA production. MDHA is either directly reduced to AsA by MDHAR or is interconverted nonenzymatically to AsA (Szarka et al. 2012; Farooq et al. 2016). Because the reducing agent DHA may undergo irreversible hydrolysis to 2,3-diketogulonic acid when it is not reduced by DHAR, the next stage of the cycle is DHAR-mediated reduction of DHA into AsA utilizing GSH. As a result, DHAR contributes to AsA regeneration and also plays a key role in maintaining the AsA pool (Gallie 2012). GSH regeneration is just as fundamental as AsA regeneration. The GSH is regenerated by NADPH-dependent action of GR on oxidized glutathione (GSSG) (Gill and Tuteja 2010).

Therefore, the AsA-GSH cycle has a significant role in balancing ascorbate redox homeostasis in plants, which further governs the growth and development of plants. The redox state of ascorbate can be controlled by employing exogenous substances the AsA-GSH cycle (Jung et al. 2019). Enzymes such as MDHAR and DHAR are responsible for maintaining the reduced form of the AsA pool, along with regulating the metabolic ratio of AsA to its oxidation state, i.e. DHA (Anjum et al. 2014). This reduced AsA pool is utilized by APX during ROS metabolism as the specific electron donor. As a result, AsA, its redox coupling (AsA/DHA), and associated enzymes like APX, DHAR, and MDHAR collectively produce an AsA redox system that effectively protects plants against the negative effects of ROS and its derivatives.

8.2.1 Components of AsA-GSH Cycle

8.2.1.1 APX

In the AsA-GSH pathway APX is the first enzyme that utilizes AsA as the basic electron donor to convert H_2O_2 into H_2O and O_2 (Asada 1999; Conklin and Barth 2004). APX scavenges H_2O_2, which is predominantly involved in the AsA-GSH cycle; it also catalyzes MDHA reactions in plant cells and generates substantial DHA from MDHA (Ullah et al. 2016). There are various isoforms of APX that are present among various subcellular compartments of the cells like mitochondria, chloroplasts, peroxisome, and cytosol. APX is the key ROS-regulating enzyme acting in these subcellular compartments (Caverzan et al. 2012). The expression of APX genes is regulated against different biotic and abiotic environmental stresses, and also throughout plant growth/development. The APX response is directly engaged in plant cell defense against unfavorable environmental conditions (Caverzan et al. 2012). There are seven APX genes present in *Lycopersicum esculentum* in which three are cytosolic, two peroxisomal, and two chloroplast localized (Najami et al. 2008; Anjum et al. 2016; del Rio and Lopez-Huertas 2016). Eight members of the APX gene family have also been identified in *Oryza sativa* (Texeira et al. 2006).

Some of these APX isoforms are resistant to stress and are affected by almost all types of environmental stresses (Akram et al. 2017). For example, three *APX* genes of *O. sativa*, namely *OsAPX2*, *OsAPX7*, and *OsAPX8*, showed alteration in transcript levels against NaCl-imposed salinity stress. The *OsAPX2* and *OsAPX7* genes showed enhanced expression where *OsAPX8* accumulation showed strong suppression against salinity stress (Teixeira et al. 2006). Similarly, during salt stress, the transcriptional level of the peroxisomal APX gene (*HvAPX1*) increased significantly in *Hordeum vulgare* (Shi et al. 2001). On the other hand, *Arabidopsis thaliana apx3* knockout mutants showed no abnormal growth or development when subjected to both normal and stressed environments. This lack of the peroxisomal isoform among these plants has been possibly compensated by other antioxidant enzymes (Narendra et al. 2006). *Arabidopsis thaliana* plants exposed to combined heat and drought stress accumulate APX1 protein and mRNA. In comparison with wild type (WT) plants, cytosolic APX1-deficient mutants accumulated higher H_2O_2 and were more susceptible to stress combinations. In contrast, thylakoid APX-deficient plants are less susceptible than APX1-deficient mutants and WT plants to this combination of stress. The cytosolic *APX1* gene may also play a key role in plant adaptation to such stressful conditions (Koussevitzky et al. 2008). Thus APX isoforms play an important role as protective elements against adverse environmental conditions.

8.2.1.2 MDHAR

MDHAR helps to regenerate AsA and several isoforms are found in different organisms (Hasanuzzaman et al. 2018). The susceptibility of plants to various stress conditions, such as intense light, causes rapid oxidation of AsA to MDHA in the chloroplast (Polle 2001; Pandey et al. 2015). MDHA is rapidly reduced to AsA in chloroplasts by photoreduced ferredoxin, and this pathway is usually the main source of AsA regeneration in the region of the thylakoid membrane (Kavitha et al. 2010). In the AsA-GSH pathway, reduction of MDHA occurs through two enzymes, i.e. DHAR and MDHAR. Using reduced nicotinamide adenine dinucleotide phosphate (NAD(P)H) as an electron donor, MDHAR immediately reduces MDHA (Banerjee and Roychoudhury 2018. Conversely, to obtain AsA and DHA, two MDHA molecules may interact nonenzymatically. Nevertheless, MDHAR has been found to reduce the majority of MDHA (Haroldsen et al. 2011). MDHAR enzyme activity is present in the whole kingdom of plants and animals. MDHARs of plants exhibit a high degree of sequence similarity with prokaryotic flavoenzymes (Pandey et al. 2015). There are 3 MDHAR genes in tomatoes, 5 genes in *A. thaliana*, 6 isoforms in *O. sativa*, and as many as 19 genes in *Triticum aestivum*, according to studies (Leterrier and Cagnac 2018).

Both MDHAR and DHAR are essential constituents in regenerating important antioxidant molecule ascorbate in plants for the maintenance of ROS-scavenging ability. There are various reports available that have shown the role of MDHAR in stress tolerance in many test species; for example, by regulating redox homeostasis, proteostasis, and the ascorbate pool, *O. sativa* (OsMDHAR)-expressing yeast cells showed improved tolerance to H_2O_2 followed by increased functioning of enzymatic antioxidants in comparison with the WT cells. In addition, it also improves the survival rate of transgenic yeast during fermentation and the fermentative capacity (Kim et al. 2016). Hasanuzzaman et al. (2019) revealed that MDHAR was found to reduce phenoxyl radicals like quercetin, ferulic acid, chlorogenic acid, and coniferyl alcohol. In addition, co-expression of MDHAR and DHAR from transgenic *Brassica rapa* (*Br*MDHAR and *Br*DHAR) effectively confers tolerance to stress by reducing freezing-induced oxidative damage (Shin et al. 2013).

8.2.1.3 DHAR

DHAR expression in cells is responsible for AsA regeneration from its oxidized state, as well as regulating AsA cellular redox status, thereby influencing the cell response and tolerance to stress-induced ROS (Bielen et al. 2013). AsA, the major antioxidant in plants, is oxidized as a free radical intermediate to DHA by a series of reversible electron transfers with MDHA (Ushimaru et al. 2006). With GSH as an electron donor, DHAR reduces DHA to AsA reflecting the importance of DHAR in its generation (Bhuyan et al. 2019). DHAR plays an important part in plant adaptation to abiotic stresses and a variety of abiotic stress factors activate its expression as well (Fan et al. 2014).

Five different DHARs have been identified in *A. thaliana*, namely At1g19550, At1g19570, At1g75270, At5g36270, and At5g16710, which are found in mitochondria, chloroplasts, or cytosol (Pandey et al. 2015). In *Spinacia oleracea* leaves, two distinct DHAR isoforms have been found; one isoform is localized in chloroplasts and another in a subcellular compartment other than chloroplasts (Shimaoka et al. 2000). In *Eucalyptus*, two DHAR genes responsible for cytosolic and chloroplastic DHARs have also been reported (Teixeira et al. 2005; Padey et al. 2015). Various researchers have observed that overexpression of DHAR ameliorates the oxidative damage in plants induced by different environmental stresses. Kim et al. (2014) reported that overexpression of *OsDHAR1* in transgenic *O. sativa* L. increases salinity stress tolerance through regulation of the AsA pool and redox homeostasis. Similarly, in transgenic tomato, overexpression of *PbDHAR2* from *Pyrus sinkiangensis* reduces stress-induced deleterious effects by increasing the expression of AsA-GSH cycle genes, which demonstrates that the regulation of *PbDHAR2* during AsA regeneration contributes to improved salt and chilling tolerance in tomato (Qin et al. 2015). The expression of *JcDHAR* gene from *Jatropha curcas* displayed improved tolerance to salt/Polyethylene glycol-induced oxidative stress in transgenic tobacco (Chang et al. 2017).

8.2.1.4 GR

By regenerating reduced glutathione, GR retains the cellular redox state thus retaining the equilibrium between both reduced GSH and the AsA pool (Rao et al. 2006). GR is one of the important enzymes in maintaining the GSH pool via the AsA-GSH pathway and plays an important role in maintaining the sulfhydryl (-SH) group. Glutathione-S-transferases utilize GR as a substrate, and it is also used in transgenics to develop stress-tolerant plants (Yousuf et al. 2012). In plants GR proteins have been classified into two classes/isoforms (GR1 and GR2) on the basis of the presence of an *N*-terminal (Trivedi et al. 2013). It has been reported that modulation in the expression profile of different GR isoforms occurs under different stress conditions (Gill et al. 2013). Drought stress induces the upregulation of cytosolic GR gene expression directly related to the severity of the stress in *Vigna unguiculata* (Contour-Ansel et al. 2006). Pilon-Smits et al. (2000) reported two times higher activity of GR in cytosolic transgenic plants (cytGR) in comparison with the WT under metal stress in *Brassica juncea*; however, in plastidic transgenics (cpGR) the activity was 50 times higher.

8.2.1.5 AsA

AsA is the primary antioxidant component of the AsA-GSH cycle. Ascorbate is an active and multifaceted biomolecule that is present in most plant species. AsA can be produced from four different pathways: the L-gulose, L-galactose, D-galacturonic acid, and myoinositol pathways (Mellidou et al. 2012; Ishikawa et al. 2018). The decisive role of

AsA is scavenging excess ROS produced in plants that are exposed to various environmental stresses, which severely disturbs growth and development by damaging cell composition (Choudhury et al. 2017). As a component of the antioxidant defense system, AsA can efficiently scavenge the accumulated ROS via direct or indirect pathways, thus protecting the plants by eliminating the stress-imposed oxidative damage and enhancing stress tolerance (Akram et al. 2017). It is also essential for the mitigation of ROS-induced damage caused in hypoxic and anoxic roots (Bai et al. 2010). From the remarkably high concentrations present in many tissues, the importance of this antioxidant in plants is noticeable. It is identified in different subcellular organelles of *A. thaliana*, such as the cell wall, cytosol, mitochondria, vacuoles, chloroplasts, and peroxisomes (Dabrowska et al. 2007).

In mitochondria, the final stage of AsA synthesis takes place and then it is transferred to the remaining compartments of the cells. AsA is able to traverse the plasma membrane by transporters/exchangers particular to AsA/DHA that have demonstrated a strong affinity for DHA (Horemans et al. 1997). It also acts as a significant co-factor in the production of certain phytohormones. As a result, it is indeed thought that AsA influences phytohormone-mediated signaling pathways throughout the shift from the vegetative to the reproductive phase, as well as during senescence (Barth et al. 2006). AsA is also involved in several other biological mechanisms, such as growth and development modulation, photosynthesis, hormone biosynthesis, and plant acclimation to extreme environmental conditions (Bartoli et al. 2017). Hemavathi et al. (2010) reported overexpression of genes related to the AsA biosynthetic pathway enhances the vitamin C content. Co-expression of *NCED* and *ALO* in transgenic *Nicotiana tabacum* plants increases the vitamin C content and tolerance to abiotic stresses (Bao et al. 2016). In addition, overexpression of *AtOxR* gene localized to the endoplasmic reticulum (ER) resulted in AsA accumulation, decreased H_2O_2 content, and improved stress tolerance in transgenic *A. thaliana* (Bu et al. 2016). Therefore, increased AsA content enhances stress tolerance and nutritional quality in crops.

8.2.1.6 GSH

GSH is categorized into three different amino acids, namely glutamine (Glu), cysteine (Cys), and glycine (Gly), but certain plants may possess GSH homologues (Noctor et al. 2012). GSH plays a decisive role in reserving Cys, maintaining redox equilibrium, including antioxidant defense, detoxification of xenobiotics, regulation of cell cycle, and apoptosis (Gill et al. 2013). It has been suggested that GSH acts as a signal transported from the shoot through the phloem, inhibiting root sulfate uptake, therefore controlling plant sulfur homeostasis (Rennenberg and Herschbach 2014). Cells typically have three major GSH reservoirs: the cytosol; mitochondria; and ER (Nahar et al. 2016). The thiol group is precise to γ-glutamyltranspeptidase (GGT) and offers greater stability to GSH (Zagorchev et al. 2013; Nahar et al. 2016). The direct involvement of GSH in detoxification of ROS and methylglyoxal (MG) and their reaction products has been reported in plant cells, which provides protection to plants against stress-imposed oxidative damage. Glutathione serves as an important molecule for a number of physiological processes, particularly during different abiotic stresses. Glutathione is oxidized to GSSG in association with AsA turnover. Other thiol-dependent enzymes, including GPX and

GST, utilize GSH as a co-factor, hence it is oxidized to GSSG, which is then reduced back to GSH using GR catalysis (Hasanuzzaman et al. 2019). It has also been reported that exposure of plants to various abiotic stresses considerably increases the activities of GSH utilizing and regenerating enzymes that are important for stress tolerance. The significance of GSH against stress tolerance has been confirmed by use of transgenic plants, in which plants overexpressing genes of GSH biosynthesis and genes allied with regulating GSH levels, had enhanced GSH levels and showed augmented tolerance to individual stressors (Gaber et al. 2017).

8.3 NO

NO is an active molecule in signaling cascades linked to core plant biology roles, such as development and growth, stomatal movement (Scuffi et al. 2016), leaf senescence (Kong et al. 2016), and responses to different environmental stresses (Fatma et al. 2016; Begara-Morales et al. 2018). NO transmits its activity primarily through posttranslational modification, for example both *S*-nitrosylation and tyrosine nitration, which may control the role of target proteins (Begara-Morales et al. 2014). Under stressful environments, these NO posttranslational modifications (NO-PTMs) could be involved in cell signaling (Corpas et al. 2015). Many other ROS-scavenging enzymes can also be affected by NO-based PTMs and some of these have been reported to be inversely regulated by *S*-nitrosylation and nitration, such as APX and SOD (Yang et al. 2015; Kolbert and Feigl 2017). Thus, for antioxidant enzyme activities, NO-related PTMs can function as an on–off switch.

8.3.1 S-nitrosylation

S-nitrosylation contains *S*-nitrosothiols (SNOs) in addition to a cysteine thiol of an NO group, which can modify the role of a wide range of proteins (Hess et al. 2005). *S*-nitrosylation of GSH and the predominantly low molecular weight of *S*-nitrosothyl generate *S*-nitrosoglutathione (GSNO). It is referred to as an NO cell reservoir (Durner et al. 1999; Leitner et al. 2009), which is active in signaling pathways due to its phloem mobility. Transnitrosylation processes, in which a new *S*-nitrosothiol is produced by changing its NO group to a new cysteine thiol group, can be mediated by GSNO (Hess et al. 2005). Thiol-based protein modulation is related to alterations in the redox homeostasis of cells and there are several main players involved in the alteration of thiol. Thioredoxins (TRXs) are ubiquitous protein molecules that, by reversibly oxidizing their redox active motif Cys(x)2Cys to disulfide, catalyze the significant reduction of thiol-disulfide in target proteins. Oxidized TRX reduction requires NADPH-dependent thioredoxin reductase (NTR) activity in the mitochondria and cytosol; therefore, assessing the status of protein thiol-disulfide is a key factor in understanding the redox status of the cell. In *A. thaliana* there are three different NTR proteins: NTRA and NTRB in mitochondria and cytosol; and NTRC in the plastids (Alkhalfioui et al. 2007). The NTR-TRX system (NTS) is involved in monitoring the reduced or oxidized state of the Cys residues in proteins and the TRX pool in the cell. The NTS, particularly *S*-nitrosothiols, is capable of cleaving reversible oxidations.

8.3.2 Tyrosine Nitrate

Primarily tyrosine nitration involves the mediation of reactive nitrogen species (RNS), for instance, peroxynitrite ($ONOO^-$) and nitrogen dioxide (NO_2), which are generated as secondary products of NO metabolism in the presence of oxidants like H_2O_2, superoxide radicals ($\bullet O_2^-$), and transition metal centers (Radi 2004). There are two important mechanisms of biological nitration, the peroxynitrite and hemeperoxidase pathways, which lead to the formation of tyrosyl radicals and NO_2, which further combine to form 3-nitrotyrosine (Gow et al. 2004; Radi 2004). This transition changes tyrosine to a negatively charged residue, resulting in a significant change in the pKa of the hydroxyl group (Turko and Murad 2002; Abello et al. 2009), which may influence target proteins (Souza et al. 2008; Radi 2013). In mammalian cells, tyrosine nitration is an irreversible process and is a nitrosative stress signal in which the presence of tyrosine denitrase activity reduces 3-nitrotyrosine (Gorg et al. 2007; Deeb et al. 2013), pointing to a role of tyrosine nitration in signal-mediated processes.

8.4 AsA-GSH Cycle and NO

The AsA-GSH cycle defends the cellular components against oxidative damage, but its components, specifically proteins like APX, DAR, HMDHAR, and GR, are also susceptible to oxidative damage, which can change their function, thus breaking down the antioxidant protection. In plants, multiple oxidative and reductive biochemical pathways can generate NO (Hasanuzzaman et al. 2018). NO interacts with the reduced form of GSH in the GSH pool and generates GSNO, which is further catalyzed by GSNO reductase (GSNOR) action that releases NO and GSSG, thus maintaining equilibrium between NO and nitrosothiols and regulating the redox state in the cell (Romero-Puertas and Sandalio 2016). In addition, *S*-nitrosylation can alter protein interactions, consequently tinkering with the antioxidant response (Yang et al. 2015). Studies indicate that all proteins of the AsA-GSH cycle are influenced by *S*-nitrosylation and/or nitration, which result from interaction with NO (see Figure 8.1). APX enzymes are the most studied enzymes of the AsA-GSH cycle and are specifically regulated by NO metabolism. The inactivation of APX1 is induced by Cys32 oxidation (Yang et al. 2015). In comparison, under salt stress the APX1 active site (Cys32) increases its activity by nitrosylation, which is followed by PTM that increases both oxidative stress and *S*-nitrosothiols (Begara-Morales et al. 2015). MDHAR is negatively modified by nitration that induces enzymatic inactivity by alteration in the location of the co-factor binding site, which thereby disrupts the recycling mechanism of AsA (Begara-Morales et al. 2015). Information on the nitration and activity regulation of DHAR proteins is available (Tanou et al. 2012). Nevertheless, nitration is also targeted against GR, which is also stated to inhibit mammalian cell development, but *Pisum sativum* chloroplastic and cytosolic GR are not influenced by nitration.

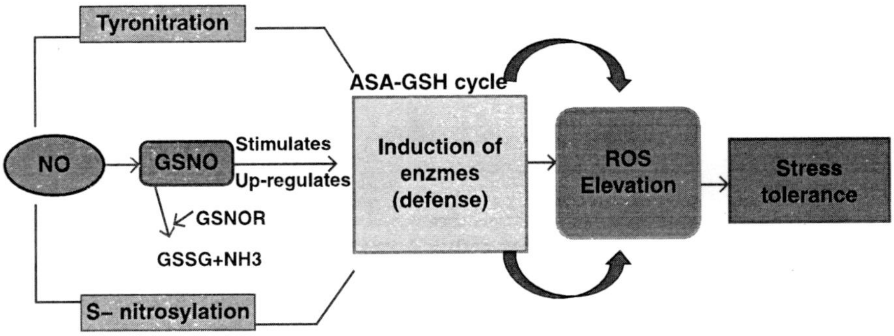

Figure 8.1 Diagram showing how NO imparts stress tolerance to plants by upregulating the enzymes of the AsA-GSH cycle for enhanced ROS detoxification.

8.5 AsA-GSH Cycle Interaction with Phytohormones

In plant cells the AsA-GSH cycle is the main pathway that links H_2O_2 scavenging to redox signaling and, because of its association with phytohormones, the AsA-GSH cycle has recently gained a lot of interest. AsA modulates plant growth by regulating phytohormone biosynthesis and subsequently hormone-mediated signal transduction under adverse environmental circumstances (Pastori et al. 2003; Kim et al. 2017). For example, improved levels of gibberellic acid (GA), indole acetic acid (IAA), zeatin, and brassinosteroids (BRs), and reduced contents of ABA have been reported by Sadak et al. (2013) upon supplementation of AsA and citric acid (2:1) ratio. Similarly, Ye et al. (2012) reported in *O. sativa* plants that there is a possible interaction between ascorbate and phytohormones. Since AsA is a co-factor for the biosynthesis of phytohormones like GA, ET, and ABA (Ye et al. 2012; Zhang 2013), through their interplay with phytohormones cellular ascorbate levels influence different signal transduction pathways in order to regulate growth and development (see Akram et al. 2017).

Hormonal signaling influences cell growth and modulates successful signaling processes (Zhou et al. 2017). Association of plant growth regulators with glutathione is important in the establishment of plants under normal as well as extreme environments. Phytohormones including SA, ABA, AT, AUX, CK (Marrs 1996), methyl jasmonate (Moons 2003), and BTs promote GST production in plants (Deng et al. 2007). Glutathione stimulates the production of ABA, which regulates seed germination, stomatal conductance, and the rate of transpiration (Chen et al. 2012). It has been observed that the enzymes of the AsA-GSH cycle and GSH levels are related to ROS detoxification. Glutathione is believed to have growth-regulating effects; increased endogenous GSH facilitates cell division in the meristematic zone of root (Vernoux et al. 2000). This elongation is a crucial morphological adaptation to drought. Exogenously introduced selenium decreases the stress-imposed oxidative damage and increases drought resistance by enhancing endogenous GSH levels, which is beneficial. The enzyme activity of the AsA-GSH cycle and the ratio of GSH to GSSG are all increased (Hasanuzzaman et al. 2011). On the other hand, GSH activates ethylene biosynthetic enzymes. GSH also increases the level of serine acetyl transferase and confers tolerance to Ni toxicity (Datta et al. 2015).

Akter et al. (2012) reported in *A. thaliana* that GSH is involved in closure of stomata triggered by ABA and methyl jasmonate (MeJA). Under oxidative stress, overexpression of genes related to SA metabolism in rice increased SA and GSH levels (Singh et al. 2015).

AUX regulates the architecture of the *A. thaliana* root system in a way that is likely independent of melatonin signaling (Yin et al. 2019). SA increases Pb-S tolerance in pepper seedlings, as well as activates the AsA-GSH cycle-related enzymes APX, GR, DHAR, and MDHAR (Kaya 2020). Similarly, salicylic acid improves the enzymatic antioxidant activity in *Artemisia aucheri* under in vitro conditions and reduces the negative effects of PEG-induced water stress. Furthermore, SA's protective role has been linked to the promotion of the AsA-GSH cycle (Abbaspour and Ehsanpour 2016). There are various other reports linked, which are shown in Table 8.1.

8.6 NO and Phytohormone Interaction

NO often associates with phytohormones and other endogenous chemicals during signaling processes that lead to plant responses (Santner et al. 2009). This suggests that almost all major plant hormones can influence endogenous levels of NO, at least to some extent. There is no longer any ambiguity about the fact that NO acts both downstream and upstream of

Table 8.1 Impact of nitric oxide (NO) on the ascorbate–glutathione (AsA-GSH) cycle under various environmental stresses.

Plant	Stress	NO-mediated effects on AsA-GSH cycle	References
Capsicum annuum	Salinity	NO induced upregulation of AsA-GSHpathway in pepper plants under salt stress	Kaya et al. (2020)
Solanum lycopersicum	Cadmium	Upregulated antioxidant enzymes of AsA-GSH cycle; enhanced the AsA and GSH concentrations and reduced Cd stress	Ahmad et al. (2018)
Brassica pekinensis	Low light	Induced and promoted the activities of AsA-GSH cycle-related enzymes (GR, MDHAR, APX, DHAR, AAO); appropriate ammonium induced NO biosynthesis	Hu et al. (2019)
Agropyron cristatum	Water	Regulated the AsA-GSH cycle	Shan et al. (2012)
Glycine max	Arsenic	Triggered upregulation of AsA-GSH cycle components; NO mitigates arsenic toxicity	Singh et al. (2021)
Triticum aestivum	Aluminum	NO enhanced DHAR and GR activity; increased GSH and GSSG ratios; activated enzymes of GSH metabolism	Sun et al. (2015)
Nitraria tangutorum	NaCl	NO enhanced the APX enzymes in the AsA-GSH pathway	Yan et al. (2018)
T. aestivum	Saline, arsenic, drought	Upregulation of antioxidant enzymes of the AsA-GSH cycle by NO treatment	Hasanuzzaman et al. (2011); Hasanuzzaman and Fujita (2013); Shan et al. (2015)

AAO, ascorbic acid oxidase; APX, ascorbate peroxidase; DHAR, dehydroascorbate reductase; GR, glutathione reductase; MDHAR, monodehydroascorbate reductase.

phytohormones (Terrile et al. 2012; Freschi 2013). It has already been confirmed that NO can influence the biosynthesis, catabolism, cognition, transmission, and transduction of various phytohormones such as AUX, GA, CK, ABA, ET, SA, JA, and BR during the induction of different plant responses (Freschi 2013; Egamberdieva et al. 2017). Since these plant hormones have a the ubiquitous role in improving resistance of plants to different environmental stresses, it has been presumed that the synthesis of NO is connected to hormone-mediated stress responses (Nawaz et al. 2017). In this section we describe different examples highlighting the role of NO-mediated hormone-regulated processes in plants, and also the influence of NO in regulation of different phytohormones and vice versa (see Figure 8.2).

Clearly, placing NO in a signaling pathway downstream of hormonal stimulus implies that the mechanism of NO synthesis, its degradation, conjugation, and deconjugation, can be influenced by some stage between the detection of hormone stimulation and triggering of the plant response (Manjunatha et al. 2010). As a result, the time interval between the transmission of the hormonal signal and recognition of variations in the endogenous levels of NO is important information. It has been identified in certain situations that lag phases have been correlated to alterations in the amount of transcripts or NO-synthesizing protein abundance or extraction of enzymes (Freschi et al. 2010). But a lag period has been identified between treatment with phytohormones and higher levels of endogenous NO in certain specific conditions, indicating that posttranslational regulation of NO

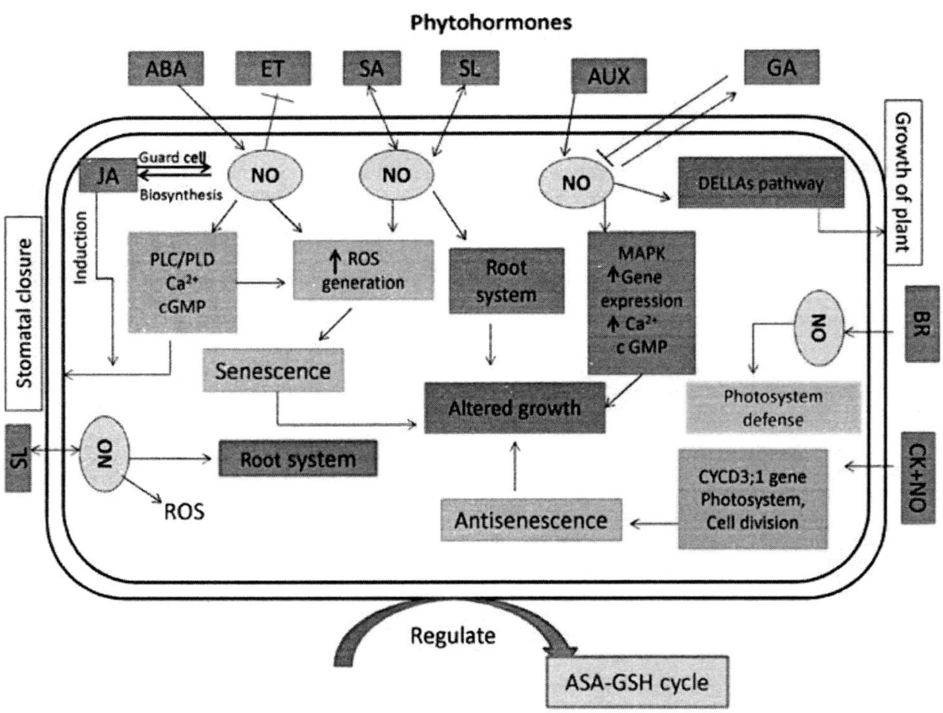

Figure 8.2 Diagrammatic representation of NO acting as a regulatory molecule in hormone-mediated plant responses. Plant hormones trigger NO levels, which regulate growth, development, and metabolism in plants including the AsA-GSH cycle.

metabolism-associated proteins, rather than their de novo synthesis, may be involved in some specific conditions (Tun et al. 2006; Sun et al. 2010). When phytohormones act upstream, NO either regulates phytohormone level distribution, for example their biosynthesis, degradation, and conjugation of enzymes; or signaling either at the transcriptional or posttranslational stages – this modification has been shown to occur (Bethke et al. 2007; Feng et al. 2013). The interactions of plant hormones with NO in different plants are shown in Table 8.2.

8.6.1 NO and AUX

AUX and NO have synergistic effects in the regulation of plant responses, which include cell division initiation and creation embryogenic cells, as well as gravitropic responses, root organogenesis, root nodule formation, and response of roots to iron deficit (Pagnussat et al. 2002, 2004; Lanteri et al. 2006; Pii et al. 2007; Chen et al. 2010). NO was found to act downstream of AUXs in substantially all of these instances, presumably via linear signaling pathways. Exogenous AUX treatment or AUX overproducer mutants typically lead to increased NO generation, which is especially noticeable in plant cells or tissues undergoing physiological AUX-dependent reactions (Pagnussat et al. 2002; Lombardo et al. 2006; Chen et al. 2010). On the contrary, other investigations have found that AUXs have no or just a little effect on NO production (Tun et al. 2001; Guo et al. 2003). Hu et al. (2005) reported formation of AUX-dependent NO can only happen under certain spatial or temporal circumstances. Presently, ample information is available on NO and AUX interactions; however, most of this has focused on the response of plant roots, with fewer reports on crosstalk among the two signaling molecules in shoot and reproductive tissues.

Almost all of these investigations found that removing NO by scavengers reduced normal AUX-dependent root responses including initiation of mitogen-activated protein kinases (MAPKs) during the development of adventitious roots (Pagnussat et al. 2004) and regulation of cell cycle genes during the formation of lateral roots (Correa-Aragunde et al. 2006). Furthermore, a direct effect of NO has been proposed on AUX perception and signal transduction, based on new evidence that AUX receptor protein (TIR1) undergoes *S*-nitrosylation at two specific cysteine residues (Terrile et al. 2012). *S*-nitrosylation of TIR1 improves its interaction with AUX/IAA, a transcriptional repressor of AUX-associated genes.

8.6.2 ABA and NO

ABA is important for seed dormancy induction and maintenance, germination rate inhibition, and subsequent postgermination developmental control points (Finkelstein 2013). NO acts downstream of ABA because reduced production of NO reduces ABA-induced responses in stressed plant tissues (Tossi et al. 2012). Due to environmental obstacles such as UV-B radiation and water scarcity, NO and ABA both crosstalk significantly during certain signaling cascades, resulting in the activation of adaptive plant responses like closure of stomata and antioxidant defense (Tossi et al. 2009; Hancock et al. 2011). Moreover, NO has been reported to interact with ABA during stress adaptation events such as seed dormancy breaking (Lozano-Juste and Leon 2010). In NO–ABA crosstalk processes, there is a certain degree of specificity that appears to be dependent on the kind of plant cells, tissues,

Table 8.2 Interactions of different plant hormones with nitric oxide (NO) in different plants.

Phytohormones involved	Plant species	NO-mediated effects	References
Auxin (AUX)	*Arabidopsis thaliana*	Induced adventitious and lateral root	Kolbert et al. (2008)
Ethylene	*A. thaliana*	Inhibition of methionine adenosyltransferase (MAT1) activity, reduced pool of the ethylene precursor *S*-adenosylmethionine (SAM)	Lindermayr et al. (2006)
Abscisic acid (ABA)	*Vicia faba*	Increased stomatal closing	Garcia-Mata and Lamattina (2001)
Cytokinins	*A. thaliana* and *Lactuca sativa*	Reduction in hypocotyl elongation and NO level increased	Beligni and Lamattina 2000
Cytokinins	*A. thaliana*	Reduced endogenous NO levels	Liu et al. 2013
AUX	*Medicago sativa*	AUX efflux carrier expression was greater in roots bearing nodules with higher indole acetic acid (IAA)	Pii et al. 2007
indole-3-butyric acid	*A. thaliana*	Induced NO production is nitrate reductase-associated	Kolbert et al. 2008
Ethylene (ET)	*Pisum sativum*	ET reduced NO level; enhanced Cd induced senescence processes	Rodríguez-serrano et al. 2006
ABA	*A. thaliana*	Higher nitrate reductase (NR) regulates stomatal mobility and is linked to increased NO generation	Garcıa-Mata and Lamattina, 2002
AUX	*Tagetes erecta*, *A. thaliana*, *Oryza sativa*	Influence the development of root hair and root meristem size	Hunt et al. 2002; Liao et al. 2012; Chen et al. 2013
ABA	*Zea mays*	Increased generation of NO in mesophyll cells of leaves	Zhang et al. 2011
Salicylate (SA)	*Triticum aestivum*	Downstream of SA signaling in the reduction of induced oxidative damage	Alavi et al. 2014
AUX	*Medicago truncatula*	Reduced AUX degradation in the roots	Xu et al. 2010
SA	*Arachis hypogaea*	Decreased oxidative stress	Xu et al. 2015
SA	*Spinacia oleracea*	Induced NR activity and reduced chilling injury	Aydin and Nalbantoğlu, 2011
Gibberellic acid (GA)	*A. thaliana*	Inhibitory effects of PR growth dependent on GA-DELLA-SLY pathway	Wu et al. 2014
Jasmonate (JA)	*T. aestivum*	Induction of NO; upregulated AsA-GSH cycle	Shan et al. 2015

or organs examined, or the essence of the physiological occurrence being analyzed. Under adverse environmental circumstances, ROS generation (H_2O_2) activates responses of NO-mediated ABA, such as induction of closure of stomata, antioxidant enzyme activation, and transcription factor upregulation (Bright et al. 2006; Lu et al. 2009). It has been observed that cyclic guanosine monophosphate (cGMP) and type 2C protein phosphatases (PP2Cs) contribute to endogenous NO-mediated ABA signaling and upstream cytosolic Ca^{2+} during the regulation of stomatal apparatus (Mioto et al. 2013). Furthermore, during plant antioxidant defense it has been shown that the calcium/calmodulin system and MAPKs act as important downstream components in NO signaling induced by ABA or H_2O_2 (Sang et al. 2008). In the ABA-dependent signaling network, crosstalk between NO and ABA is upregulated by cytosolic Ca^{2+} in order to regulate the expression of CAM in bromeliads, which augments stress tolerance in plants facing an inadequate supply of water (Freschi et al. 2010; Mioto et al. 2013). ABA-induced closure of stomatal apertures in turgid leaves of several plants may be reduced by NO scavengers (Neill et al. 2002). Hence, it has been demonstrated that NO activates and deactivates plasma membrane Ca^{2+}-dependent anion channels and inward rectifying K^+ channels, respectively, as a result of variations in the amounts of guard cell cytoplasmatic Ca^{2+} caused by NO-triggered release of this anion from intercellular storage.

8.6.3 RT and NO

An antagonistic relationship between NO and ET is indicated by the interaction between these two molecules. NO may influence ET production, which is important for plant tissue development and senescence (Arasimowicz and Floryszak-Wieczorek 2007). ET production in plants is regulated by exogenous NO application (Zhu and Zhou 2007). Only NO has been shown to combat the production of ET, fruit ripening control, and leaf senescence (Leshem et al. 1998; Manjunatha et al. 2010). ET has recently been recognized as a main promoting signal for these responses and some research suggests that the mechanisms for producing and detecting ET are highly regulated, and that they are influenced not just by plant growth and development but also by a variety of external variables. Recent research has found that NO can impede fruit ET formation, resulting in reduction of transcription levels or activity of major ET biosynthesis enzymes (Manjunatha et al. 2010). The activity of 1-aminocyclopropane 1-carboxylic acid (ACC) synthase converts SAM, which is produced from the "Yang cycle," into ethylene precursor ACC in vegetative and reproductive plant tissues (ACS). ACC can be subsequently converted to ET by ACC oxidase (ACO) (Yang and Hoffman 1984). Because the availability of ACC, ACS, and ACO in plant tissues represents the important factor in determining production of ET rates (Barry et al. 2000), it can be concluded that an inhibitory effect of NO on all of these components is an important mechanism for downregulating ET synthesis.

NO inhibits ET production in *A. thaliana* plants by *S*-nitrosylating the enzyme methionine adenosyltransferase (MAT1). NO addition leads to the inhibition of MAT1 activity, which results in a decrease in ET precursor SAM levels (Lindermayr et al. 2006). To understand the mechanisms of plant responses to heavy metal (HM)-induced oxidative stress, analysis of the crosstalk between ET and stress signaling molecules such as NO is essential.

In plants exposed to HM stress Thao et al. (2015) indicated a potential relation between NO and ET via MAPKs. As a result, recent research identifies ET as a stress signaling hormone that interacts with other phytohormone signaling pathways to enhance resistance against various environmental stresses.

6.4 CKs and NO

CK is the main hormone that regulates root growth, and its vascular differentiation and gravitropism. Thus, NO can act either downstream mediating CK responses (Tun et al. 2008) or independently (Romanov et al. 2008) depending on the condition of plant cells. The synergistic function of CKs with NO has been clearly demonstrated in demolishing the expression of the *CYCD3;1* gene (transcriptional gene for cell cycle proliferation) induced by CKs in plants treated with NO scavengers/NOS inhibitors or mutant plants deprived of NO production. Leaf senescence, apoptosis, photosynthetic adaptation to water stress, cell division, and differentiation are instances of synergistic interactions between CKs and NO (Carimi et al. 2005; Mishina et al. 2007; Shao et al. 2010; Shen et al. 2012). Further suggesting the defensive and antisenescence function of NO and CKS, it has been demonstrated that higher levels of NO during CK induced photosynthetic adaptation to water stress and a close relationship between generation of NO and activity of NR in such drought-adaptive plant responses has been shown (Shao et al. 2010).

6.5 NO and GAs

GAs play a key role in the germination of seed, hypocotyl elongation, primary root development, re-orientation, and pollen tube growth (Beligni and Lamattina 2000; Leon and Lozano-Juste 2011). The underlying relation between GAs and NO is limited to the regulation, germination, and suppression of hypocotyl elongation during seedling de-etiolation (Bethke et al. 2007). It has also been identified that NO acts upstream of GA in the regulation of all these responses (Bethke et al. 2007), governing both synthesis of GA and perception/transduction, and acts as a hypocotyl elongation inhibitor. Phytochrome-interacting factors (PIF) PIF and DELLA proteins appear to include a potential pathway fundamental to the crosstalk between NO, light, and GAs. NO coordinates the suppression of PIF genes that facilitate growth and an increase in DELLA protein content (Lozano-Juste and León 2011).

8.6.6 JA and NO

JA is an intrinsic hormone, controls growth, and was first found in higher plants as a stress-related hormone (Wang et al. 2020). As a natural plant growth regulator, JA is widely dispersed in plants (Ghasemi Pirbalouti et al. 2014; Wasternack and Strand 2016) and (like ABA) mediates stomatal closure through a ROS-dependent signaling pathway during response to wounding or attacks by pathogens (Suhita et al. 2004; Devoto and Turner 2005). It belongs to the group of compounds produced by the oxygenation of the fatty acids (Bartoli et al. 2013). JA may not act as a single receptor, but rather as part of a complex signaling network that involves other plant hormone signaling pathways. In the course of resisting environmental stress, ABA, ET, SA, and other phytohormones have both synergistic and antagonistic relationships with JA.

NO stimulates JA production and expression of genes responding to JA as well as affecting JA signaling in *A. thaliana* (Huang et al. 2004). NO and JA can work together to stimulate matrine production. Furthermore, NO treatment activates lipoxygenase (LOX) activity and increases JA levels in the cells, suggesting that NO can activate *Sophora flavescens* cells' endogenous JA biosynthesis. External application of JA stimulates NO production in *S. flavescens* cells and induces NO synthase-like activities, indicating that JA can cause NO production in the cells (Xu and Dong 2008). NO production is increased in guard cells by JA, and both JA and NO cause closure of stomatal apparatus (Liu et al. 2005). Mur et al. (2013) found interactions of NO with both SA and JA signaling pathways in various places and importantly it modulates them, though it also stimulates independent responses through posttranscriptional modification of key regulator proteins (Kovacs and Lindermayr 2013).

8.6.7 SA and NO

SA has a role in plant growth and development, as well as vital physiological activities including enhancing the plant response to stress (Hassoon and Abduljabbar 2019). Exogenous application of SA improves drought resistance by enhancing both enzymatic and nonenzymatic antioxidant systems under water stress. The triggering of each other's defensive responses has been shown to be antagonistic, and only SA can inhibit JA biosynthesis (Van Loon et al. 1998; Glazebrook 2001). NO stimulates the SA production in *N. tabacum* as has been reported by Durner et al. (1998) and Durner and Klessig (1999). NO demonstrated substantial increase of SA biosynthesis in *A. thaliana* (Huang et al. 2004). In contrast, the interaction of SA and NO improved physiological and biochemical reactions to reduce Cd toxicity more effectively than either SA or NO separately (Mostofa et al. 2019). NO signaling downstream of SA signaling can reduce oxidative damage in *T. aestivum* seedlings (Naser Alavi et al. 2014).

8.6.8 BRs and NO

BRs are steroid hormones that are essential for growth and development of plants. This hormone regulates a variety of plant activities, such as cell elongation, cell division, photomorphogenesis, xylem differentiation, and reproduction, as well as abiotic and biotic stress (Ahanger et al. 2018, 2020; Planas-Riverola et al. 2019; Nolan et al. 2020). In recent research, NO accumulation was revealed to be significantly enhanced by BR treatments in cucumber leaves and also found to be essential in the BR-induced alternative respiratory pathways in the response of *Nicotiana benthamiana* seedlings to salinity stress (Cui et al. 2011; Zhu et al. 2016). In *A. thaliana*, BR treatment increased NO accumulation and greatly decreased virus accumulation, and an interactive impact of NO and BR was observed on photosynthesis and the antioxidant system of *Lycopersicon esculentum* (Hayat et al. 2010). Arfan and co-workers also reported that H_2O_2 was involved in the generation of NO, while NO was considered to be important in BR-induced alternative oxidase efficiency in *M. truncatula* under cold stress conditions, contributing to the protection of the photosystem (Arfan et al. 2019).

8.6.9 SLs and NO

SLs were discovered to be germination inducers for parasitic plants in the late 1960s (Cook et al. 1966). Because of their various roles in controlling growth and development of plants, SLs are classified as phytohormones (Umehara et al. 2008; Bouwmeester et al. 2019). SLs have been shown to be synthesized in both the roots and the shoots of plants (Foo et al. 2001). SL-induced gene expression acts in physiological effects like shoot branching inhibition, root system, and suppressing leaf senescence (Pandey et al. 2016; Marzec and Melzer 2018). NO and SL are both significant growth controlling signals in plants. SL and NO focus on the root system of crops like *Zea mays* (Manoli et al. 2016), *Helianthus annuus* (Bharti and Bhatla 2015), and *O. sativa* (Sun et al. 2014), which were grown with multiple nutrient sources. NO is an upstream SL signaling regulator, according to the results of these trials, but the existence of the NO–SL interaction depends on the availability of nutrients. NO decreases salicylates / salicylic acid biosynthesis during nitrate-induced root elongation resulting in pin-formed-mediated auxin transport changes that contribute to cell elongation. Exogenously supplied SL induces NO generation, showing that SL levels are negatively regulated (Manoli et al. 2016). Low levels of N and P induce NO production, which leads to proteasomal destruction of the D53 repressor protein, increases SL signaling, and contributes to root elongation (Sun et al. 2016). Marzec and Melzer (2018) conducted plant experiments under stress-free environments to understand the role of SLs in root growth. Other research used related pharmacological, molecular, and biological methods in model *A. thaliana* cultivated in a stress-free environment to generate new information about the signal interactions between NO and SL in root system architecture development (Olah et al. 2020).

8.7 Conclusion

The AsA-GSH cycle plays an important role in mediating against the harmful effects of ROS and, acting as an antioxidant defense system, detoxifies ROS molecules at greater levels and decreases the oxidative stress in plants, thereby fulfilling other functions related to redox sensing and signaling. Nevertheless, in plants NO is the key signaling molecule that takes part in AsA-GSH cycle regulation by enhancing the activity of its different components. NO serves as a crossroad to activate metabolic and physiological responses in hormone signaling, which needs further confirmation at the genetic and molecular levels. Proteins of the AsA-GSH pathway are affected by *S*-nitrosylation and/or nitration, which are obtained by interactions with NO. In addition, GSH modulates plant growth by regulating phytohormone biosynthesis.

References

Abbaspour, J. and Ehsanpour, A.A. (2016). Physiological targets of salicylic acid on *Artemisia aucheri* BOISS as a medicinal and aromatic plant grown under in vitro drought stress. *Botanical Studies* 57 (1): 1–12.

Abello, N., Kerstjens, H.A., Postma, D.S. et al. (2009). Protein tyrosine nitration: selectivity, physicochemical and biological consequences, denitration, and proteomics methods for the identification of tyrosine-nitrated proteins. *Journal of Proteome Research* 8 (7): 3222–3238.

Ahanger, M.A., Ashraf, M., Bajguz, A. et al. (2018). Brassinosteroids regulate growth in plants under stressful environments and crosstalk with other potential phytohormones. *Journal of Plant Growth Regulation* 37 (4): 1007–1024.

Ahanger, M.A., Mir, R.A., Alyemeni, M.N. et al. (2020). Combined effects of brassinosteroid and kinetin mitigates salinity stress in tomato through the modulation of antioxidant and osmolyte metabolism. *Plant Physiology and Biochemistry* 147: 31–42.

Ahanger, M.A., Qi, M., Huang, Z. et al. (2021). Improving growth and photosynthetic performance of drought stressed tomato by application of nano-organic fertilizer involves up-regulation of nitrogen, antioxidant and osmolyte metabolism. *Ecotoxicology and Environmental Safety* 216. doi:10.1016/j.ecoenv.2021.112195.

Ahanger, M.A., Qin, C., Begum, N. et al. (2019). Nitrogen availability prevents oxidative effects of salinity on wheat growth and photosynthesis by up-regulating the antioxidants and osmolytes metabolism, and secondary metabolite accumulation. *BMC Plant Biology* 19: 479. doi: 10.1186/s12870-019-2085-3.

Ahanger, M.A., Tomar, N.S., Tittal, M. et al. (2017). Plant growth under water/ salt stress: ROS production; antioxidants and significance of added potassium under such conditions. *Physiology and Molecular Biology of Plants* 23 (4): 731–744.

Ahmad, P., Ahanger, M.A., Alyemeni, M.N. et al. (2018). Exogenous application of nitric oxide modulates osmolyte metabolism, antioxidants, enzymes of ascorbate-glutathione cycle and promotes growth under cadmium stress in tomato. *Protoplasma* 255 (1): 79–93.

Akram, N.A., Shafiq, F., and Ashraf, M. (2017). Ascorbic acid-a potential oxidant scavenger and its role in plant development and abiotic stress tolerance. *Frontiers of Plant Science* 8: 613. doi: 10.3389/fpls.2017.00613.

Akter, N., Sobahan, M.A., Uraji, M. et al. (2012). Effects of depletion of glutathione on abscisic acid- and methyl jasmonate-induced stomatal closure in *Arabidopsis thaliana*. *Bioscience, Biotechnology and Biochemistry* 76: 2032–2037.

Alavi, S.M., Arvin, M.J., and Manoochehri Kalantari, K. (2014). Salicylic acid and nitric oxide alleviate osmotic stress in wheat (*Triticum aestivum* L.) seedlings. *Journal of Plant Interactions* 9: 683–688.

Alkhalfioui, F., Renard, M., and Montrichard, F. (2007). Unique properties of NADP-thioredoxin reductase C in legumes. *Journal of Experimental Botany* 58 (5): 969–978.

Anjum, N.A., Gill, S.S., Gill, R. et al. (2014). Metal/metalloid stress tolerance in plants: role of ascorbate, its redox couple, and associated enzymes. *Protoplasma* 251 (6): 1265–1283.

Anjum, N.A., Sharma, P., Gill, S.S. et al. (2016). Catalase and ascorbate peroxidase—Representative H_2O_2-detoxifying heme enzymes in plants. *Environmental Science and Pollution Research* 23: 19002–19029.

Arasimowicz, M. and Floryszak-Wieczorek, J. (2007). Nitric oxide as a bioactive signalling molecule in plant stress responses. *Plant Science* 172 (5): 876–887.

Arfan, M., Zhang, D.W., Zou, L.J. et al. (2019). Hydrogen peroxide and nitric oxide crosstalk mediates brassinosteroids induced cold stress tolerance in *Medicago truncatula*. *International Journal of Molecular Sciences* 20: 1–15.

Asada, K. (1999). The water-water cycle in chloroplasts: scavenging of active oxygens and dissipation of excess photons. *Annual Review of Plant Biology* 50 (1): 601–639.

Asgher, M., Per, T.S., Masood, A. et al. (2017). Nitric oxide signaling and its crosstalk with other plant growth regulators in plant responses to abiotic stress. *Environmental Science and Pollution Research* 24 (3): 2273–2285.

Aydın, B. and Nalbantoglu, B. (2011). Effects of cold and salicylic acid treatments on nitrate reductase activity in spinach leaves. *Turkish Journal of Biology* 35: 443–448.

Bai, T., Li, C., Ma, F. et al. (2010). Responses of growth and antioxidant system to root-zone hypoxia stress in two *Malus* species. *Plant Soil* 327 (1): 95–105.

Bao, G., Zhuo, C., Qian, C. et al. (2016). Co-expression of NCED and ALO improves vitamin C level and tolerance to drought and chilling in transgenic tobacco and stylo plants. *Plant Biotechnology Journal* 1: 206–214.

Barry, C.S., Llop-Tous, M.I., and Grierson, D. (2000). The regulation of 1-aminocyclopropane-1-carboxylic acid synthase gene expression during the transition from system-1 to system-2 ethylene synthesis in tomato. *Plant Physiology* 123 (3): 979–986.

Barth, C., De Tullio, M., and Conklin, P.L. (2006). The role of ascorbic acid in the control of flowering time and the onset of senescence. *Journal of Experimental Botany* 57 (8): 1657–1665.

Bartoli, C.G., Buet, A., Gergoff, G.G. et al. (2017). Ascorbate-glutathione cycle and abiotic stress tolerance in plants. In: *Ascorbic Acid in Plant Growth, Development and Stress Tolerance* (eds. M. Hossain, S. Munné-Bosch, D. Burritt, P. Diaz-Vivancos, M. Fujita, and A. Lorence), 177–200. Cham, Switzerland: Springer.

Bartoli, C.G., Casalongue, C.A., Simontacchi, M. et al. (2013). Interactions between hormone and redox signalling pathways in the control of growth and cross tolerance to stress. *Environmental and Experimental Botany* 94: 73–88.

Begara-Morales, J.C., Chaki, M., Valderrama, R. et al. (2018). Nitric oxide buffering and conditional nitric oxide release in stress response. *Journal of Experimental Botany* 69: 3425–3438.

Begara-Morales, J.C., Sánchez-Calvo, B., Chaki, M. et al. (2015). Differential molecular response of monodehydroascorbate reductase and glutathione reductase by nitration and S-nitrosylation. *Journal of Experimental Botany* 66: 5983–5996.

Begara-Morales, J.C., Sánchez-Calvo, B., Chaki, M. et al. (2014). Dual regulation of cytosolic ascorbate peroxidase (APX) by tyrosine nitration and S-nitrosylation. *Journal of Experimental Botany* 65 (2): 527–538.

Beligni, M.V. and Lamattina, L. (2000). Nitric oxide stimulates seed germination and de-etiolation, and inhibits hypocotyl elongation, three light-inducible responses in plants. *Planta* 210: 215–221.

Bethke, P.C., Libourel, I.G., Aoyama, N. et al. (2007). The Arabidopsis aleurone layer responds to nitric oxide, gibberellin, and abscisic acid and is sufficient and necessary for seed dormancy. *Plant Physiology* 143 (3): 1173–1188.

Bharti, N. and Bhatla, S.C. (2015). Nitric oxide mediates strigolactone signaling in auxin and ethylene-sensitive lateral root formation in sunflower seedlings. *Plant Signaling and Behavior* 10: e1054087.

Bhuyan, M.H.M.B., Hasanuzzaman, M., Mahmud, J.A. et al. (2019). Unraveling morphophysiological and biochemical responses of *Triticum aestivum* L. to extreme pH: coordinated actions of antioxidant defense and glyoxalase systems. *Plants* 8: 1–18.

Bielen, A., Remans, T., Vangronsveld, J. et al. (2013). The influence of metal stress on the availability and redox state of ascorbate, and possible interference with its cellular functions. *International Journal of Molecular Sciences* 14 (3): 6382–6413.

Bouwmeester, H.J., Fonne-Pfister, R., Screpanti, C. et al. (2019). Strigolactones: plant hormones with promising features. *Angewandte Chemie International Edition* 58 (37): 12778–12786.

Bright, J., Desikan, R., Hancock, J.T. et al. (2006). ABA-induced NO generation and stomatal closure in *Arabidopsis* are dependent on H_2O_2 synthesis. *Plant Journal* 45: 113–122. doi:10.1111/j.1365-313X.2005.02615.x.

Bu, Y., Sun, B., Zhou, A. et al. (2016). Overexpression of *AtOxR* gene improves abiotic stresses tolerance and vitamin C content in *Arabidopsis thaliana. BMC Biotechnology* 16: 69. doi:10.1186/s12896-016-0299-0.

Campos, F.V., Oliveira, J.A., Pereira, M.G. et al. (2019). Nitric oxide and phytohormone interactions in the response of *Lactuca sativa* to salinity stress. *Planta* 250 (5): 1475–1489.

Carimi, F., Zottini, M., Costa, A. et al. (2005). NO signalling in cytokinin-induced programmed cell death. *Plant, Cell and Environment* 28: 1171–1178. doi:10.1111/j.1365-3040.2005.01355.x.

Caverzan, A., Passaia, G., Rosa, S.B. et al. (2012). Plant responses to stresses: role of ascorbate peroxidase in the antioxidant protection. *Genetics and Molecular Biology* 35 (4, suppl. 1). doi:10.1590/S1415-47572012000600016.

Chang, L., Sun, H., Yang, H. et al. (2017). Over-expression of dehydroascorbate reductase enhances oxidative stress tolerance in tobacco. *Electronic Journal of Biotechnology* 25: 1–8.

Chen, J.H., Jiang, H.W., Hsieh, E.J. et al. (2012). Drought and salt stress tolerance of an Arabidopsis glutathione S-transferase U17 knockout mutant are attributed to the combined effect of glutathione and abscisic acid. *Plant Physiology* 158: 340–351.

Chen, W.W., Yang, J.L., Qin, C. et al. (2010). Nitric oxide acts downstream of auxin to trigger root ferric-chelate reductase activity in response to iron deficiency in *Arabidopsis. Plant Physiology* 154: 810–819. doi:10.1104/pp.110.161109.

Choudhury, F.K., Rivero, R.M., Blumwald, E. et al. (2017). Reactive oxygen species, abiotic stress and stress combination. *Plant Journal* 90: 856–867.

Conklin, P.L. and Barth, C. (2004). Ascorbic acid, a familiar small molecule intertwined in the response of plants to ozone, pathogens, and the onset of senescence. *Plant, Cell and Environment* 27: 959–970.

Contour-Ansel, D., Torres-Franklin, M.L., Cruz de Carvalho, M.H. et al. (2006). Glutathione reductase in leaves of cowpea: cloning of two cDNAs, expression and enzymatic activity under progressive drought stress, desiccation and abscisic acid treatment. *Annals of Botany* 98: 1279–1287.

Cook, C.E., Whichard, L.P., Turner, B. et al. (1966). Germination of witchweed (*Striga lutea* Lour.): isolation and properties of a potent stimulant. *Science* 154 (3753): 1189–1190.

Corpas, F.J., Begara-Morales, J.C., Sánchez-Calvo, B. et al. (2015). Nitration and S-nitrosylation: two post-translational modifications (PTMs) mediated by reactive nitrogen species (RNS) and their role in signalling processes of plant cells. In: *Reactive Oxygen and Nitrogen Species Signaling and Communication in Plants* (eds. J. Kagapunti, K.J. Gupta, and A.U. Igamberdiev), 267–281. Berlin, Heidelberg: Springer-Verlag.

Corpas, F.J. and Palma, J.M. (2018). Nitric oxide on/off in fruit ripening. *Plant Biology* 20 (5): 805–807.

Correa-Aragunde, N., Graziano, M., Chevalier, C. et al. (2006). Nitric oxide modulates the expression of cell cycle regulatory genes during lateral root formation in tomato. *Journal of Experimental Botany* 57 (3): 581–588.

Cui, J.X., Zhou, Y.H., Ding, J.G. et al. (2011). Role of nitric oxide in hydrogen peroxide dependent induction of abiotic stress tolerance by brassinosteroids in cucumber. *Plant, Cell and Environment* 34: 347–358.

Dąbrowska, G., Kata, A., Goc, A. et al. (2007). Characteristics of the plant ascorbate peroxidase family. *Acta Biologica Cracoviensia Series Botanica* 49 (1): 7–17.

Datta, R., Kumar, D., Sultana, A. et al. (2015). Glutathione regulates 1-aminocyclopropane-1-carboxylate synthase transcription via WRKY33 and 1-aminocyclopropane-1-carboxylate oxidase by modulating messenger RNA stability to induce ethylene synthesis during stress. *Plant Physiology* 169: 2963–2981.

Davies, P. (2004). *Plant Hormones: Biosynthesis, Signal Transduction, Action!* New York: Springer Science & Business Media Press.

Deeb, R.S., Nuriel, T., Cheung, C. et al. (2013). Characterization of a cellular denitrase activity that reverses nitration of cyclooxygenase. *American Journal of Physiology – Heart and Circulatory Physiology* 305: H687–H698. doi:10.1152/ajpheart.00876.2012.

del Rio, L.A. and Lopez-Huertas, E. (2016). ROS generation in peroxisomes and its role in cell signaling. *Plant and Cell Physiology* 57: 1364–1376.

Devoto, A. and Turner, J.G. (2005). Jasmonate-regulated Arabidopsis stress signalling network. *Physiology Plant* 123: 161–172.

Domingos, P., Prado, A.M., Wong, A. et al. (2015). Nitric oxide: a multitasked signaling gas in plants. *Molecular Plant* 8 (4): 506–520.

Durner, J., Gow, A.J., Stamler, J.S. et al. (1999). Ancient origins of nitric oxide signaling in biological systems. *Proceedings of the National Academy of Sciences* 96 (25): 14206–14207.

Durner, J. and Klessig, D.F. (1999). Nitric oxide as a signal in plants. *Current Opinion in Plant Biology* 2: 369–372.

Durner, J., Wendehenne, D., and Klessig, D.F. (1998). Defense gene induction in tobacco by nitric oxide, cyclic GMP, and cyclic ADP-ribose. *Proceedings of the National Academy of Sciences* 95: 10328–10333.

Egamberdieva, D., Wirth, S.J., Alqarawi, A.A. et al. (2017). Phytohormones and beneficial microbes: essential components for plants to balance stress and fitness. *Frontiers in Microbiology* 8: 2104. doi:10.3389/fmicb.2017.02104.

Fan, H.F., Ding, L., Du, C.X. et al. (2014). Effect of short-term water deficit stress on antioxidative systems in cucumber seedling roots. *Botanical Studies* 55 (1): 1–7.

Farooq, M.A., Gill, R.A., Islam, F. et al. (2016). Methyl jasmonate regulates antioxidant defense and suppresses arsenic uptake in *Brassica napus* L. *Frontiers of Plant Science* 7: 468. doi:10.3389/fpls.2016.00468.

Fatma, M., Masood, A., Per, T.S. et al. (2016). Interplay between nitric oxide and sulfur assimilation in salt tolerance in plants. *Crop Journal* 4 (3): 153–161.

Feng, J., Wang, C., Chen, Q. et al. (2013). S-nitrosylation of phosphotransfer proteins represses cytokinin signaling. *Nature Communications* 4 (1): 1–9.

Finkelstein, R. (2013). Abscisic acid synthesis and response. In: *The Arabidopsis Book*, vol. 11. American Society of Plant Biologists.

Foo, E., Turnbull, C.G., and Beveridge, C.A. (2001). Long-distance signaling and the control of branching in therms1 mutant of pea. *Plant Physiology* 126 (1): 203–209.

Freschi, L. (2013). Nitric oxide and phytohormone interactions: current status and perspectives. *Frontiers of Plant Science* 4: 398. doi:10.3389/fpls.2013.00398.

Freschi, L., Rodrigues, M.A., Domingues, D.S. et al. (2010). Nitric oxide mediates the hormonal control of Crassulacean acid metabolism expression in young pineapple plants. *Plant Physiology* 152: 1971–1985. doi:10.1104/pp.109.151613.

Gaber, A., Burritt, D.J., and Hossain, M.A. (2017). Transgenic plants over-expressing glutathione biosynthetic genes and abiotic stress tolerance. In: *Glutathione in Plant Growth, Development, and Stress Tolerance* (eds. M. Hossain, M. Mostofa, P. Diaz-Vivancos, D. Burritt, M. Fujita, and L.S. Tran). Cham, Switzerland: Springer. 397–412. doi:10.1007/978-3-319-66682-2_18.

Gallie, D.R. (2012). The role of L-ascorbic acid recycling in responding to environmental stress and in promoting plant growth. *Journal of Experimental Botany* 64: 433–443.

Garcia-Mata, C. and Lamattina, L. (2001). Nitric oxide induces stomatal closure and enhances the adaptive plant responses against drought stress. *Plant Physiology* 126: 1196–1204.

García-Mata, C. and Lamattina, L. (2002). Nitric oxide and abscisic acid cross talk in guard cells. *Plant Physiology* 128 (3): 790–792.

Ghasemi Pirbalouti, A., Sajjadi, S.E., and Parang, K. (2014). A review (research and patents) on jasmonic acid and its derivatives. *Archiv der Pharmazie* 347 (4): 229–239.

Gill, S.S., Anjum, N.A., Hasanuzzaman, M. et al. (2013). Glutathione and glutathione reductase: a boon in disguise for plant abiotic stress defense operations. *Plant Physiology and Biochemistry* 70: 204–212.

Gill, S.S. and Tuteja, N. (2010). Reactive oxygen species and antioxidant machinery in abiotic stress tolerance in crop plants. *Plant Physiology and Biochemistry* 48: 909–930.

Glazebrook, J. (2001). Genes controlling expression of defense responses in Arabidopsis – 2001 status. *Current Opinion in Plant Biology* 4: 301–308.

Görg, B., Qvartskhava, N., Voss, P. et al. (2007). Reversible inhibition of mammalian glutamine synthetase by tyrosine nitration. *FEBS Letters* 581: 84–90. doi:10.1016/j.febslet.2006.11.081.

Gow, A.J., Farkouh, C.R., Munson, D.A. et al. (2004). Biological significance of nitric oxide-mediated protein modifications. *American Journal of Physiology – Lung Cellular and Molecular Physiology* 287: L262–L268. doi:10.1152/ajplung.00295.2003.

Guo, F.Q., Okamoto, M., and Crawford, N.M. (2003). Identification of a plant nitric oxide synthase gene involved in hormonal signaling. *Science* 302: 100–103. doi:10.1126/science.1086770.

Hancock, J.T., Neill, S.J., and Wilson, I.D. (2011). Nitric oxide and ABA in the control of plant function. *Plant Science* 181: 555–559. doi:10.1016/j.plantsci.2011.03.017.

Haroldsen, V.M., Chi-Ham, C.L., Kulkarni, S. et al. (2011). Constitutively expressed DHAR and MDHAR influence fruit, but not foliar ascorbate levels in tomato. *Plant Physiology and Biochemistry* 49 (10): 1244–1249.

Hasanuzzaman, M., Bhuyan, M.H.M., Anee, T.I. et al. (2019). Regulation of ascorbate-glutathione pathway in mitigating oxidative damage in plants under abiotic stress. *Antioxidants* 8 (9): 384.

Hasanuzzaman, M. and Fujita, M. (2013). Exogenous sodium nitroprusside alleviates arsenic induced oxidative stress in wheat (*Triticum aestivum* L.) seedlings by enhancing antioxidant defense and glyoxalase system. *Ecotoxicology* 22 (3): 584–596.

Hasanuzzaman, M., Hossain, M.A., and Fujita, M. (2011). Nitric oxide modulates antioxidant defense and the methylglyoxal detoxification system and reduces salinity-induced damage of wheat seedlings. *Plant Biotechnology Reports* 5 (4): 353–565.

Hasanuzzaman, M., Hossain, M.A., Teixeira da Silva, J.A. et al. (2012). Plant responses and tolerance to abiotic oxidative stress: antioxidant defense is a key factor. In: *Crop Stress and Its Management: Perspectives and Strategies* (eds. V. Bandi, A.K. Shanker, C. Shanker, and M. Mandapaka), 261–315. Dordrecht, the Netherlands: Springer.

Hasanuzzaman, M., Nahar, K., Anee, T.I. et al. (2017). Exogenous silicon attenuates cadmium-induced oxidative stress in *Brassica napus* L. by modulating AsA-GSH pathway and glyoxalase system. *Frontiers of Plant Science* 8: 1061. doi:10.3389/fpls.2017.01061.

Hasanuzzaman, M., Nahar, K., Rahman, A. et al. (2018). Exogenous glutathione attenuates lead-induced oxidative stress in wheat by improving antioxidant defense and physiological mechanisms. *Journal of Plant Interactions* 13: 203–212.

Hassoon, A.S. and Abduljabbar, I.A. (2019). Review on the role of salicylic acid in plants. In: *Sustainable Crop Production*. (eds. M. Hasanuzzaman, M.C.M.T. Filho, M. Fujita, and T.A.R. Nogueira). Sustainable Crop Production, London. Intech Open 10.5772/intechopen.83521.

Hayat, S., Yadav, S., Ali, B. et al. (2010). Interactive effect of nitric oxide and brassinosteroids on photosynthesis and the antioxidant system of *Lycopersicon esculentum*. *Russian Journal of Plant Physiology* 57 (2): 212–221.

He, H., He, L., and Gu, M. (2012). Interactions between nitric oxide and plant hormones in aluminum tolerance. *Plant Signaling and Behavior* 7 (4): 469–471.

Hemavathi, et al. (2010). Enhanced ascorbic acid accumulation in transgenic potato confers tolerance to various abiotic stresses. *Biotechnology Letters* 32: 321–330.

Hess, D.T., Matsumoto, A., Kim, S.O. et al. (2005). Protein S-nitrosylation: purview and parameters. *Nature Reviews Molecular Cell Biology* 6 (2): 150–166.

Horemans, N., Asard, H., and Caubergs, R.J. (1997). The ascorbate carrier of higher plant plasma membranes preferentially translocates the fully oxidized (dehydroascorbate) molecule. *Plant Physiology* 114 (4): 1247–1253.

Hu, L., Li, Y., Wu, Y. et al. (2019). Nitric oxide is involved in the regulation of the ascorbate–glutathione cycle induced by the appropriate ammonium: nitrate to mitigate low light stress in *Brassica pekinensis*. *Plants* 8 (11): 489. doi:10.3390/plants8110489.

Hu, X.Y., Neill, S.J., Tang, Z.C. et al. (2005). Nitric oxide mediates gravitropic bending in soybean roots. *Plant Physiology* 137: 663–670. doi:10.1104/pp.104.054494.

Huang, X., Stettmaier, K., Michel, C. et al. (2004). Nitric oxide is induced by wounding and influences jasmonic acid signaling in *Arabidopsis thaliana*. *Planta* 218 (6): 938–946.

Hunt, P.W., Klok, E.J., Trevaskis, B. et al. (2002). Increased level of hemoglobin 1 enhances survival of hypoxic stress and promotes early growth in *Arabidopsis thaliana*. *Proceedings of the National Academy of Sciences* 99: 17197–17202.

Innocenti, G., Pucciariello, C., Le Gleuher, M. et al. (2007). Glutathione synthesis is regulated by nitric oxide in *Medicago truncatula* roots. *Planta* 225: 1597–1602.

Iqbal, N., Umar, S., Khan, N.A. et al. (2021). Nitric oxide and hydrogen sulfide coordinately reduce glucose sensitivity and decrease oxidative stress via ascorbate-glutathione cycle in heat-stressed wheat (*Triticum aestivum* L.) plants. *Antioxidants* 10 (1): 108.

Ishikawa, T., Maruta, T., Yoshimura, K. et al. (2018). Biosynthesis and regulation of ascorbic acid in plants. In: *Antioxidants and Antioxidant Enzymes in Higher Plants* (eds. D. Gupta, J. Palma, and F. Corpas), 163–179. Cham, Switzerland: Springer.

Jung, H.I., Kong, M.S., Lee, B.R. et al. (2019). Exogenous glutathione increases arsenic translocation into shoots and alleviates arsenic-induced oxidative stress by sustaining ascorbate–glutathione homeostasis in rice seedlings. *Frontiers of Plant Science* 10: 1089. doi:10.3389/fpls.2019.01089.

Kavitha, K., George, S., Venkataraman, G. et al. (2010). A salt-inducible chloroplastic monodehydroascorbate reductase from halophyte *Avicennia marina* confers salt stress tolerance on transgenic plants. *Biochimie* 92 (10): 1321–1329.

Kaya, C. (2020). Salicylic acid-induced hydrogen sulphide improves lead stress tolerance in pepper plants by upraising the ascorbate-glutathione cycle. *Physiology Plant.* doi: 10.1111/ppl.13159.

Kaya, C., Ashraf, M., Alyemeni, M.N. et al. (2020). The role of endogenous nitric oxide in salicylic acid-induced up-regulation of ascorbate-glutathione cycle involved in salinity tolerance of pepper (*Capsicum annuum* L.) plants. *Plant Physiology and Biochemistry* 147: 10–20.

Kim, I.S., Kim, Y.S., Kim, Y.H. et al. (2016). Potential application of the *Oryza sativa* monodehydroascorbate reductase gene (OsMDHAR) to improve the stress tolerance and fermentative capacity of *Saccharomyces cerevisiae. PLoS ONE* 11 (7): e0158841.

Kim, Y.O., Bae, H.J., Cho, E. et al. (2017). Exogenous glutathione enhances mercury tolerance by inhibiting mercury entry into plant cells. *Frontiers of Plant Science* 8: 683. doi:10.3389/fpls.2017.00683.

Kim, Y.S., Kim, I.S., Shin, S.Y. et al. (2014). Overexpression of dehydroascorbate reductase confers enhanced tolerance to salt stress in rice plants (*Oryza sativa* L. japonica). *Journal of Agronomy and Crop Science* 200 (6): 444–456.

Kolbert, Z., Bartha, B., and Erder, L. (2008). Exogenous auxin-induced NO synthesis is nitrate reductase-associated in *Arabidopsis thaliana* root primordia. *Journal of Plant Physiology* 165: 967–975.

Kolbert, Z. and Feigl, G. (2017). Cross-talk of reactive oxygen species and nitric oxide in various processes of plant development: past and present. In: *Reactive Oxygen Species in Plants: Boon or Bane – Revisiting the Role of ROS* (eds. V.P. Singh, S. Singh, D.K. Tripathi, S.M. Prasad, and D.K. Chauhan), 261–289. Chichester, UK: John Wiley & Sons, Ltd.

Kong, X., Wang, T., Li, W. et al. (2016). Exogenous nitric oxide delays salt-induced leaf senescence in cotton (*Gossypium hirsutum* L.). *Acta Physiologiae Plantarum* 38: 61. doi:10.1007/s11738-016-2079-9.

Koussevitzky, S., Suzuki, N., Huntington, S. et al. (2008). Ascorbate peroxidase 1 plays a key role in the response of *Arabidopsis thaliana* to stress combination. *Journal of Biological Chemistry* 283: 34197–34203.

Kovacs, I. and Lindermayr, C. (2013). Nitric oxide-based protein modification: formation and site-specificity of protein S-nitrosylation. *Frontiers of Plant Science* 4: 137. doi:10.3389/fpls.2013.00137.

Kushwaha, B.K., Singh, S., Tripathi, D.K. et al. (2019). New adventitious root formation and primary root biomass accumulation are regulated by nitric oxide and reactive oxygen species in rice seedlings under arsenate stress. *Journal of Hazardous Materials* 361: 134–140.

Lanteri, M.L., Pagnussat, G.C., and Lamattina, L. (2006). Calcium and calcium-dependent protein kinases are involved in nitric oxide- and auxin-induced adventitious root formation in cucumber. *Journal of Experimental Botany* 57 (6): 1341–1351.

Leitner, M., Vandelle, E., Gaupels, F. et al. (2009). NO signals in the haze: nitric oxide signalling in plant defence. *Current Opinion in Plant Biology* 12 (4): 451–458.

Leshem, Y.Y., Wills, R.B.H., and Ku, V.V.V. (1998). Evidence for the function of the free radical gas – nitric oxide (NO) – as an endogenous maturation and senescence regulating factor in higher plants. *Plant Physiology and Biochemistry* 36: 825–833. doi:10.1016/S0981-9428(99)80020-5.

Leterrier, M. and Cagnac, O. (2018). Function of the various MDAR isoforms in higher plants. In: *Antioxidants and Antioxidant Enzymes in Higher Plants* (eds. D. Gupta, J. Palma, and F. Corpas), 83–94. Cham, Switzerland: Springer.

Liao, W.B., Huang, G.B., Yu, J.H. et al. (2012). Nitric oxide and hydrogen peroxide alleviate drought stress in marigold explants and promote its adventitious root development. *Plant Physiology and Biochemistry* 58: 6–15.

Lindermayr, C., Saalbach, G., Bahnweg, G. et al. (2006). Differential inhibition of *Arabidopsis* methionine adenosyltransferases by protein S-nitrosylation. *Journal of Biological Chemistry* 281: 4285–4291.

Liu, X., Shi, W., Zhang, S. et al. (2005). Nitric oxide involved in signal transduction of jasmonic acid-induced stomatal closure of *Vicia faba* L. *Chinese Science Bulletin* 50 (6): 520–525.

Liu, Y., Lai, N., Gao, K. et al. (2013). Ammonium inhibits primary root growth by reducing the length of meristem and elongation zone and decreasing elemental expansion rate in the root apex in *Arabidopsis thaliana*. *PLoS ONE* 8 (4): e61031.

Lombardo, M.C., Graziano, M., Polacco, J. et al. (2006). Nitric oxide functions as a positive regulator of root hair development. *Plant Signaling and Behavior* 1: 28–33.

Lozano-Juste, J. and Leon, J. (2010). Enhanced abscisic acid-mediated responses in *nia1nia2noa1-2* triple mutant impaired in NIA/NR- and AtNOA1-dependent nitric oxide biosynthesis in *Arabidopsis*. *Plant Physiology* 152: 891–903. doi:10.1104/pp.109.148023.

Lozano-Juste, J. and León, J. (2011). Nitric oxide regulates DELLA content and PIF expression to promote photomorphogenesis in Arabidopsis. *Plant Physiology* 156: 1410–1423.

Lu, S.Y., Su, W., Li, H.H. et al. (2009). Abscisic acid improves drought tolerance of triploid bermudagrass and involves H_2O_2- and NO-induced antioxidant enzyme activities. *Plant Physiology and Biochemistry* 47: 132–138. doi:10.1016/j.plaphy.2008.10.006.

Manjunatha, G., Lokesh, V., and Neelwarne, B. (2010). Nitric oxide in fruit ripening: trends and opportunities. *Biotechnology Advances* 28: 489–499. doi:10.1016/j.biotechadv.2010.03.001.

Manoli, A., Trevisan, S., Voigt, B. et al. (2016). Nitric oxide-mediated maize root apex responses to nitrate are regulated by auxin and strigolactones. *Frontiers of Plant Science* 6: 1269. doi:10.3389/fpls.2015.01269.

Marrs, K.A. (1996). The functions and regulation of glutathione transferases in plants. *Annual Review of Plant Physiology and Plant Molecular Biology* 47: 127–158.

Marzec, M. and Melzer, M. (2018). Regulation of root development and architecture by strigolactones under optimal and nutrient deficiency conditions. *International Journal of Molecular Sciences* 19: 1887. doi:10.3390/ijms19071887.

Mellidou, I., Keulemans, J., Kanellis, A.K. et al. (2012). Regulation of fruit ascorbic acid concentrations during ripening in high and low vitamin C tomato cultivars. *BMC Plant Biology* 12: 1–19.

Mishina, T.E., Lamb, C., and Zeier, J. (2007). Expression of a nitric oxide degrading enzyme induces a senescence programme in *Arabidopsis*. *Plant, Cell and Environment* 30: 39–52. doi:10.1111/j.1365-3040.2006.01604.x.

Mittler, R., Vanderauwera, S., Gollery, M. et al. (2004). The reactive oxygen gene network of plants. *Trends in Plant Science* 9: 490–498.

Moons, A. (2003). Osgstu3 and osgtu4, encoding tau class glutathione S-transferases, are heavy metal- and hypoxic stress-induced and differentially salt stress responsive in rice roots. *FEBS Letters* 553: 427–432.

Mostofa, M.G., Rahman, M., Ansary, M. et al. (2019). Interactive effects of salicylic acid and nitric oxide in enhancing rice tolerance to cadmium stress. *International Journal of Molecular Sciences* 20 (22): 5798.

Mur, L.A., Prats, E., Pierre, S. et al. (2013). Integrating nitric oxide into salicylic acid and jasmonic acid/ethylene plant defense pathways. *Frontiers of Plant Science* 4: 215. doi:10.3389/fpls.2013.00215.

Nahar, K., Hasanuzzaman, M., and Fujita, M. (2016). Physiological roles of glutathione in conferring abiotic stress tolerance to plants. In: *Abiotic Stress Response in Plants* (eds. S.S. Gill and N. Tuteja), 151–179. Weinheim, Germany: Wiley.

Najami, N., Janda, T., Barriah, W. et al. (2008). Ascorbate peroxidase gene family in tomato: its identification and characterization. *Molecular Genetics and Genomics* 279: 171–182.

Narendra, S., Venkataramani, S., Shen, G. et al. (2006). The Arabidopsis ascorbate peroxidase 3 is a peroxisomal membrane-bound antioxidant enzyme and is dispensable for Arabidopsis growth and development. *Journal of Experimental Botany* 57: 3033–3042.

Naser Alavi, S.M., Arvin, M.J., and Manoochehri Kalantari, K. (2014). Salicylic acid and nitric oxide alleviate osmotic stress in wheat (*Triticum aestivum* L.) seedlings. *Journal of Plant Interactions* 9 (1): 683–688.

Nawaz, F., Shabbir, R.N., Shahbaz, M. et al. (2017). Cross talk between nitric oxide and phytohormones regulate plant development during abiotic stresses. In: *Phytohormones: Signaling Mechanisms and Crosstalk in Plant Development and Stress Responses* (ed. M. EL-Esawai), 117–141. Rijeka, Croatia: InTech.

Neill, S.J., Desikan, R., Clarke, A. et al. (2002). Hydrogen peroxide and nitric oxide as signalling molecules in plants. *Journal of Experimental Botany* 53: 1237–1247. doi:10.1093/jexbot/53.372.1237.

Noctor, G. and Foyer, C.H. (1998). Ascorbate and glutathione: keeping active oxygen under control. *Annual Review of Plant Biology* 49 (1): 249–279.

Noctor, G., Mhamdi, A., Chaouch, S. et al. (2012). Glutathione in plants: an integrated overview. *Plant, Cell and Environment* 35: 454–484.

Nolan, T.M., Vukasinovic, N., Liu, D. et al. (2020). Brassinosteroids: multidimensional regulators of plant growth, development, and stress responses. *Plant Cell* 32 (2): 295–318.

Olah, D., Feigl, G., Molnar, A. et al. (2020). Strigolactones interact with nitric oxide in regulating root system architecture of *Arabidopsis thaliana*. *Frontiers of Plant Science* 11: 1019. doi:10.3389/fpls.2020.01019.

Pagnussat, G.C., Lanteri, M.L., Lombardo, M.C. et al. (2004). Nitric oxide mediates the indole acetic acid induction activation of a mitogen-activated protein kinase cascade involved in adventitious root development. *Plant Physiology* 135 (1): 279–286.

Pagnussat, G.C., Simontacchi, M., Puntarulo, S. et al. (2002). Nitric oxide is required for root organogenesis. *Plant Physiology* 129 (3): 954–956.

Pandey, A., Sharma, M., and Pandey, G.K. (2016). Emerging roles of strigolactones in plant responses to stress and development. *Frontiers of Plant Science* 7: 434. doi:10.3389/fpls.2016.00434.

Pandey, P., Singh, J., Achary, V. et al. (2015). Redox homeostasis via gene families of ascorbate-glutathione pathway. *Frontiers of Plant Science* 3: 25. doi:10.3389/fenvs.2015.00025.

Pastori, G.M., Kiddle, G., Antoniw, J. et al. (2003). Leaf vitamin C contents modulate plant defense transcripts and regulate genes controlling development through hormone signalling. *Plant Cell* 15: 1212–1226.

Pii, Y., Crimi, M., Cremonese, G. et al. (2007). Auxin and nitric oxide control indeterminate nodule formation. *BMC Plant Biology* 7 (1): 1–11.

Pilon-Smits, E.A., Zhu, Y.L., Sears, T. et al. (2000). Overexpression of glutathione reductase in *Brassica juncea*: effects on cadmium accumulation and tolerance. *Physiologia Plantarum* 110 (4): 455–460.

Planas-Riverola, A., Gupta, A., Betegón-Putze, I. et al. (2019). Brassinosteroid signaling in plant development and adaptation to stress. *Development* 146: dev151894. doi:10.1242/dev.151894.

Polle, A. (2001). Dissecting the superoxide dismutase–ascorbate–glutathione-pathway in chloroplasts by metabolic modeling: computer simulations as a step towards flux analysis. *Plant Physiology* 126: 445–462.

Qin, A., Huang, X., Zhang, H. et al. (2015). Overexpression of PbDHAR2 from *Pyrus sinkiangensis* in transgenic tomato confers enhanced tolerance to salt and chilling stresses. *Horticultural Science* 50 (6): 789–796.

Radi, R. (2004). Nitric oxide, oxidants, and protein tyrosine nitration. *Proceedings of the National Academy of Sciences* 101: 4003–4008. doi:10.1073/pnas.0307446101.

Radi, R. (2013). Peroxynitrite, a stealthy biological oxidant. *Journal of Biological Chemistry* 288: 26464–26472. doi:10.1074/jbc.R113.472936.

Rao, K., Raghavendra, A., and Reddy, K. (2006). *Physiology and Molecular Biology of Stress Tolerance*, 1–14. Dordrecht, the Netherlands: Springer.

Rennenberg, H. and Herschbach, C. (2014). A detailed view on sulphur metabolism at the cellular and whole-plant level illustrates challenges in metabolite flux analyses. *Journal of Experimental Botany* 65 (20): 5711–5724.

Rodríguez-Serrano, M., Romero-Puertas, M.C., Zabalza, ANA et al. (2006). Cadmium effect on oxidative metabolism of pea (*Pisum sativum* L.) roots. Imaging of reactive oxygen species and nitric oxide accumulation in vivo. *Plant, Cell and Environment* 29 (8): 1532–1544.

Romanov, G.A., Lomin, S.N., Rakova, N.Y. et al. (2008). Does NO play a role in cytokinin signal transduction? *FEBS Letters* 582: 874–880.

Romero-Puertas, M.C. and Sandalio, L.M. (2016). Nitric oxide level is self-regulating and also regulates its ROS partners. *Frontiers of Plant Science* 7: 316. doi:10.3389/fpls.2016.00316.

Sadak, M.S., Elhamid, E.M.A., and Mostafa, H.M. (2013). Alleviation of adverse effects of salt stress in wheat cultivars by foliar treatment with antioxidants I. changes in growth, some biochemical aspects and yield quantity and quality. *American-Eurasian Journal of Agricultural and Environmental Sciences* 13: 1476–1487.

Sang, J.R., Zhang, A., Lin, F. et al. (2008). Cross-talk between calcium-calmodulin and nitric oxide in abscisic acid signaling in leaves of maize plants. *Cell Research* 18: 577–588. doi:10.1038/cr.2008.39.

Santner, A., Calderón-Villalobos, L.I., and Estelle, M. (2009). Plant hormones are versatile chemical regulators of plant growth. *Nature Chemical Biology* 5: 301–307. doi:10.1038/nchembio.165.

Scuffi, D., Lamattina, L, and García-Mata, C. (2016). Decoding the interaction between nitric oxide and hydrogen sulfide in stomatal movement. In: *Gasotransmitters in Plants*, 271–287. Cham, Switzerland: Springer.

Sehar, Z., Masood, A., and Khan, N.A. (2019). Nitric oxide reverses glucose-mediated photosynthetic repression in wheat (*Triticum aestivum* L.) under salt stress. *Environmental and Experimental Botany* 161: 277–289.

Shafi, A. and Zahoor, I. (2020). Abiotic and biotic stress-induced alterations in the micronutrient status of plants. In: *Plant Micronutrients*, (eds. T. Aftab, and K.R. Hakeem). 285–309. Cham, Switzerland: Springer.

Shan, C., He, F., Xu, G. et al. (2012). Nitric oxide is involved in the regulation of ascorbate and glutathione metabolism in *Agropyron cristatum* leaves under water stress. *Biology Plant* 56 (1): 187–191.

Shan, C., Zhou, Y., and Liu, M. (2015). Nitric oxide participates in the regulation of the ascorbate-glutathione cycle by exogenous jasmonic acid in the leaves of wheat seedlings under drought stress. *Protoplasma* 252 (5): 1397–1405.

Shao, R.X., Wang, K.B., and Shangguan, Z.P. (2010). Cytokinin-induced photosynthetic adaptability of *Zea mays* L. to drought stress associated with nitric oxide signal: probed by ESR spectroscopy and fast OJIP fluorescence rise. *Journal of Plant Physiology* 167: 472–479. doi:10.1016/j.jplph.2009.10.020.

Shen, Q., Wang, Y.T., Tian, H. et al. (2012). Nitric oxide mediates cytokinin functions in cell proliferation and meristem maintenance in *Arabidopsis*. *Molecular Plant* 10. 1093/mp/sss148.

Shi, W.M., Muramoto, Y., Ueda, A. et al. (2001). Cloning of peroxisomal ascorbate peroxidase gene from barley and enhanced thermotolerance by overexpressing in *Arabidopsis thaliana*. *Gene* 273: 23–27.

Shimaoka, T., Yokota, A., and Miyake, C. (2000). Purification and characterization of chloroplast dehydroascorbate reductase from spinach leaves. *Plant and Cell Physiology* 41 (10): 1110–1118.

Shin, S.Y., Kim, M.H., Kim, Y.H. et al. (2013). Co-expression of monodehydroascorbate reductase and dehydroascorbate reductase from *Brassica rapa* effectively confers tolerance to freezing-induced oxidative stress. *Molecular Cells* 36: 304–315.

Singh, A.P., Dixit, G., Kumar, A. et al. (2017). A protective role for nitric oxide and salicylic acid for arsenite phytotoxicity in rice (*Oryza sativa* L.). *Plant Physiology and Biochemistry* 115: 163–173.

Singh, A.P., Dixit, G., Mishra, S. et al. (2015). Salicylic acid modulates arsenic toxicity by reducing its root to shoot translocation in rice (*Oryza sativa* L.). *Frontiers of Plant Science* 6: 340.

Singh, S., Husain, T., Kushwaha, B.K. et al. (2021). Regulation of ascorbate-glutathione cycle by exogenous nitric oxide and hydrogen peroxide in soybean roots under arsenate stress. *Journal of Hazardous Materials* 409: 123686.

Souza, J.M., Peluffo, G., and Radi, R. (2008). Protein tyrosine nitration – functional alteration or just a biomarker? *Free Radical Biology and Medicine* 45 (4): 357–366.

Su, Y., Xia, S., Wang, R. et al. (2017). Phytohormonal quantification based on biological principles. *Hormone and Metabolic Signaling in Plants* 13: 431–470.

Suhita, D., Raghavendra, A.S., Kwak, J.M. et al. (2004). Cytoplasmic alkalization precedes reactive oxygen species production during methyl jasmonate- and abscisic acid-induced stomatal closure 1. *Plant Physiology* 134: 1536–1545.

Sun, C., Liu, L., Yu, Y. et al. (2015). Nitric oxide alleviates aluminum-induced oxidative damage through regulating the ascorbate-glutathione cycle in roots of wheat. *Journal of Integrative Plant Biology* 57 (6): 550–561.

Sun, H., Bi, Y., Tao, J. et al. (2016). Strigolactones are required for nitric oxide to induce root elongation in response to nitrogen and phosphate deficiencies in rice. *Plant, Cell and Environment* 39: 1473–1484.

Sun, H., Tao, J., Liu, S. et al. (2014). Strigolactones are involved in phosphate- and nitrate-deficiency-induced root development and auxin transport in rice. *Journal of Experimental Botany* 65: 6735–6746.

Sun, L.R., Hao, F.S., Lu, B.S. et al. (2010). AtNOA1 modulates nitric oxide accumulation and stomatal closure induced by salicylic acid in *Arabidopsis*. *Plant Signaling and Behavior* 5: 1022–1024. doi:10.4161/psb.5.8.12293.

Szarka, A., Tomasskovics, B., and Bánhegyi, G. (2012). The ascorbate-glutathione-tocopherol triad in abiotic stress response. *International Journal of Molecular Sciences* 13: 4458–4483.

Tanou, G., Filippou, P., Belghazi, M. et al. (2012). Oxidative and nitrosative-based signaling and associated post-translational modifications orchestrate the acclimation of citrus plants to salinity stress. *Plant Journal* 72: 585–599.

Teixeira, F.K., Menezes-Benavente, L., Galvão, V.C. et al. (2006). Rice ascorbate peroxidase gene family encodes functionally diverse isoforms localized in different subcellular compartments. *Planta* 224 (2): 300–314.

Teixeira, F.K., Menezes-Benavente, L., Galvão, V.C. et al. (2005). Multigene families encode the major enzymes of antioxidant metabolism in *Eucalyptus grandis* L. *Genetics and Molecular Biology* 28 (3): 529–538.

Terrile, M.C., Parïs, R., Calderón-Villalobos, L.I.A. et al. (2012). Nitric oxide influences auxin signaling through *S*-nitrosylation of the *Arabidopsis* TRANSPORT INHIBITOR RESPONSE 1 auxin receptor. *Plant Journal* 70: 492–500. doi:10.1111/j.1365-313X.2011.04885.x.

Thao, N.P., Khan, M.I.R., Thu, N.B.A. (2015). Role of ethylene and its cross talk with other signaling molecules in plant responses to heavy metal stress. *Plant Physiology* 169 (1): 73–84.

Tossi, V., Cassia, R., Bruzzone, S. et al. (2012). ABA says NO to UV-B: a universal response? *Trends in Plant Science* 17 (9): 510–517.

Tossi, V., Lamattina, L., and Cassia, R. (2009). An increase in the concentration of abscisic acid is critical for nitric oxide-mediated plant adaptive responses to UV-B irradiation. *New Phytologist* 181: 871–879. doi:10.1111/j.1469-8137.2008.02722.x.

Trivedi, D.K., Gill, S.S., Yadav, S. et al. (2013). Genome-wide analysis of glutathione reductase (GR) genes from rice and Arabidopsis. *Plant Signaling and Behavior* 8 (2): e2301–e2307.

Tun, N.N., Holk, A., and Scherer, G.F.E. (2001). Rapid increase of NO release in plant cell cultures induced by cytokinin. *FEBS Letters* 509: 174–176. doi:10.1016/S0014-5793(01)03164-7.

Tun, N.N., Livaja, M., Kieber, J.J. et al. (2008). Zeatin-induced nitric oxide (NO) biosynthesis in *Arabidopsis thaliana* mutants of NO biosynthesis and of two-component signaling genes. *New Phytologist* 178 (3): 515–531.

Tun, N.N., Santa-Catarina, C., Begum, T. et al. (2006). Polyamines induce rapid biosynthesis of nitric oxide (NO) in *Arabidopsis thaliana* seedlings. *Plant and Cell Physiology* 47: 346–354. doi:10.1093/pcp/pci252.

Turko, I.V. and Murad, F. (2002). Protein nitration in cardiovascular diseases. *Pharmacological Reviews* 54 (4): 619–634.

Ullah, S., Kolo, Z., Egbichi, I. et al. (2016). Nitric oxide influences glycine betaine content and ascorbate peroxidase activity in maize. *South African Journal of Botany* 105: 218–225.

Umehara, M., Hanada, A., Yoshida, S. et al. (2008). Inhibition of shoot branching by new terpenoid plant hormones. *Nature* 455 (7210): 195–200.

Ushimaru, T., Nakagawa, T., Fujioka, Y. et al. (2006). Transgenic Arabidopsis plants expressing the rice dehydroascorbate reductase gene are resistant to salt stress. *Journal of Plant Physiology* 163 (11): 1179–1184.

Van Loon, L.C., Bakker, P.A.H.M., and Pieterse, C.M.J. (1998). Systemic resistance induced by rhizosphere bacteria. *Annual Review of Phytopathology* 36: 453–483.

Vernoux, T., Wilson, R.C., Seeley, K.A. et al. (2000). The root meristemless1/cadmium sensitive2 gene defines a glutathione-dependent pathway involved in initiation and maintenance of cell division during postembryonic root development. *Plant Cell* 12: 97–110.

Wang, J., Song, L., Gong, X. et al. (2020). Functions of jasmonic acid in plant regulation and response to abiotic stress. *International Journal of Molecular Sciences* 21 (4): 1446. doi:10.3390/ijms21041446.

Wasternack, C. and Strand, M. (2016). Jasmonate signaling in plant stress responses and development – active and inactive compounds. *New Biotechnology* 33 (5): 604–613.

Wendehenne, D., Durner, J., and Klessig, D.F. (2004). Nitric oxide: a new player in plant signaling and defence responses. *Current Opinion in Plant Biology* 7: 449–455.

Wu, A.P., Gong, L., Chen, X. et al. (2014). Interactions between nitric oxide, gibberellic acid, and phosphorus regulate primary root growth in Arabidopsis. *Biology Plant* 58 (2): 335–340.

Wu, Y., Hu, L., Liao, W. et al. (2019). Foliar application of 5-aminolevulinic acid (ALA) alleviates NaCl stress in cucumber (*Cucumis sativus* L.) seedlings through the enhancement of ascorbate-glutathione cycle. *Scientia Horticulturae* 257: 108761. doi:10.1016/j.scienta.2019.108761.

Xu, J., Wang, W., Yin, H. et al. (2010). Exogenous nitric oxide improves antioxidative capacity and reduces auxin degradation in roots of *Medicago truncatula* seedlings under cadmium stress. *Plant Soil* 326 (1): 321–330.

Xu, L.L., Fan, Z.Y., Dong, Y.J. et al. (2015). Effects of exogenous salicylic acid and nitric oxide on physiological characteristics of two peanut cultivars under cadmium stress. *Biology Plant* 59 (1): 171–182.

Xu, M.J. and Dong, J.F. (2008). Synergistic action between jasmonic acid and nitric oxide in inducing matrine accumulation of *Sophora flavescens* suspension cells. *Journal of Integrative Plant Biology* 50 (1): 92–101.

Yan, Y., Pan, C., Du, Y. et al. (2018). Exogenous salicylic acid regulates reactive oxygen species metabolism and ascorbate–glutathione cycle in *Nitraria tangutorum* Bobr. under salinity stress. *Physiology and Molecular Biology of Plants* 24 (4): 577–589.

Yang, H., Mu, J., Chen, L. et al. (2015). S-nitrosylation positively regulates ascorbate peroxidase activity during plant stress responses. *Plant Physiology* 167 (4): 1604–1615.

Yang, S.F. and Hoffman, N.E. (1984). Ethylene biosynthesis and its regulation in higher plants. *Annual Review of Plant Physiology* 35 (1): 155–189.

Ye, N., Zhu, G., Liu, Y. et al. (2012). Ascorbic acid and reactive oxygen species are involved in the inhibition of seed germination by abscisic acid in rice seeds. *Journal of Experimental Botany* 63: 1809–1822.

Yin, Z., Lu, J., Meng, S. et al. (2019). Exogenous melatonin improves salt tolerance in tomato by regulating photosynthetic electron flux and the ascorbate–glutathione cycle. *Journal of Plant Interactions* 14 (1): 453–463.

Yousuf, P.Y., Hakeem, K.U.R., Chandna et al. (2012). Role of glutathione reductase in plant abiotic stress. In: *Abiotic Stress Responses in Plants*, (eds. P. Ahmad, and M.N.V. Prasad). 149–158. New York: Springer.

Zagorchev, L., Seal, C., Kranner, I. et al. (2013). A central role for thiols in plant tolerance to abiotic stress. *International Journal of Molecular Sciences* 14: 7405–7432.

Zhang, A., Zhang, J., Zhang, J. et al. (2011). Nitric oxide mediates brassinosteroid-induced ABA biosynthesis involved in oxidative stress tolerance in maize leaves. *Plant and Cell Physiology* 52: 181–192.

Zhang, Y. (ed.) (2013). Biological role of ascorbate in plants. In: *Ascorbic Acid in Plants: Biosynthesis, Regulation and Enhancement*, (eds. Springer Nature Switzerland AG). 7–33. Springer.

Zhou, Y., Wen, Z., Zhang, J. et al. (2017). Exogenous glutathione alleviates salt-induced oxidative stress in tomato seedlings by regulating glutathione metabolism, redox status, and the antioxidant system. *Scientia Horticulturae* 220: 90–101.

Zhu, S.H. and Zhou, J. (2007). Effect of nitric oxide on ethylene production in strawberry fruit during storage. *Food Chemistry* 100 (4): 1517–1522.

Zhu, T., Deng, X.G., Tan, W.R. et al. (2016). Nitric oxide is involved in brassinosteroid-induced alternative respiratory pathway in *Nicotiana benthamiana* seedlings' response to salt stress. *Physiology Plant* 156 (2): 150–163.

9

Phytohormones and Nitric Oxide Cross-talk in Regulation of Stress Tolerance in Plants

Debjani Dutta[1], *Sananda Mondal*[2], *and Snehashis Karmakar*[1]

[1]*Department of Plant Physiology, Faculty of Agriculture, Bidhan Chandra Krishi Viswavidyalaya, Mohanpur, Nadia, West Bengal, India*
[2]*Department of Crop Physiology, Institute of Agriculture, Visva-Bharati, Sriniketan, Birbhum, West Bengal, India*

9.1 Introduction

Plants grown in natural ecosystems usually face different kinds of biotic (i.e. fungi, bacteria, and pathogenic viruses) and abiotic stresses (oxidative stress, drought, chilling and freezing injury, salt, flood, heavy metals, high light, and temperature intensity), which hamper the normal growth and yield potential of plants, and accordingly global food security is also affected. To understand how plants will sense, perceive, and respond to stress signals is a vital matter for researchers, especially with the current changing climatic conditions where food security for every nation is an issue. Plants undergo different kinds of changes in terms of regulation of genes, metabolic, and physiological homeostasis in response to various environmental stresses as compared with normal conditions. In stressful situations plants will respond by perceiving the exogenous and endogenous stress stimuli via the hormonal signaling network, which coordinates some downstream signaling components like cyclic nucleotides (cyclic guanosine monophosphate, cGMP), Ca^{2+} ions, and reactive oxygen (superoxides, O_2^-; hydrogen peroxide, H_2O_2; singlet oxygen, O_2^1; hydroxyl radicals, etc.) or nitrogen (nitric oxide, i.e. NO) species (Jaspers and Kangasjärvi 2010). Based on the concentrations, reactive oxygen species (ROS) and/or reactive nitrogen species (RNS) are potentially toxic and may be detrimental or beneficial to the plant cells (Miller et al. 2010). According to Asada (2006), synthesis of ROS alters the redox homeostasis of the cells and ultimately oxidative damage takes place within the plants cell; as a whole it creates necrotic lesions across plant species (Mano et al. 2002; Petrov et al. 2015). During the onset of stress, ROS production is regulated by NO, which interacts with different signaling components thus maintaining balanced concentrations of ROS (Vranova et al. 2002). Acclimatization is a process in which plants can achieve some metabolic adjustments like accumulation of osmolytes (Shabbir et al. 2016), maintenance of turgor (Waraich et al. 2011), photosynthetic regulation, and rate of transpiration alongside stimulation of antioxidant mechanisms (Nawaz et al. 2015) to withstand different abiotic stresses. In addition to this, gene expression is altered due to stress leading to synthesis of different antistress compounds, which helps to modify plant physiology, growth, development, phenology, and reproduction in response to exposure to various adverse climatic conditions in plants (Penuelas et al. 2013).

In living cells including plants, NO, a gaseous, redox, highly reactive nitrogen species (RNS), is produced under stressful and normal conditions. It takes part in signaling cascades, interacts with different stress regulators, and confers plant responses to abiotic stress conditions. In recent years, NO has been proven to provide extraordinary potential in engineering of crop plants for abiotic stress tolerance (Zhou et al. 2021). When ROS concentration levels become toxic for a plant, NO may act as a detoxifier reducing the detrimental effects caused by ROS (Lipton et al. 1993; Zhou et al. 2021). Respiratory function, mainly the electron transport pathway of mitochondria, is regulated by NO, which modulates ROS and activates defense mechanisms by enhancing the production of antioxidants in plants, in the presence of different abiotic stresses (Zottini et al. 2002). Moreover, exogenously applied NO may play an important role in antioxidant enzyme activation, likely for superoxide dismutase (SOD), which restricts O_2^-, organic radicals (R), and lipid O_2 (Shi et al. 2007). NO acts as an endogenous regulator of several phytohormones, as well as inhibiting programmed cell death, which is aided by stomatal function in various plant species like wheat, *Arabidopsis*, and pea (Leshem and Haramaty 1996; Garcia-Mata and Lamattina 2001; Bright et al. 2006). During unfavorable/multiple stressful conditions, NO acts as a signaling molecule, enabled by its status as a small redox molecule and a neutrally charged free radical; together with its ability to diffuse easily through membranes these properties render it a significantly dynamic molecule (Domingos et al. 2015).

To explore the multiple, diverse, interactive mechanisms between NO and phytohormones, several widespread investigations have been carried out during the past few years. There is no doubt that NO possibly acts both downstream and upstream of phytohormone signaling pathways (Hancock et al. 2011; Simontacchi et al. 2013). Moreover, the transport and signaling of phytohormones involves synthesis, distribution, conjugation, and degradation of elements, modulated by NO (Manjunatha et al. 2010; Lozano-Juste and León 2011; Terrile et al. 2012; Feng et al. 2013; Freschi 2013). However, to understand the mechanism of how NO forms a link with phytohormone-related proteins at translational or post-transcriptional level, further extensive studies are required. Similarly, the cross-talk mechanisms between NO and phytohormone signaling cascades, i.e. auxins (AUXs), cytokinins (CKs), gibberellins (GBs), abscisic acid (ABA), ethylene (ET), jasmonic acid (JA), brassinosteroids (BRs), salicylic acid (SA), polyamines (PAs), and strigolactones (SLs) during abiotic stress conditions still remain elusive. Sometimes, transcript level changes or an abundance of proteins for NO synthesis or removal of enzymes has been observed during the lag phases (Pagnussat et al. 2002; Freschi et al. 2010).

By evaluating the cross-talk mechanisms of different phytohormones and NO, it has been observed that AUX and NO act synergistically to regulate a sequence of plant responses, such as initiation of cell division and formation of embryogenic cells (Ötvös et al. 2005), root organogenesis (Pagnussat et al. 2002, 2003; Lanteri et al. 2006), responses to gravity (Hu et al. 2005), root responses to iron deficiency (Chen et al. 2010), formation of root nodules (Pii et al. 2007), and, finally, stimulation of nitrate reductase (NR) activity (Du et al. 2008). For instance, CKs and NO were noted to regulate various physiological processes synergistically in a positive direction, such as division and differentiation of cells (Shen et al. 2012), photosynthetic adaptation to drought stress (Shao et al. 2010), programmed cell death (Carimi et al. 2005), and senescence of leaves (Mishina et al. 2007). The vital "stress-related" biomolecules, i.e. ABA and NO, having been intensively involved in various cross-talk mechanisms as part of different signaling cascades instigated by environmental challenges like UV-B radiation and water limitation, induce the plant adaptive responses, i.e. closing of stomata and antioxidant defense responses (Neill et al. 2008; Tossi et al. 2009; Hancock et al. 2011). Depending on the environmental condition, plants respond in different ways by activating various pathways. One such pathway is ABA signaling where NO acts as a downstream element. However, the inhibition of ABA production does not have any effect on these induced responses when NO is applied exogenously. Regarding the GA and NO cross-talk mechanism it has been found that it is only limited with two key roles and these are the modulation of seed germination (Beligni et al. 2002; Bethke et al. 2007) and hindrance of the elongation of the hypocotyl at the time of de-etiolation of seedlings (Lozano-Juste and León 2011). To balance such kinds of plant responses, NO plays an important role as an upstream modulator of GA biosynthesis and perception/transduction (Bethke et al. 2007; Lozano-Juste and León 2011).

9.2 Cross-talk Between Phytohormones and NO in Drought Stress Tolerance

Drought stress, being a major limiting factor, affects multiple features of growth, development, and yield of plants (Shabbir et al. 2016). The closing of stomata to conserve water within cells is the typical mechanism of plants with regard to water deficit conditions, which are mainly caused by drought stress. The concrete participation of NO in response to drought exposure at the whole plant level is still not well known. To date, researchers have found that NO acts in drought stress, both alone and in association with phytohormones and other signaling molecules. Here we intend to establish the details of the cross-talk relationship between phytohormones and NO to cope with drought stress in plants (Figure 9.1).

The interactions of NO with phytohormones under drought stress may be categorized in three different ways (Adimulam et al. 2017):

1) Phytohormones and NO are reciprocally induced under drought conditions.
2) Phytohormones help recruit NO under drought conditions.
3) NO acts as a cross-talk switch for interacting phytohormones in stress situations.

Several studies have highlighted synergistic as well as antagonistic cross-talk between NO and ABA in drought conditions (Freschi 2013; Santisree et al. 2015; Prakash et al. 2019). According to Garcia-Mata and Lamattina (2001), NO takes part in drought-induced stomatal closure by interacting with ABA signaling cascades in *Vicia faba*. Synergistic cross-talk between ABA and NO as part of stomatal closure under stress conditions was also reported earlier in *Pisum sativum* and *Arabidopsis thaliana* (Neill et al. 2002). Many components, i.e. Ca^{2+}, cGMP, cyclic adenosine 5′-diphosphate ribose (cADPR), and G proteins, are shared as the common signaling elements for NO and ABA under drought conditions (Xing et al. 2004). For example, in ABA-regulated stomatal closure in guard cells of faba bean, NO is involved by selectively triggering the intracellular Ca^{2+} channels through a signaling cascade mediated by cGMP/cADPR (Mioto and Mercier 2013). On the other hand, through the cross-talk between ABA and NO in an ABA-mediated signaling network, cytosolic Ca^{2+}

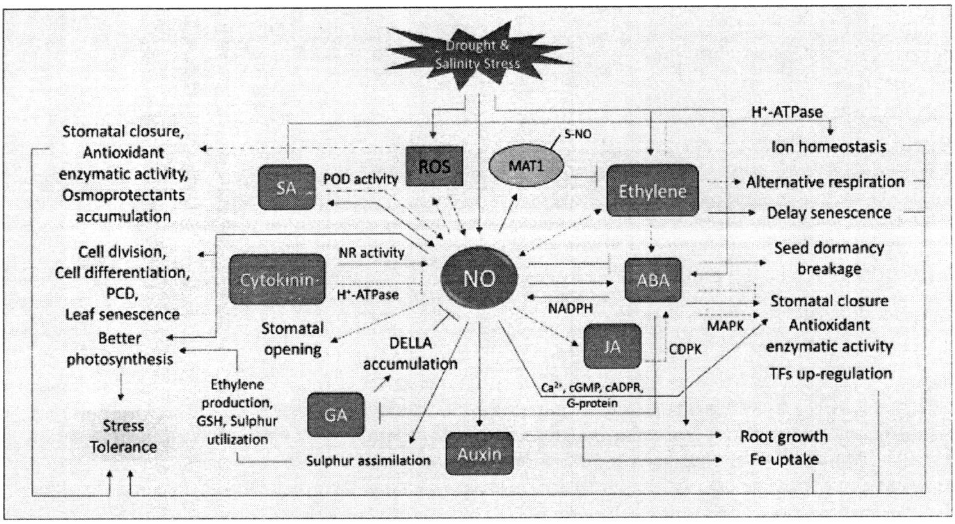

Figure 9.1 A schematic representation portraying the crosstalk between NO and phytohormones in drought and salinity stress tolerance. NO and phytohormones show either synergistic and/or antagonistic interactions under drought and salinity stress conditions. NO and ABA are mutually regulated under both types of stress, triggering Ca^{2+} channels through the cGMP/cADPR/G-protein pathway, thereby inducing stomatal closure. ROS-mediated increased NO content stimulates JA biosynthesis, consequently the CDPK-induced ABA response results in closing of stomata, activation of the antioxidant defense system, and upregulation of many TFs, leading to improved drought tolerance. CK-induced NR activity increases NO accumulation, improving photosynthesis, thereby enhancing drought endurance. Drought-induced NO increases AUX levels, which enhances root growth and iron uptake, mediated by CDPK signaling. NO induces S-nitrosylation of MAT1, which helps reduce ET levels, causing a delay in the senescence process. Drought-induced ET production inhibits stomatal closure by interacting with either NO or ABA. NO antagonistically regulates GB signaling by inducing DELLA repressor accumulation. Under salinity stress, ABA increases H_2O_2-mediated NO accumulation that activates antioxidant enzymatic defenses through MAPK signaling. Under saline conditions, NO and ET are reciprocally regulated, thus maintaining ion homeostasis through activation of H^+-ATPase and causing a bypass to alternative respiration, promoting salinity stress tolerance. SA-induced NO also helps maintain ion homeostasis by activating H^+-ATPase activity. SA triggers POD activity, which induces stomatal closure, antioxidant enzyme activation, and osmoprotectant accumulation, mediated by the NO/ROS pathway. Salinity-induced NO signaling increases sulfur assimilation, which enhances ET production, GSH, and sulfur utilization, consequently improving photosynthesis and stress endurance capability. NO, nitric oxide; ABA, abscisic acid; cGMP, cyclic guanosine monophosphate; cADPR, cyclic adenosine 5′-diphosphate ribose; ROS, reactive oxygen species; JA, jasmonic acid; CDPK, Ca-dependent protein kinase; TFs, transcription factors; CK, cytokinin; NR, nitrate reductase; AUX, auxin; MAT1, methionine adenosyltransferase 1; ET, ethylene; GB, gibberellin; MAPK, mitogen-activated protein kinase; SA, salicylic acid; POD, peroxidase; GSH, reduced glutathione.

ions were induced to regulate Crassulacean acid metabolism (CAM) expression, which significantly improved stress tolerance in plants under water deficit stress (Freschi et al. 2010; Mioto and Mercier 2013; Prakash et al. 2019). Likewise, ABA and NO are mutually regulated under exposure to drought conditions. For instance, via nitric oxide synthase (NOS)-like activity, NO induces the accumulation of ABA in root tips of wheat during stress conditions. Similarly, ABA-induced synthesis of NO through the elevation of nicotinamide adenine dinucleotide phosphate (NADPH) in maize during drought stress proves the functional relationship between NO and ABA (Adimulam et al. 2017). Under adverse environmental conditions, the generation of ROS enhances NO-facilitated ABA responses such as activation of the enzymatic antioxidant defense system (Lu et al. 2009) and upregulation of transcription factors (TFs) (Zhang et al. 2007), as well as the induction of stomatal closure. During seed germination, after being induced by NO synthesis, the expression of *CYP707A2* gene initiated ABA catabolism and consequently an increased level of ABA, which aided in improvement of drought tolerance in plants (Freschi 2013; León et al. 2014). Contrary to this, ABA and NO are also well known to interact antagonistically by offsetting each other in some cases, such as in breaking of seed dormancy exposed to drought conditions (Bethke et al. 2006; Liu et al. 2010, 2011).

A synergistic interaction between CKs and NO under water deficit conditions was found by Shao et al. (2010). During water stress, CK stimulated NO signaling cascades probably through a NR source. Interaction of CK with NO signaling networks regulated photosynthesis in *Zea mays*, leading to an improvement in drought tolerance capacity in plants (Shao et al. 2010). In addition, the synergistic relationship between CK and NO to control cell division, differentiation, programmed cell death, and leaf senescence has also been studied (Shen et al. 2012; Simontacchi et al. 2013). An antagonistic interaction has also been reported. For instance, CK induced reduced levels of NO promoted the opening of stomata in dark-grown seedlings of *V. faba* (Xiao-Ping and Xi-Gui 2006). Most probably, by reducing NO concentrations in guard cells, CKs stimulate plasma membrane H^+-ATPase and simultaneously induce opening of stomata in darkness.

NO is known to serve as a secondary messenger in the signaling cross-talk of many phytohormones, e.g. AUX, ET, GBs, and JA, in regulation of a number of plant growth processes under stress situations (Saito et al. 2009; Liu et al. 2010). It has been well documented that the stimulation of iron uptake and root growth in plants by AUX is mediated via the endogenous levels of NO (Liao et al. 2011; Yadav et al. 2011; Simontacchi et al. 2013). Association of AUX and ET for the regulation of root development and morphology is one of the key aspects of plant tolerance to drought (Kohli et al. 2013). The Ca^{2+}-dependent protein kinase (CDPK)-activated signaling cross-talk between NO and AUX is involved in the formation of adventitious roots in cucumber (Pagnussat et al. 2002). As NO is known to be involved in the formation of lateral roots in drought conditions (Liao et al. 2012), it is assumed that the NO and AUX signaling network interacts in the development and architecture of root meristem and root hair (Hunt et al. 2002; Chen et al. 2013) to extract more water under water deficit stress. It is worth highlighting that NO has been confirmed to have a role in inducing AUX accumulation in plants by regulating biosynthesis, transport, and degradation of AUX (Xu et al. 2010; Fernandez-Marcos et al. 2011). To date, a few AUX transporters have been identified as playing vital roles in mitigating against drought stress (Remy et al. 2013). However, further study is needed on NO-facilitated regulation of synthesis and transport of AUX under drought stress.

Antagonistic cross-talk between NO and ET has been reported in most cases (Desikan et al. 2004; Lindermayr et al. 2005; Freschi 2013). Both the endogenous and exogenous levels of NO can affect ET biosynthesis in plants during senescence and fruit ripening (Manjunatha et al. 2012; Freschi 2013). A previous report stated that through the *S*-nitrosylation of methionine adenosyltransferase (MAT1), NO can downregulate the ET precursor *S*-adenosylmethionine (AdoMet) in *Arabidopsis* (Lindermayr et al. 2005); thus, as a result of decreasing ET synthesis, senescence can be delayed and plants show better tolerance to stress. The same type of antagonistic relationship is also formed when ET inhibits closing of stomata by combining with either NO or ABA (Desikan et al. 2004). On the contrary, NO has also been reported to promote ET production in cucumber and *A. thaliana* by upregulating ET biosynthesis genes, while NO production is also stimulated in senescing leaves by ET precursor according to Manjunatha et al. (2012). A similar kind of antagonism was described between NO and GB during photomorphogenesis and root growth in *A. thaliana* (Lozano-Juste and León 2011; Freschi 2013). GBs antagonistically controlled NO production by stimulating the accumulation of DELLA repressors, which are identified to act as key modulators in the signaling cascades of GB and NO interactions (Hui et al. 2009).

On the other hand, a synergistic interaction for JA and SA with NO has been documented (Palmieri et al. 2008; Alavi et al. 2014). NO was shown to upregulate many genes involved in JA biosynthesis, which potentially regulates JA signaling via the pathway mediated by ROS/NO (Palmieri et al. 2008). Simultaneously, JA induces ABA-regulated stomatal closure through the signaling of Ca^{2+} influx-induced CDPK production. Equally, in water-stressed plants, NO was noted to act downstream of SA signaling to alleviate ROS-induced oxidative stress. The SA and NO were observed to act together to reduce the adverse effects of water stress in wheat seedlings as induced by polyethylene glycol (PEG) (Alavi et al. 2014; Prakash et al. 2021).

9.3 Cross-talk Between Phytohormones and NO in Salinity Stress Tolerance

Salinity stress is thought of as one of the most severe forms of stress because of its worldwide distribution and high magnitude. Cross-talk between phytohormones and NO under salinity stress is very interesting in plants. NO, an important signaling element, helps significantly in alleviating adverse effects of salinity stress in plants. It stimulates plant growth, improves photosynthesis and stomatal conductance, sustains ion homeostasis, accumulates osmoprotectants, and activates antioxidant defense mechanisms, thus enhancing plant ability to withstand salinity stress (Bhardwaj et al. 2021). In the case of salinity stress, phytohormones play key roles in improving stress tolerance as well as adaptability in plants (Hussain et al. 2016). Various studies have supported the theory that NO interacts with the signaling cascades of various phytohormones and induces salinity stress tolerance (Poór and Tari 2011b; Fatima et al. 2016) (Figure 9.1). Very often in plants, phytohormones, NO, and ROS initiate different responses under abiotic stress (Chen et al. 2021). In the case of salinity-exposed plants, phytohormones such as AUX, ET, ABA, SA, etc. are transported from roots to leaves or throughout the plant body to promote NO synthesis (Molassiotis et al. 2010). In salinity-affected plants, NO-induced amelioration of oxidative stress is associated with a decrease in thiobarbituric acid (TBA) reactive substance content and an increase in antioxidant enzymatic activities (Xu et al. 2011).

In *Arabidopsis*, participation of both NO and AUX in growth retardation of root apical meristem in salt-affected plants was reported by Liu et al. (2015). It was noted that saline conditions inhibit the translation of *PIN-FORMED* (*PIN*) genes, thereby a reduction in AUX levels occurs and ultimately root meristem growth inhibition takes place. Stabilization of the genes *INDOLE-3-ACETIC ACID17 (IAA17)/AUXIN RESISTANT3 (AXR3)* inhibits AUX signaling via accumulation of NO.

Salinity-induced leaf senescence of cotton is mitigated by exogenous NO application. Here, sodium nitroprusside (SNP) acts as an NO donor. NO decreases ABA content by downregulating the genes *NCED2* and *NCED9*, which are responsible for ABA biosynthesis (Kong et al. 2016). ABA promotes H_2O_2 accumulation in plant cells, which increases NO production. It triggers the activation of mitogen-activated protein kinase (MAPK) and expression of several genes related to antioxidant enzyme activities in plants under salinity stress (Zhang et al. 2007; Lu et al. 2009).

However, Fatima et al. (2016) reported that ET acts as a stress signaling hormone. It enhances stress tolerance ability in plants against many abiotic and biotic stresses by interacting with the signaling cascade of the other phytohormones. Transcriptionally activated ethylene response factor (ERF) in the ET signaling pathway improves tolerance to salinity stress by reducing ROS accumulation in tobacco seedlings (Wu et al. 2014). Wang et al. (2009) stated that *Arabidopsis* callus treated with 100 mM NaCl had increased NO accumulation and ultimately caused ET emission. This resulted in upregulation in the expression of the genes that encode plasmalemma H^+-ATPase. Both ET and NO take part in upregulating the expression of H^+-ATPase genes, which help maintain ion homeostasis and tolerance to salinity stress. NO and ET also participate in triggering alternative respiration during salinity stress (Wang et al. 2010). A contradictory report of Poór and Tari (2011b) stated an antagonistic association between ET and NO in NaCl-treated (100 and 250 mM) tomato suspension culture. Increased ET biosynthesis stimulated ROS generation and increased the dead cell ratio in plant cell culture exposed to salinity stress. However, NO production reduced ROS concentrations and blocked ET biosynthesis, which reduced the dead cell ratio. Poór et al. (2011a) reported that absence of NO and ET caused ionic (Na^+/K^+) imbalance in cell suspension culture of apical root segment, which in turn increased sensitivity to salinity stress.

An important role for both NO and ROS in SA-induced closing of stomata was reported by Khokon et al. (2011). An induction of peroxidase activity by SA accumulation stimulated accumulation of NO and generation of ROS in guard cells, which caused stomatal closure. According to Liu et al. (2014) and Dong et al. (2015), combined application of SA and SNP can mitigate salinity-induced stress by activating antioxidant enzymes (ascorbate peroxidase [APX], glutathione peroxidase [GPX], and catalase [CAT]) and increasing proline accumulation in *Gossypium hirsutum* and *Fagur sylvatica*. Interaction of SA with the signaling pathway of NO decreased H_2O_2 accumulation, which led to H^+-ATPase influx in plasma membrane. Synergistic interactions between NO and SA decreased Na^+ uptake and increased Ca^{2+}/Mg^{2+} absorption under salinity stress (Dong et al. 2015).

Sulfur (S), an important element in various metabolites (e.g. coenzyme A, reduced glutathione (GSH), cysteine, methionine, Fe-S cluster, thioredoxin etc.), plays a vital role in regulation of various physiological processes under salinity stress (Khan, Ahmad et al. 2013). According to Fatima et al. (2016), NO encourages sulfur assimilation and it is linked to ET production by cysteine biosynthesis. In salinity-affected mustard plants, the combined effect of S and NO controlled the utilization of GSH and S, resulting in improvement in photosynthesis and plant growth. NO is the main regulatory signal that triggers many biochemical processes. It interacts with nitro and sulfhydryl groups in the nitration process and improves salt tolerance.

According to Kong et al. (2016), application of SNP as foliar spray delays salinity-induced senescence of leaves in cotton seedlings. Here, NO upregulates the expression of isopentyl transferase (IPT) gene, which is a key enzyme of CK biosynthetic pathways.

Fan et al. (2013) suggested that exogenously applied NO reduced the activity of polyamine oxidase (PAO) enzyme, the content of total free PAs like spermidine (Spd) and putrescine (Put) under salinity stress, and alleviated salt-induced damage in cucumber seedlings.

All these reports are strong evidence of the NO and phytohormone crosstalk to enhance salt tolerance under high-salinity conditions.

9.4 Cross-talk Between Phytohormones and NO in Heavy Metal Stress Tolerance

Heavy metal (HM) or metalloid toxicity, an alarming threat to crop production, creates adverse impacts on plant growth and development. To cope with this HM stress, plants regulate the endogenous levels of NO via interacting with the signaling pathways of different phytohormones (Figure 9.2). Here, we are going to discuss the key roles of NO in plants against HM toxicity with a specific focus on recent studies that implicate the interaction of NO with different phytohormones in alleviation of HM stress in plants.

NO interacts with plant hormones, such as AUXs, ET, ABA, GB, and CK, through various signaling networks during HM toxicity to induce plant responses against this stress condition. Studies done so far have emphasized the downstream and upstream interactions between AUX and NO concerning metal tolerance in plants. Under arsenate (As^V) stress, both IAA and NO play key roles in Si-induced adventitious root development in rice (Tripathi et al. 2021). In *A. thaliana* seedlings treated with Cd, NO enhanced lateral root formation by acting as a downstream messenger in AUX signaling cascades (Xu et al. 2011). In Cu-intoxicated seedlings of *A. thaliana*, NO and AUX have been observed to affect the growth of primary root and stem because of the negative regulation between AUX and NO (Pető et al., 2011). NO promoted plant growth mainly by regulating AUX signaling through intervention in the interaction between transport inhibitor response 1 (TIR1) and AUX/IAA and the *S*-nitrosylation of TIR1 (Wei et al. 2020). Additionally, NO supplementation in Cd-exposed *Medicago truncatula* seedlings resulted in recovery of IAA levels, which had been reduced by Cd-induced IAA oxidase activity, and thereby helped plant adaptation to Cd toxicity (Xu et al. 2010). Likewise, in As-stressed rice seedlings, NO was proven to lessen metal toxicity by inducing the expression of PIN proteins (Praveen and Gupta 2018). Further, it was noted that Cd toxicity elevated the endogenous levels of AUX, whereas the exogenous levels of AUX alleviated the accumulation of Cd in *A. thaliana*, thus relieving the symptoms of Cd stress (Zhu et al. 2013).

ET, a gaseous phytohormone, influences plant growth and development by regulating seed germination, adventitious root formation, leaf and flower senescence, and fruit ripening. However, recent investigations have demonstrated an antagonistic interaction between ET and NO under HM toxicity (Liu et al. 2015a, 2015b). SNP suppressed the expression of ET, reduced the accumulation of Cd, improved the uptake of minerals, and thus lessened Cd stress in *Trifolium repens* (Liu et al. 2015b). Cross-talk between NO and

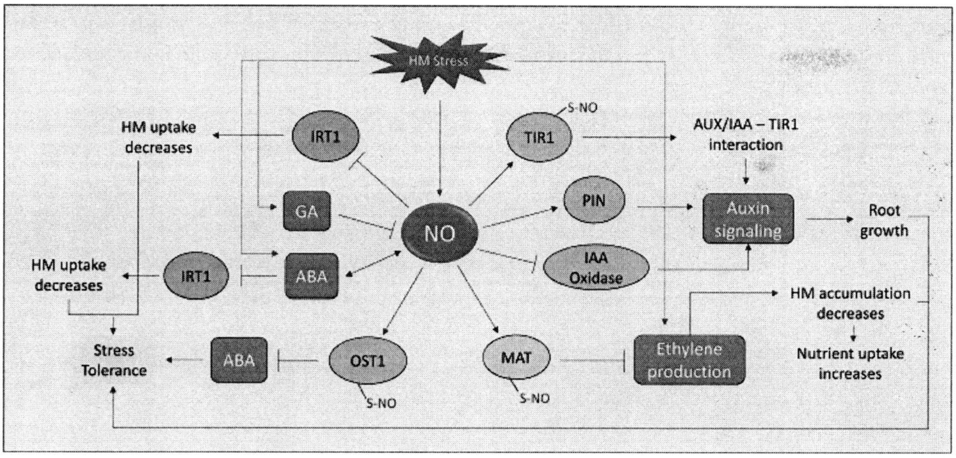

Figure 9.2 A schematic representation portraying the cross-talk between NO and phytohormones in heavy metal (HM) stress tolerance. Heavy metal stress negatively affects the level of auxin (AUX), whereas, ethylene (ET) and abscisic acid (ABA) levels are affected positively. HM-induced NO upregulates the genes responsible for PIN-FORMED (PIN) protein synthesis and downregulates the genes for indole-3-acetic acid (IAA) oxidase. The NO-induced S-nitrosylation of transport inhibitor response 1 (TIR1) and its interaction with AUX/IAA regulates AUX signaling. These upregulate AUX signaling cascades, thus stimulating root growth and improving stress tolerance. NO decreases ET level by triggering S-nitrosylation of methionine adenosyltransferase (MAT), thereby decreasing HM accumulation and increasing nutrient uptake, resulting in enhanced stress endurance ability. Stress-induced NO and ABA are mutually regulated, downregulating iron-regulated transporter 1 (IRT1) gene expression, leading to a decrease in HM uptake. HM-induced NO decreases ABA level by stimulating S-nitrosylation of open stomata 1 (OST1), causing HM stress tolerance. Exogenous gibberellic acid decreases HM-induced NO accumulation, inhibiting IRT1 gene expression, reducing HM uptake.

ET is associated with cell responses to Cd stress, which suggests a deeper understanding of the mechanisms of cellular responses to metal stress. Furthermore, NO application in *A. thaliana* reduced ET biosynthesis via *S*-nitrosylation of methionine adenosyltransferase (MAT) (Lindermayr et al. 2006).

The interaction of ABA and NO in signaling pathways was known as one of the key modulators of plants growth under HM toxicity. Recent investigations have proved that ABA helps reduce the accumulation of Cd in *A. thaliana* by inhibiting iron-regulated transporter 1 (IRT1)-mediated uptake of Cd, thereby enhancing plant tolerance to Cd toxicity (Luo et al. 2012; Fan et al. 2014). In the same way, NO interacts with ABA during HM toxicity to lower the degree of HM stress. For instance, NO was observed to decrease As toxicity in *V. faba* through the reduction of As-induced increase in ABA concentration (Mohamed et al. 2016). Consistent with this, it was also reported in Pb-exposed *Vigna unguiculata* that NO significantly reduced ABA content when applied exogenously, and thus enhanced plant tolerance to Pb toxicity (Sadeghipour 2017). Fascinatingly, NO was noted to control ABA signaling antagonistically both in vitro and in vivo in guard cells via *S*-nitrosylation of open stomata 1 (OST1) (Wang, Liu et al. 2015). Hence, this information suggests that in response to HM toxicity, NO is involved in ABA signaling by the use of *S*-nitrosylation of different proteins.

Recently, several studies have recounted that GBs and NO interact with each other both synergistically and antagonistically in response to HM stress (Zhu et al. 2012; Mohamed et al. 2016). According to Zhu et al. (2012), Cd-intoxicated *Arabidopsis* seedlings induce increased levels of endogenous GA and NO, whereas exogenous application of GA reduced the accumulation of Cd-induced NO and also downregulated the *IRT1* gene. Simultaneously, the uptake of Cd was inhibited and accordingly Cd toxicity was reduced in *Arabidopsis*. In addition, NO could promote the upregulation of *IRT1* transporter gene, thus the stimulation of Cd uptake resulted in an increase in Cd toxicity (Besson-Bard et al. 2009). Alternatively, exogenously applied NO could increase GA content and participate in the mitigating role of GA on As toxicity in faba bean (Mohamed et al. 2016). Thus, these results suggest that NO could act downstream to GB in signaling cascades, inhibiting uptake of excess HM, and taking part in the alleviating role of GB on HM stress.

9.5 Cross-talk Between Phytohormones and NO in Temperature Stress Tolerance

In the modern era, temperature stress (whether heat or cold stress) is one of the biggest challenges for crop production. Several researchers have reported that heat stress has adverse effects on the molecular, biochemical, and physiological characteristics of plants. It can cause protein denaturation, oxidative damage, lipid peroxidation in cell membranes, macromolecular damage, and enzyme deactivation in plants (Khan, Asgher et al. 2013; Brestic et al. 2016; Prasad et al. 2016; Santisree et al. 2017). According to Leshem (2001) and Yu et al. (2014), NO accumulation rises in many plant species in response to heat stress. NO-mediated high-temperature stress signaling and tolerance mechanisms in various plant species have been reported (Uchida et al. 2002; Song et al. 2006). The cross-talk between phytohormones (JA, ABA, SA, PAs) and NO (Figure 9.3) induces thermotolerance responses in plants by upregulating the gene expression of many heat-shock proteins (HSPs) and activating antioxidant defense systems (Zandalinas et al. 2016).

Mutant studies on ABA biosynthesis (abal-1) and ABA signaling (aba I-1) of *A. thaliana* proposed that stomatal closure induced by heat or drought stress was regulated through JA and H_2O_2 signaling networks, which in turn induced endogenous levels of NO (Zandalinas et al. 2016), and functions of Ca^{2+} and guard cell S-type anion channel (SLAC1) (Daszkowska-Golec and Szarejko 2013; Murata et al. 2015). ABA was further reported to boost the activity of NOS and to increase the levels of NO, which helped to improve the thermotolerance capacity in *Phragmites communis* (Song et al. 2008). The exogenous application of ABA in seedlings of *Stylosanthes guianensis* enhanced the activities of many antioxidant enzymes, for example SOD, APX, and CAT, by stimulating the synthesis of NO (Zhou et al. 2005). Likewise, an increase in SA-induced glutathione-S-transferase (GST) activity was witnessed to increase heat tolerance in maize (Cetinkya et al. 2014) possibly due to NO production in plants as triggered by SA-induced H_2O_2 accumulation. However, direct evidence related to cross-talk between NO and phytohormones in improving heat tolerance to plants is lacking and further research is needed.

Low-temperature stress also causes serious damage to organelle structures and hampers cellular metabolic activities, leading to a restriction in plant growth and development

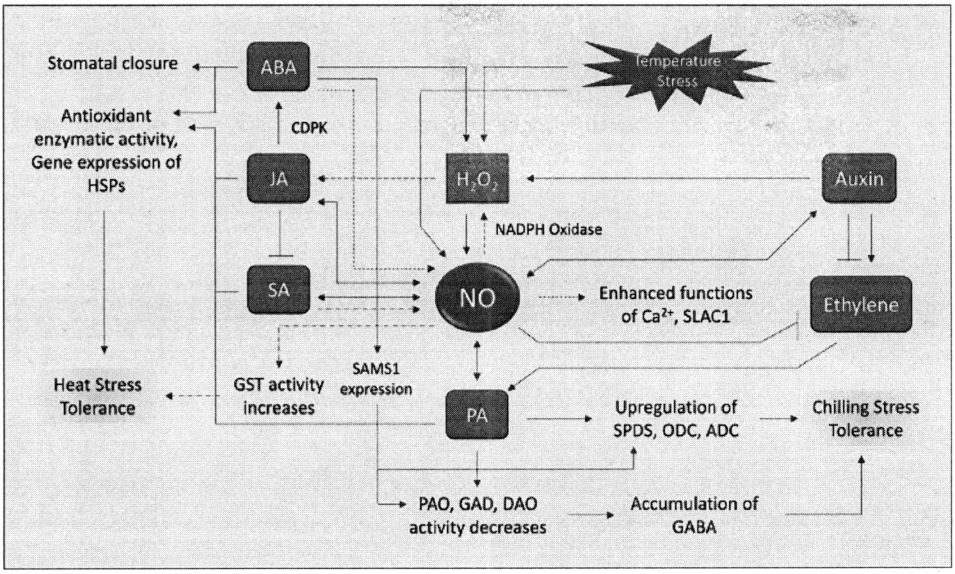

Figure 9.3 A schematic representation portraying the cross-talk between nitric oxide (NO) and phytohormones in temperature stress tolerance. Temperature stress (i.e. heat stress, chilling stress) stimulates accumulation of NO and H_2O_2, which in turn trigger (represented by normal-ended arrow) or inhibit (represented by blunt-ended arrow) phytohormone-mediated stress responses. The mutual interaction between NO and polyamines (PAs) upregulates the expression of spermidine synthase (SPDS), ornithine decarboxylase (ODC), arginine decarboxylase (ADC), and decreases the activities of polyamine oxidase (PAO), glutamate decarboxylase (GAD), and diamine oxidase (DAO), thus improving chilling stress tolerance. Salicylic acid (SA)-induced increased glutathione-S-transferase (GST) activity, abscisic acid (ABA)-induced enhanced functions of Ca^{2+}-channels, guard cell S-type anion channel (SLAC1), and cross-talk of other phytohormones with NO involve stomatal closure, activation of antioxidant defense systems, and gene expression of heat-shock proteins (HSPs), leading to improved heat tolerance in plants.

(Atıcı and Nalbantoğlu, 2003). Plants exposed to low-temperature stress face oxidative and nitrosative stress, thus stimulating the biosynthesis of NO (Corpas et al. 2011), which plays a crucial role in interactions between ABA and PAs (e.g. spermine [Spm], Spd, and Put) to induce plant responses to stress conditions (Wimalasekera et al. 2011). Earlier studies reported coordinated actions among NO, PAs, ABA, and H_2O_2 that aided in modulation of many physiological as well as temperature stress responses under such stressful environments (Zhou et al. 2005; Diao et al. 2017). Esim and Atici (2014) noted that the cross-talk among ABA, SA, and NO signaling resulted in enhanced antioxidant enzymatic activities (SOD, CAT, and peroxidase [POD]) in seedlings of *Z. mays*, consequently contributing to better stress tolerance under chilling stress. NO and SA act as very efficient tools to mitigate the adverse effects of abiotic stress. Their interaction involves various signaling networks and different regulatory elements that modulate plant responses to adverse stress conditions (Prakash et al. 2021). In another study by Guo et al. (2014), the interplay between ABA and NO was noted to upregulate the expression of cold-induced *MfSAMS1* gene, thereby resulting in improved acclimatization to cold stress in *M. sativa*. Additionally, *MfSAMS1* expression aided in altering the activities of Cu-containing amine oxidase, PAO, and the levels of Spm, Spd, and Put, which in turn regulate the actions of the antioxidant system during acclimatization to cold. Exogenous supply of NO improved tolerance to chilling stress in *Lycopersicon esculentum* by increasing the levels of Spd and Put and stimulating the gene expression of Spd synthase (*LeSPDS*), ornithine decarboxylase (*LeODC*), and arginine decarboxylase (*LeADC. LeADC1*). According to Li et al. (2014), NO application in *Zingiber officinale* seedlings was noted to convert Put into Spm or Spd, thus conferring tolerance to cold stress. A recent investigation by Wang et al. (2016) also proposed that cross-talk between PAs and NO enhanced tolerance to chilling stress in cold-stored banana as the activities of PAO, glutamate decarboxylase (GAD), and diamine oxidase (DAO) were increased in response to NO treatment, leading to an accumulation of gamma-aminobutyric acid (GABA) and thus preventing chilling injury in bananas.

More relevant studies concerning the effect of NO on CK, BR, ET, and JA signaling are needed to gather more key information about the signaling cross-talk of these modulatory components in relation to temperature stress tolerance in plants.

9.6 Cross-talk Between Phytohormones and NO in Other Stress Tolerance

It is recognized now that interactions of NO with various phytohormones like ET, ABA, and JA control the signaling pathways in guard cells under various altered environmental conditions (Gonugunta et al. 2009; Hasanuzzaman et al. 2018).

Ozone (O_3), a highly toxic air pollutant, severely restricts growth and development of plants (Ahlfors et al. 2009). According to Wilkinson and Davies (2010), for the purpose of sensing O_3 various transportable chemical compounds are used as signaling molecules by plants from the guard cells to adjacent mesophyll and epidermal cells, one of which is NO. Kwak et al. (2006) described how NO production due to ozone-induced stress triggers ET and ABA biosynthesis and affects ABA-mediated stomatal responses by impeding K^+ ion efflux at stomatal guard cells. To inhibit ozone toxicity, alternative oxidase (AOX) is also involved both ET- and NO-reliant pathways in plants (Ederli et al. 2006). Application of NO causes leaf injury owing to an increased level of ozone-induced ET production. Both ozone and SNP treatments upregulate the expression of the genes that are involved in ET biosynthesis (1-aminocyclopropane-1-carboxylate oxidase [ACC oxidase] and 1-aminocyclopropane-1-carboxylate synthase 6 [ACS6]) (Overmyer et al. 2000; Rao and Davis 2001).

Esringu et al. (2016) stated that in *Lactuca sativa* seedlings, exogenous application of NO prevented the inhibition of IAA and GA biosynthesis that is induced by ultraviolet-B (UV-B). Combined action of SA and NO reduces UV-B-induced stress in wheat plants (Yan et al. 2016).

There are many reports available regarding cross-talk between the signaling cascades of phytohormones and NO as the earliest response to plant wounding (Lamattina et al. 2003). París et al. (2009) stated that both AUX and NO have active participation in wound healing responses in plants. In sweet potatoes, Imanishi et al. (1997) presented evidence for the participation of methyl jasmonate and the expression of ipomoelin (IPO) gene in mechanical wounding. In wounded epidermis of *Arabidopsis* leaves, NO application leads to elevation of expression of the key enzymes (e.g. OPR, OPR3, and LOX2) that are involved in the octadecanoid pathway (Hung and Kao 2004). According to Grün et al. (2006), both H_2O_2 accumulation and NO-induced wound response may act as the modulators of initiation of cell death. All these studies very clearly established that initiation of different wound responses comprises cross-talk among various stress signaling elements such as phytohormones and NO.

9.7 Conclusion and Future Perspectives

In the past few years, much exploratory work has been done on NO cross-talk with plant hormones in the regulation of various stress tolerances, but still some information and evidence are lacking. It is clear that NO interacts with different phytohormones to regulate the activities of different enzymatic antioxidants and expression of genes. For production of phytohormones, perception of signals by the receptor, and signally the requirement of NO-dependent posttranslational modifications of various biosynthetic enzymes, receptors, and transcriptional factors have been identified. However, during various abiotic stress responses, the physiological significance of these types of modifications is not clearly understood. Likewise, the impact of NO on transcriptional regulation of genes encoding phytohormone-related proteins has not been entirely revealed until now. Since, a number of proteins partake in phytohormone biosynthesis, transport, perception, catabolism, and regulation of signaling mechanisms, the huge number of possible targets of NO interaction may represent a challenge for further study on NO–phytohormone cross-talk under stress conditions. Most of the research concerning NO–phytohormone interactions is conducted in controlled laboratory conditions, but plants grown under open field conditions often face multiple stresses. Further research is very much needed to postulate how NO participates in the signaling networks of different phytohormones in field-grown plants that are exposed to multiple stresses.

References

Adimulam, S.S., Bhatnagar-Mathur, P., and Santisree, P. (2017). Interaction of nitric oxide with phytohormones under drought stress. *Journal of Plant Studies* 6: 58–61. doi: 10.5539/jps.v6n1p58.

Ahlfors, R., Brosché, M., Kollist, H. et al. (2009). Nitric oxide modulates ozone-induced cell death, hormone biosynthesis and gene expression in *Arabidopsis thaliana. The Plant Journal* 58: 1–2.

Alavi, S.M., Arvin, M.J., and Manoochehri Kalantari, K. (2014). Salicylic acid and nitric oxide alleviate osmotic stress in wheat (*T. aestivum* L.) seedlings. *Journal of Plant Interactions* 9: 683–688.

Asada, K. (2006). Production and scavenging of reactive oxygen species in chloroplasts and their functions. *Plant Physiology* 141: 391–396. doi: 10.1104/pp.106.082040.

Atıcı, Ö. and Nalbantoğlu, B. (2003). Antifreeze proteins in higher plants. *Phytochemistry* 64: 1187–1196.

Beligni, M.V., Fath, A., Bethke, P.C. et al. (2002). Nitric oxide acts as an antioxidant and delays programmed cell death in barley aleurone layers. *Plant Physiology* 129: 1642–1650. doi: 10.1104/pp.002337.

Besson-Bard, A., Gravot, A., Richaud, P. et al. (2009). Nitric oxide contributes to cadmium toxicity in *Arabidopsis* by promoting cadmium accumulation in roots and by upregulating genes related to iron uptake. *Plant Physiology* 149: 1302e1315. doi: 10.1104/pp.108.133348.

Bethke, P.C., Libourel, I.G., Reinöhl, V. et al. (2006). Sodium nitroprusside, cyanide, nitrite, and nitrate break *Arabidopsis* seed dormancy in a nitric oxide-dependent manner. *Planta* 223: 805–812. doi: 10.1007/s00425-005-0116-9.

Bethke, P.C., Libourel, I.G.L., Aoyama, N. et al. (2007). The *Arabidopsis* aleurone layer responds to nitric oxide, gibberellin, and abscisic acid and is sufficient and necessary for seed dormancy. *Plant Physiology* 143: 1173–1188. doi: 10.1104/pp.106.093435.

Bhardwaj, S., Kapoor, D., Singh, S. et al. (2021). Nitric oxide: a ubiquitous signal molecule for enhancing plant tolerance to salinity stress and their molecular mechanisms. *Journal of Plant Growth Regulation*. doi: 10.1007/s00344-021-10394-3.

Brestic, M., Zivcak, M., Kunderlikova et al. (2016). High temperature specifically affects the photoprotective responses of chlorophyll b deficient wheat mutant lines. *Photosynthesis Research* 130: 251–266. doi: 10.1007/s11120-016-0249-7.

Bright, J., Desikan, R., Hancock, J.T. et al. (2006). ABA-induced NO generation and stomatal closure in *Arabidopsis* are dependent on H_2O_2 synthesis. *The Plant Journal* 45: 113–122. doi: 10.1111/j.1365-313X.2005.02615.x.

Carimi, F., Zottini, M., Costa, A. et al. (2005). NO signaling in cytokinin-induced programmed cell death. *Plant Cell and Environment* 28 1171–1178. doi: 10.1111/j.1365-3040.2005.01355.x.

Cetinkya, H., Dinler, B.S., and Tasci, E. (2014). Investigation of comparative regulation on antioxidant enzyme system under copper treatment and drought stress in maize. *Notulae Botanicae Horti Agrobotanici Cluj-Napoca* 42: 363.

Chen, H., Qin, Y., Pu, J. et al. (2021). Phytotoxicity of the chiral herbicide dichlorprop: cross-talk between nitric oxide, reactive oxygen species and phytohormones. *The Science of the Total Environment* 788: 1–9. doi: 10.1016/j.scitotenv.2021.147866.

Chen, J., Zhang, H.Q., Hu, L.B. et al. (2013). Microcystin-LR-induced phytotoxicity in rice crown root is associated with the cross-talk between auxin and nitric oxide. *Chemosphere* 93: 283–293.

Chen, W.W., Yang, J.L., Qin, C. et al. (2010). Nitric oxide acts downstream of auxin to trigger root ferric chelate reductase activity in response to iron deficiency in *Arabidopsis*. *Plant Physiology* 154: 810–819. doi: 10.1104/pp.110.161109.

Corpas, F.J., Leterrier, M., Valderrama, R. et al. (2011). Nitric oxide imbalance provokes a nitrosative response in plants under abiotic stress. *Plant Science* 181: 604–611.

Daszkowska-Golec, A. and Szarejko, I. (2013). Open or close the gate – stomata action under the control of phytohormones in drought stress conditions. *Frontiers in Plant Science* 4: 138. doi: 10.3389/fpls.2013.00138.

Desikan, R., Cheung, M.K., Bright, J. et al. (2004). ABA, hydrogen peroxide and nitric oxide signaling in stomatal guard cells. *Journal of Experimental Botany* 55: 205–212. doi: 10.1093/jxb/erh033.

Diao, Q., Song, Y., Shi, D. et al. (2017). Interaction of polyamines, abscisic acid, nitric oxide, and hydrogen peroxide under chilling stress in tomato (*Lycopersicon esculentum* Mill.) seedlings. *Frontiers in Plant Science* 8: 203. doi: 10.3389/fpls.2017.00203.

Domingos, P., Prado, A.M., Wong, A. et al. (2015). Nitric oxide: a multitasked signaling gas in plants. *Molecular Plant* 8: 506–520. doi: 10.1016/j.molp.2014.12.010

Dong, F., Simon, J., Rienks, M. et al. (2015). Effects of rhizopheric nitric oxide (NO) on N uptake in *Fagus sylvatica* seedlings depend on soil CO_2 concentration, soil N availability and N source. *Tree Physiology* 35: 910–920.

Du, S.T., Zhang, Y.S., Lin, X.Y. et al. (2008). Regulation of nitrate reductase by nitric oxide in Chinese cabbage pakchoi (*Brassica chinensis* L.). *Plant, Cell & Environment* 31: 195–204. doi: 10.1111/j.1365-3040.2007.01750.x.

Ederli, L., Morettini, R., Borgogni, A. et al. (2006). Interaction between nitric oxide and ethylene in the induction of alternative oxidase in ozone-treated tobacco plants. *Plant Physiology* 142: 595–608.

Esim, N. and Atici, O. (2014). Nitric oxide improves chilling tolerance of maize by affecting apoplastic antioxidative enzymes in leaves. *Plant Growth Regulation* 72: 29–38.

Esringu, A., Aksakal, O., Tabay, D. et al. (2016). Effects of sodium nitroprusside (SNP) pretreatment on UV-B stress tolerance in lettuce (*Lactuca sativa* L.) seedlings. *Environmental Science and Pollution Research* 23: 589–597.

Fan, H.F., Du, C.X., and Guo, S.R. (2013). Nitric oxide enhances salt tolerance in cucumber seedlings by regulating free polyamine content. *Environmental and Experimental Botany* 86: 52–59.

Fan, S.K., Fang, X.Z., Guan, M.Y. et al. (2014). Exogenous abscisic acid application decreases cadmium accumulation in *Arabidopsis* plants, which is associated with the inhibition of IRT1-mediated cadmium uptake. *Frontiers in Plant Science* 5: 721. doi: 10.3389/fpls.2014.00721.

Fatima, M., Masood, A., Per, T.S. et al. (2016). Nitric oxide alleviates salt stress inhibited photosynthetic performance by interacting with sulfur assimilation in mustard. *Frontiers in Plant Science* 7: 521. doi: 10.3389/fpls.2016.00521.

Feng, J., Wang, C., Chen, Q. et al. (2013). S-nitrosylation of phosphotransfer proteins represses cytokinin signaling. *Nature Communications* 4: 1529.

Fernandez-Marcos, M., Sanz Lewis, D.R., Muday, G.K. et al. (2011). Nitric oxide causes root apical meristem defects and growth inhibition while reducing PIN-FORMED 1 (PIN1)-dependent acropetal auxin transport. *Proceedings of the National Academy of Sciences of the United States of America* 108: 18506–18511. doi: 10.1073/pnas.1108644108.

Freschi, L. (2013). Nitric oxide and phytohormone interactions: current status and perspectives. *Frontiers in Plant Science* 4: 398. doi: 10.3389/fpls.2013.00398.

Freschi, L., Rodrigues, M.A., Domingues, D.S. et al. (2010). Nitric oxide mediates the hormonal control of Crassulacean acid metabolism expression in young pineapple plants. *Plant Physiology* 152: 1971–1985. doi: 10.1104/pp.109.151613.

García-Mata, C. and Lamattina, L. (2001). Nitric oxide induces stomatal closure and enhances the adaptive plant responses against drought stress. *Plant Physiology* 126: 1196–1204. doi: 10.1104/pp.126.3.1196.

Gonugunta, V.K., Srivastava, N., and Raghavendra, A.S. (2009). Cytosolic alkalinization is a common and early messenger preceding the production of ROS and NO during stomatal closure by variable signals, including abscisic acid, methyl jasmonate and chitosan. *Plant Signaling & Behavior* 4: 561–564.

Grün, S., Lindermayr, C., Sell, S. et al. (2006). Nitric oxide and gene regulation in plants. *Journal of Experimental Botany* 57: 507–516.

Guo, Q., Wu, B., Chen, W. et al. (2014). Effects of nitric oxide treatment on the cell wall softening related enzymes and several hormones of papaya fruit during storage. *Food Science and Technology International* 20: 309–317.

Hancock, J.T., Neill, S.J., and Wilson, I.D. (2011). Nitric oxide and ABA in the control of plant function. *Plant Science* 181: 555–559. doi: 10.1016/j.plantsci.2011.03.017.

Hasanuzzaman, M., Oku, H., Nahar, K. et al. (2018). Nitric oxide-induced salt stress tolerance in plants: ROS metabolism, signaling, and molecular interactions. *Plant Biotechnology Reports* 12: 1–16.

Hu, X.Y., Neill, S.J., Tang, Z.C. et al. (2005). Nitric oxide mediates gravitropic bending in soybean roots. *Plant Physiology* 137: 663–670. doi: 10.1104/pp.104.054494.

Hui, C., Xiao-Wei, W., Yan-Mei, Z. et al. (2009). Effects of exogenous nitric oxide on the several enzymes of nitrogen metabolism in *Malus hupehensis* (Pamp.) Rehd. seedlings under water stress. *Acta Horticulturae Sinica* 36: 781–786.

Hung, K.T. and Kao, C.H. (2004). Hydrogen peroxide is necessary for abscisic acid-induced senescence of rice leaves. *Journal of Plant Physiology* 161: 1347–1357.

Hunt, P.W., Klok, E.J., Trevaskis, B. et al. (2002). Increased level of hemoglobin 1 enhances survival of hypoxic stress and promotes early growth in *Arabidopsis thaliana. Proceedings of the National Academy of Sciences of the United States of America* 99: 17197–17202.

Hussain, R.A., Ahmad, R., Nawaz, F. et al. (2016). Foliar NK application mitigates drought effects in sunflower (*Helianthus annuus* L.). *Acta Physiologiae Plantarum* 38: 1–4.

Imanishi, S., Kito-Nakamura, K., Matsuoka, K. et al. (1997). A major jasmonate-inducible protein of sweet potato, ipomoelin, is an ABA-independent wound inducible protein. *Plant & Cell Physiology* 38: 643–652.

Jaspers, P. and Kangasjärvi, J. (2010). Reactive oxygen species in abiotic stress signaling. *Physiologia Plantarum* 138: 405–413. doi: 10.1111/j.1399-3054.2009.01321.x.

Khan, A., Ahmad, I., Shah, A. et al. (2013). Amelioration of salinity stress in wheat (*Triticum aestivum* L.) by foliar application of phosphorus. *Phyton* 82: 281–287.

Khan, M.I.R., Asgher, M., and Khan, N.A. (2013). Rising temperature in the changing environment: a serious threat to plants. *Climate Change and Environmental Sustainability* 1: 25–36. doi: 10.5958/j.2320-6411.1.1.004.

Khokon, M.D., Okuma, E.I., Hossain, M.A. et al. (2011). Involvement of extracellular oxidative burst in salicylic acid-induced stomatal closure in *Arabidopsis. Plant, Cell & Environment* 34: 434–443.

Kohli, A., Sreenivasulu, N., Lakshmanan, P. et al. (2013). The phytohormone crosstalk paradigm takes center stage in understanding how plants respond to abiotic stresses. *Plant Cell Reports* 32: 945–957.

Kong, X., Wang, T., Li, W. et al. (2016). Exogenous nitric oxide delays salt induced leaf senescence in cotton (*Gossypium hirsutum* L.). *Acta Physiologiae Plantarum* 38: 1–9.

Kwak, J.M., Nguyen, V., and Schroeder, J.I. (2006). The role of reactive oxygen species in hormonal responses. *Plant Physiology* 141: 323–329.

Lamattina, L., García-Mata, C., Graziano, M. et al. (2003). Nitric oxide: The versatility of an extensive signal molecule. *Annual Review of Plant Biology* 54: 109–136.

Lanteri, M.L., Pagnussat, G.C., and Lamattina, L. (2006). Calcium and calcium-dependent protein kinases are involved in nitric oxide- and auxin-induced adventitious root formation in cucumber. *Journal of Experimental Botany* 57: 1341–1351. doi: 10.1093/jxb/erj109.

León, J., Castillo, M.C., Coego, A. et al. (2014). Diverse functional interactions between nitric oxide and abscisic acid in plant development and responses to stress. *Journal of Experimental Botany* 65: 907–921.

Leshem, Y.Y. (2001). *Nitric Oxide in Plants. Occurrence, Function and Use*. Boston, MA: Kluwer Academic Publishers.

Leshem, Y.Y. and Haramaty, E. (1996). The characterization and contrasting effects of the nitric oxide free radical in vegetative stress and senescence of *Pisum sativum* Linn. foliage. *Journal of Plant Physiology* 148: 258–263. doi: 10.1016/S0176-1617(96)80251-3.

Li, Y., Van den Ende, W., and Rolland, F. (2014). Sucrose induction of anthocyanin biosynthesis is mediated by DELLA. *Molecular Plant* 7: 570–572.

Liao, W.B., Huang, G.B., Yu, J.H. et al. (2012). Nitric oxide and hydrogen peroxide alleviate drought stress in marigold explants and promote its adventitious root development. *Plant Physiology and Biochemistry* 58: 6–15.

Liao, W.B., Huang, G.B., Yu, J.H., Zhang, M.L. et al. (2011). Nitric oxide and hydrogen peroxide are involved in indole-3-butyric acid-induced adventitious root development in marigold. *Journal of Horticultural Science and Biotechnology* 86: 159–165. doi: 10.1080/14620316.2011.11512742.

Lindermayr, C., Saalbach, G., Bahnweg, G. et al. (2006). Differential inhibition of *Arabidopsis* methionine adenosyltransferases by protein S-nitrosylation. *Journal of Biological Chemistry* 281: 4285e4291. doi: 10.1074/jbc.M511635200.

Lindermayr, C., Saalbach, G., and Durner, J. (2005). Proteomic identification of S-nitrosylated proteins in *Arabidopsis*. *Plant Physiology* 137: 921–930. doi: 10.1104/pp.104.058719.

Lipton, S.A., Choi, Y.B., Pan, Z.H. et al. (1993). A redox-based mechanism for the neuroprotective and neuro destructive effects of nitric oxide and related nitroso-compounds. *Nature* 364: 626–632. doi: 10.1038/364626a0.

Liu, N., Tu, L., Tang, W. et al. (2014). Small RNA and degradome profiling reveals a role for miRNAs and their targets in the developing fibers of *Gossypium barbadense*. *The Plant Journal* 80: 331–344.

Liu, S.L., Yang, R.J., Pan, Y.Z. et al. (2015a). Nitric oxide contributes to minerals absorption, proton pumps and hormone equilibrium under cadmium excess in *Trifolium repens* L. plants. *Ecotoxicology and Environmental Safety* 119: 35e46.

Liu, S.L., Yang, R.J., Pan, Y.Z. et al. (2015b). Exogenous NO depletes Cd-induced toxicity by eliminating oxidative damage, re-establishing ATPase activity, and maintaining stress related hormone equilibrium in white clover plants. *Environmental Science and Pollution Research* 22: 16843e16856. doi: 10.1007/s11356-015-4888-6.

Liu, W., Li, R.J., Han, T.T. et al. (2015). Salt stress reduces root meristem size by nitric oxide-mediated modulation of auxin accumulation and signaling in *Arabidopsis*. *Plant Physiology* 168: 343–356.

Liu, X., Deng, Z., Cheng, H. et al. (2011). Nitrite, sodium nitroprusside, potassium ferricyanide and hydrogen peroxide release dormancy of *Amaranthus retroflexus* seeds in a nitric oxide-dependent manner. *Plant Growth Regulation* 64: 155–161. doi: 10.1007/s10725-010-9551-0.

Liu, Y., Xu, S., Ling, T. et al. (2010). Heme oxygenase/carbon monoxide system participates in regulating wheat seed germination under osmotic stress involving the nitric oxide pathway. *Journal of Plant Physiology* 167: 1371–1379. doi: 10.1016/j.jplph.2010.05.021.

Lozano-Juste, J. and León, J. (2011). Nitric oxide regulates DELLA content and PIF expression to promote photomorphogenesis in *Arabidopsis*. *Plant Physiology* 156: 1410–1423. doi: 10.1104/pp.111.177741.

Lu, G., Gao, C., Zheng, X. et al. (2009). Identification of OsbZIP72 as a positive regulator of ABA response and drought tolerance in rice. *Planta* 229: 605–615.

Luo, B.F., Du, S.T., Lu, K.X. et al. (2012). Iron uptake system mediates nitrate-facilitated cadmium accumulation in tomato (*Solanum lycopersicum*) plants. *Journal of Experimental Botany* 63: 3127e3136. doi: 10.1093/jxb/ers036.

Manjunatha, G., Gupta, K.J., Lokesh, V. et al. (2012). Nitric oxide counters ethylene effects on ripening fruits. *Plant Signaling & Behavior* 7: 476–483. doi: 10.4161/psb.19523.

Manjunatha, G., Lokesh, V., and Neelwarne, B. (2010). Nitric oxide in fruit ripening: trends and opportunities. *Biotechnology Advances* 28: 489–499.

Mano, S., Nakamori, C., Hayashi, M. et al. (2002). Distribution and characterization of peroxisomes in *Arabidopsis* by visualization with GFP: dynamic morphology and actin-dependent movement. *Plant & Cell Physiology* 43: 331–341. doi: 10.1093/pcp/pcf037.

Miller, G., Suzuki, N., Ciftci-Yilmaz, S. et al. (2010). Reactive oxygen species homeostasis and signaling during drought and salinity stresses. *Plant, Cell & Environment* 33: 453–467. doi: 10.1111/j.1365-3040.2009.02041.x.

Mioto, P.T. and Mercier, H. (2013). Abscisic acid and nitric oxide signaling in two different portions of detached leaves of *Guzmania monostachia* with CAM up-regulated by drought. *Journal of Plant Physiology* 170: 996–1002. doi: 10.1016/j.jplph.2013.02.004.

Mishina, T.E., Lamb, C., and Zeier, J. (2007). Expression of a nitric oxide degrading enzyme induces a senescence programme in *Arabidopsis*. *Plant, Cell & Environment* 30: 39–52. doi: 10.1111/j.1365-3040.2006.01604.x.

Mohamed, H.I., Latif, H.H., and Hanafy, R.S. (2016). Influence of nitric oxide application on some biochemical aspects, endogenous hormones, minerals and phenolic compounds of *Vicia faba* plant grown under arsenic stress. *Gesunde Pflanzen* 68: 99e107. doi: 10.1007/s10343-016-0363-7.

Molassiotis, A., Tanou, G., and Diamantidis, G. (2010). NO says more than 'YES' to salt tolerance: salt priming and systemic nitric oxide signaling in plants. *Plant Signaling & Behavior* 5: 209–212.

Murata, Y., Mori, I.C., and Munemasa, S. (2015). Diverse stomatal signaling and the signal integration mechanism. *Annual Review of Plant Biology* 66: 369–392.

Nawaz, F., Ashraf, M.Y., Ahmad, R. et al. (2015). Supplemental selenium improves wheat grain yield and quality through alterations in biochemical processes under normal and water deficit conditions. *Food Chemistry* 175: 350–357.

Neill, S., Barros, R., Bright, J. et al. (2008). Nitric oxide, stomatal closure, and abiotic stress. *Journal of Experimental Botany* 59: 165–176. doi: 10.1093/jxb/erm293.

Neill, S.J., Desikan, R., Clarke, A. et al. (2002). Nitric oxide is a novel component of abscisic acid signaling in stomatal guard cells. *Plant Physiology* 128: 13–16. doi: 10.1104/pp.010707.

Ötvös, K., Pasternak, T.P., Miskolczi, P. et al. (2005). Nitric oxide is required for, and promotes auxin mediated activation of, cell division and embryogenic cell formation but does not influence cell cycle progression in alfalfa cell cultures. *The Plant Journal* 43: 849–860. doi: 10.1111/j.1365-313X.2005.02494.x.

Overmyer, K., Tuominen, H., Kettunen, R. et al. (2000). Ozone-sensitive *Arabidopsis* rcd1 mutant reveals opposite roles for ethylene and jasmonate signaling pathways in regulating superoxide-dependent cell death. *The Plant Cell* 12: 1849–1862.

Pagnussat, G.C., Lanteri, M.L., and Lamattina, L. (2003). Nitric oxide and cyclic GMP are messengers in the indole acetic acid induced adventitious rooting process. *Plant Physiology* 132: 1241–1248. doi: 10.1104/pp.103.022228.

Pagnussat, G.C., Simontacchi, M., Puntarulo, S. et al. (2002). Nitric oxide is required for root organogenesis. *Plant Physiology* 129: 954–956. doi: 10.1104/pp.004036.

Palmieri, M.C., Sell, S., Huang, X. et al. (2008). Nitric oxide-responsive genes and promoters in *Arabidopsis thaliana*: a bioinformatics approach. *Journal of Experimental Botany* 59: 177–186.

París, R., Lamattina, L., and Casalongué, C.A. (2009). Nitric oxide promotes the wound-healing response of potato leaflets. *Plant Physiology and Biochemistry* 45: 80–86.

Penuelas, J., Sardans, J., Estiarte, M. et al. (2013). Evidence of current impact of climate change on life: a walk from genes to the biosphere. *Global Change Biology* 19: 2303–2338.

Pető, A., Lehotai, N., Lozano-Juste, J. et al. (2011). Involvement of nitric oxide and auxin in signal transduction of copper induced morphological responses in *Arabidopsis* seedlings. *Annals of Botany* 108: 449e457. doi: 10.1093/aob/mcr176.

Petrov, V., Hille, J., Mueller-Roeber, B. et al. (2015). ROS-mediated abiotic stress induced programmed cell death in plants. *Frontiers in Plant Science* 6: 1–16. doi: 10.3389/fpls.2015.00069.

Pii, Y., Crimi, M., Cremonese, G. et al. (2007). Auxin and nitric oxide control indeterminate nodule formation. *BMC Plant Biology* 7: 21. doi: 10.1186/1471-2229-7-21.

Poór, P., Gémes, K., Horváth, F. et al. (2011a). Salicylic acid treatment via the rooting medium interferes with stomatal response, CO_2 fixation rate and carbohydrate metabolism in tomato, and decreases harmful effects of subsequent salt stress. *Plant Biology* 13: 105–114.

Poór, P. and Tari, I. (2011b). Ethylene-regulated reactive oxygen species and nitric oxide under salt stress in tomato cell suspension culture. *Acta Biologica Szegediensis* 55: 143–146.

Prakash, V., Singh, V.P., Tripathi, D.K. et al. (2019). Crosstalk between nitric oxide (NO) and abscisic acid (ABA) signalling molecules in higher plants. *Environmental and Experimental Botany* 161: 41–49. doi: 10.1016/j.envexpbot.2018.10.033.

Prakash, V., Singh, V.P., Tripathi, D.K. et al. (2021). Nitric oxide (NO) and salicylic acid (SA): a framework for their relationship in plant development under abiotic stress. *Plant Biology* 23: 39–49. doi: 10.1111/plb.13246.

Prasad, A., Ferretti, U., Sedlářová, M. et al. (2016). Singlet oxygen production in *Chlamydomonas reinhardtii* under heat stress. *Scientific Reports* 6: 20094. doi: 10.1038/srep20094.

Praveen, A. and Gupta, M. (2018). Nitric oxide confronts arsenic stimulated oxidative stress and root architecture through distinct gene expression of auxin transporters, nutrient related genes and modulates biochemical responses in *Oryza sativa* L. *Environmental Pollution* 240: 950e962. doi: 10.1016/j.envpol.2018.04.096.

Rao, M.V. and Davis, K.R. (2001). The physiology of ozone induced cell death. *Planta* 213: 682–690.

Remy, E., Cabrito, T.R., Baster, P. et al. (2013). A major facilitator super family transporter plays a dual role in polar auxin transport and drought stress tolerance in *Arabidopsis*. *The Plant Cell* 25: 901–926. doi: 10.1105/tpc.113.110353.

Sadeghipour, O. (2017). Nitric oxide increases Pb tolerance by lowering Pb uptake and translocation as well as phytohormonal changes in cowpea (*Vigna unguiculata* (L.) Walp.). *Sains Malays* 46: 189e195. doi: 10.17576/jsm-2017-4602-02.

Saito, N., Yoshimasa, N., Mori, I.C. et al. (2009). Nitric oxide functions in both methyl jasmonate signaling and abscisic acid signaling in *Arabidopsis* guard cells. *Plant Signaling & Behavior* 4: 119–120.

Santisree, P., Bhatnagar-Mathur, P., and Sharma, K.K. (2015). NO to drought-multifunctional role of nitric oxide in plant drought: do we have all the Answers? *Plant Science* 239: 44–55. doi: 10.1016/j.plantsci.2015.07.012.

Santisree, P., Bhatnagar-Mathur, P., and Sharma, K.K. (2017). Heat responsive proteome changes reveal molecular mechanisms underlying heat tolerance in chickpea. *Environmental and Experimental Botany* 141: 132–144. doi: 10.1016/j.envexpbot.2017.07.007.

Shabbir, R.N., Waraich, E.A., Ali, H. et al. (2016). Supplemental exogenous NPK application alters biochemical processes to improve yield and drought tolerance in wheat (*Triticum aestivum* L.). *Environmental Science and Pollution Research* 23: 2651–2662.

Shao, R.X., Wang, K.B., and Shangguan, Z.P. (2010). Cytokinin-induced photosynthetic adaptability of *Zea mays* L. to drought stress associated with nitric oxide signal: probed by ESR spectroscopy and fast OJIP fluorescence rise. *Journal of Plant Physiology* 167: 472–479. doi: 10.1016/j.jplph.2009.10.020.

Shen, Q., Wang, Y.T., Tian, H. et al. (2012). Nitric oxide mediates cytokinin functions in cell proliferation and meristem maintenance in *Arabidopsis*. *Molecular Plant* 6: 1214–1225. doi: 10.1093/mp/sss148.

Shi, Q., Ding, F., Wang, X. et al. (2007). Exogenous nitric oxide protect cucumber roots against oxidative stress induced by salt stress. *Plant Physiology and Biochemistry* 45: 542–550. doi: 10.1016/j.plaphy.2007.05.005.

Simontacchi, M., García-Mata, C., Bartoli, C.G. et al. (2013). Nitric oxide as a key component in hormone regulated processes. *Plant Cell Reports* 32: 853–866. doi: 10.1007/s00299-013-1434-1.

Song, L., Ding, W., Shen, J. et al. (2008). Nitric oxide mediates abscisic acid induced thermotolerance in the calluses from two ecotypes of reed under heat stress. *Plant Science*175: 826–832.

Song, L., Ding, W., Zhao, M. et al. (2006). Nitric oxide protects against oxidative stress under heat stress in the calluses from two ecotypes of reed. *Plant Science* 171: 449–458. doi: 10.1016/j.plantsci.2006.05.002.

Terrile, M.C., París, R., Calderón-Villalobos, L.I. et al. (2012). Nitric oxide influences auxin signaling through S-nitrosylation of the *Arabidopsis* TRANSPORT INHIBITOR RESPONSE 1 auxin receptor. *The Plant Journal* 70: 492–500.

Tossi, V., Lamattina, L., and Cassia, R. (2009). An increase in the concentration of abscisic acid is critical for nitric oxide-mediated plant adaptive responses to UV-B irradiation. *The New Phytologist* 181: 871–879. doi: 10.1111/j.1469-8137.2008.02722.x.

Tripathi, D.K., Rai, P., Guerriero, G. et al. (2021). Silicon induces adventitious root formation in rice under arsenate stress with involvement of nitric oxide and indole-3-acetic acid. *The Journal of Experimental Botany* 72. doi: 10.1093/jxb/eraa488.

Uchida, A., Jagendorf, A., Hibino, T. et al. (2002). Effects of hydrogen peroxide and nitric oxide on both salt and heat stress tolerance in rice. *Plant Science* 163: 515–523. doi: 10.1016/S0168-9452(02)00159-0.

Vranova, E., Atichartpongkul, S., Villarroel, R. et al. (2002). Comprehensive analysis of gene expression in *Nicotiana tabacum* leaves acclimated to oxidative stress. *PNAS* 99: 10870–10875. doi: 10.1073/pnas.152337999.

Wang, C.L., Wu, J., Xu, G.H. et al. (2010). S-RNase disrupts tip-localized reactive oxygen species and induces nuclear DNA degradation in incompatible pollen tubes of *Pyrus pyrifolia*. *Journal of Cell Science* 123: 4301–4309.

Wang, D., Liu, Y., Tan, X. et al. (2015). Effect of exogenous nitric oxide on antioxidative system and S-nitrosylation in leaves of *Boehmeria nivea* (L.) Gaud under cadmium stress. *Environmental Science and Pollution Research* 22: 3489e3497. doi: 10.1007/s11356-014-3581-5.

Wang, H., Liang, X., Wan, Q. et al. (2009). Ethylene and nitric oxide are involved in maintaining ion homeostasis in *Arabidopsis* callus under salt stress. *Planta* 230: 293–307.

Wang, P., Du, Y., Hou, Y.J. et al. (2015). Nitric oxide negatively regulates abscisic acid signaling in guard cells by S-nitrosylation of OST1. *Proceedings of the National Academy of Sciences of the United States of America* 112: 613e618. doi: 10.1073/pnas.1423481112.

Wang, Y., Luo, Z., Mao, L. et al. (2016). Contribution of polyamines metabolism and GABA shunt to chilling tolerance induced by nitric oxide in cold-stored banana fruit. *Food Chemistry* 197: 333–339.

Waraich, E.A., Ahmad, R., and Ashraf, M.Y. (2011). Role of mineral nutrition in alleviation of drought stress in plants. *Australian Journal of Crop Science* 5: 764–777.

Wei, L., Zhang, M., Wei, S. et al. (2020). Roles of nitric oxide in heavy metal stress in plants: cross-talk with phytohormones and protein S-nitrosylation. *Environmental Pollution* 259: 113943. doi: 10.1016/j.envpol.2020.113943.

Wilkinson, S. and Davies, W.J. (2010). Drought, ozone, ABA and ethylene: new insights from cell to plant to community. *Plant, Cell & Environment* 33: 510–525.

Wimalasekera, R., Villar, C., Begum, T. et al. (2011). COPPER AMINE OXIDASE1 (CuAO1) of *Arabidopsis thaliana* contributes to abscisic acid- and polyamine-induced nitric oxide biosynthesis and abscisic acid signal transduction. *Molecular Plant* 4: 663–678.

Wu, A.P., Gong, L., Chen, X. et al. (2014). Interactions between nitric oxide, gibberellic acid, and phosphorus regulate primary root growth in *Arabidopsis*. *Biologia Plantarum* 58: 335–340.

Xiao-Ping, S. and Xi-Gui, S. (2006). Cytokinin and auxin-induced stomatal opening is related to the change of nitric oxide levels in guard cells in broad bean. *Planta* 128: 569–579.

Xing, H., Tan, L., An, L. et al. (2004). Evidence for the involvement of nitric oxide and reactive oxygen species in osmotic stress tolerance of wheat seedlings: inverse correlation between leaf abscisic acid accumulation and leaf water loss. *Plant Growth Regulon* 42: 61–68. doi: 10.1023/B:GROW.0000014894.48683.1b.

Xu, J., Wang, W., Sun, J. et al. (2011). Involvement of auxin and nitric oxide in plant Cd-stress responses. *Plant and Soil* 346: 107. doi: 10.1007/s11104-011-0800-4.

Xu, J., Wang, W., Yin, H. et al. (2010). Exogenous nitric oxide improves antioxidative capacity and reduces auxin degradation in roots of *Medicago truncatula* seedlings under cadmium stress. *Plant and Soil* 326: 321. doi: 10.1007/s11104-009-0011-4.

Yadav, S., David, A., and Bhatla, S.C. (2011). Nitric oxide accumulation and actin distribution during auxin-induced adventitious root development in sunflower. *Scientia Horticulturae* 129: 159–166. doi: 10.1016/j.scienta.2011.03.030.

Yan, F., Liu, Y., Sheng, H. et al. (2016). Salicylic acid and nitric oxide increase photosynthesis and antioxidant defense in wheat under UV-B stress. *Biologia Plantarum* 60: 686–694.

Yu, M., Lamattina, L., Spoel, S.H., and Loake, G.J. (2014). Nitric oxide function in plant biology: a redox cue in deconvolution. *The New Phytologist* 202: 1142–1156. doi: 10.1111/nph.12739.

Zandalinas, S.I., Balfagón, D., Arbona, V. et al. (2016). ABA is required for the accumulation of APX1 and MBF1c during a combination of water deficit and heat stress. *Journal of Experimental Botany* 18: 5381–5390. doi: 10.1093/jxb/erw299.

Zhang, A., Jiang, M., Zhang, J. et al. (2007). Nitric oxide induced by hydrogen peroxide mediates abscisic acid-induced activation of the mitogen-activated protein kinase cascade involved in antioxidant defense in maize leaves. *The New Phytologist* 175: 36–50.

Zhou, B., Guo, Z., Xing, J. et al. (2005). Nitric oxide is involved in abscisic acid-induced antioxidant activities in *Stylosanthes guianensis*. *Journal of Experimental Botany* 56: 3223–3228.

Zhou, X., Joshi, S., Khare, T. et al. (2021). Nitric oxide, crosstalk with stress regulators and plant abiotic stress tolerance. *Plant Cell Reports*. doi: 10.1007/s00299-021-02705-5.

Zhu, X.F., Jiang, T., Wang, Z.W. et al. (2012). Gibberellic acid alleviates cadmium toxicity by reducing nitric oxide accumulation and expression of IRT1 in *Arabidopsis thaliana*. *Journal of Hazardous Materials* 239: 302e307. doi: 10.1016/j.jhazmat.2012.08.077.

Zhu, X.F., Wang, Z.W., Dong, F. et al. (2013). Exogenous auxin alleviates cadmium toxicity in *Arabidopsis thaliana* by stimulating synthesis of hemicellulose 1 and increasing the cadmium fixation capacity of root cell walls. *Journal of Hazardous Materials* 263: 398e403. doi: 10.1016/j.jhazmat.2013.09.018.

Zottini, M., Formentin, E., Scattolin, M. et al. (2002). Nitric oxide affects plant mitochondrial functionality in vivo. *FEBS Letters* 515: 75–78. doi: 10.1016/S0014-5793(02)02438-9.

10

Nitric Oxide Cross-talk with Phytohormones Vis-à-Vis Photosynthetic Regulation Under Extreme Environments

Yachana Jha

Natubhai V. Patel College of Pure and Applied Sciences, Charutar Vidya Mandal University, Vallabh Vidyanagar, Anand, Gujarat, India

10.1 Introduction

The growth and development of plants are directly influenced by the surrounding environmental conditions as plants remain static. Varying environmental conditions such as cold, flood, drought, salinity, and heat are responsible for restricting plant growth and for plants to survive in such varying environmental conditions, they require a certain degree of adaptation (Raza et al. 2019). With continuous global climate change crop production is badly affected leading to reduced food production with poor food quality. A clear correlation exists among changing global environment, increasing population, and reduced food production, which has resulted in severe food scarcity for the growing population.

The production of nitric oxide (NO), the most abundant reactive nitrogen species in plants, is a regular and complex process. It is a small lipid- and water-soluble gas, which moves freely in the cell cytoplasm by diffusion or is diffused through the cell membrane to other sites (Postila and Róg 2020). It is a very reactive free radical with a very short half-life and regulates several scavenging and synthesis pathways in plants under stress. It plays a critical role in central metabolic activity during the entire life of the plant, for proper growth under normal and stress conditions. The targeted production, regulation, and distribution of NO in plant are responsible for variation in metabolic activities in plants. As NO has the ability to regulate several metabolic activities, it is considered as the universal signaling molecule (Sanz-Luque et al. 2015). The NO generated in the cell rapidly reacts with important biomolecules in the cell such as DNA, RNA, enzymes, and fatty acids, and causes structural and functional modifications in these biomolecules. Such modifications cause conformational changes in biomolecules and affect their stability, change their activity, interactions, and translocation, and induce altered gene expression. In summary, this gasotransmitter acts as a signaling effector and the overaccumulation or deficiency of NO seriously impairs the growth and development of a plant (Kolbert et al. 2019).

Phytohormones are an essential component of signal transduction pathways and have a regulatory role in response to external and internal stimuli in plants. There are five classes of phytohormones, abscisic acid (ABA), ethylene, auxins, cytokinins, gibberellins, and brassinosteroids. All these phytohormones have a regulatory role in plant responses and interact with signal transducing molecules by activating a phosphorylation cascade or secondary messengers of the plant metabolic pathway. The plant hormones work in coordination for proper growth and development of the plant under normal conditions and under stress to regulate signaling pathways for the better survival of the plant.

Abiotic factors directly influence plant growth and induce different physiological activities in the plant. Photosynthesis is a key physiological activity of plants that is directly associated with growth and yield. Abiotic stress has inhibitory effects on major physiological activities of plants by limiting metabolic precursors, diffusion of carbon dioxide, and stomatal conductance, and causing change in plant pigment concentrations, mineral nutrient imbalances, and altered allocation of intermediate and end products in the plant (Sharma et al. 2020). Now it is the need of the hour to develop more efficient, environmentally friendly agricultural practices such as development of new varieties with short harvesting times and enhanced ability for environmental stress tolerance.

10.2 Sources of NO Production in Plants

NO has many metabolic and physiological activities in all living organisms including plants. Initially it was considered as an air pollutant, and only recently has it become known that it is involved in cell signaling functions in plants. In plants NO is involved in a wide range of physiological activities such as seed germination, older leaf senescence, flowering, and regulating plant response to surrounding environmental conditions. In plants it acts as a molecular messenger and regulates the expression of genes involved in cell signaling. However, despite such wide applications of NO in plants under stress, the production mechanism of NO is not very clear to date. There are two major synthesis pathways for NO in plants: enzymatic and nonenzymatic, i.e. reductive and oxidative (Nabi et al. 2019). The oxidation of aminated molecules is the basis of the oxidative pathway and the reduction of nitrites to NO is the basis of the reductive pathway. The nonenzymatic reduction of NO takes place only under specific conditions such as highly depleted environments, low pH, or in the presence of high concentrations of nitrate. Such specific conditions are very rare, but if generated this pathway is the only mechanism for the rapid and efficient production of NO, whereas enzymatic reduction of NO in plants is a regular process and takes place with the help of different enzymes like nitrate and nitrite reductase (Tejada-Jimenez et al. 2019). In addition NO is also produced enzymatically in the plant cell cytoplasm by the reduction of nitrate to nitrite by nitrate reductase enzyme (Figure 10.1). A multifunctional cytoplasmic enzyme, nitrate reductase catalyzes the first rate-limiting steps, i.e. the reduction of nitrate to nitrite by utilizing reduced nicotinamide adenine dinucleotide (NADH) as an electron donor, and is involved in nitrogen metabolism/assimilation in plant. This enzyme requires molybdopterin, flavin adenine dinucleotide (FAD), and heme as cofactors and becomes active in homodimer form. The production of NO by this enzyme in plants takes place at different stages of development for proper physiological functioning, proper stomatal movements during photosynthesis and transpiration, floral and root development, accumulation of ions under osmotic stress, and regulated hormonal responses at different growth stages or under stress, and production of NO by this enzyme drastically increases in plant under adverse circumstances such as acidic or anoxic conditions (Salgado et al. 2013).This enzyme also has nitrite reductase and nitrate reductase (NR) activity.

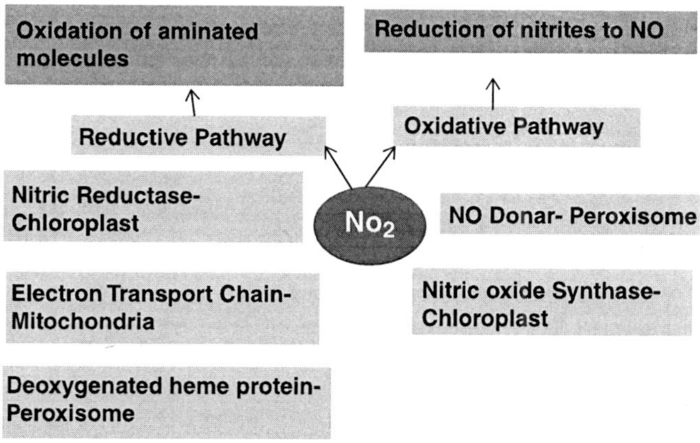

Figure 10.1 Mechanism of production of nitric oxide in the plant.

In nature the nitrification/denitrification process is responsible for the production of NO as a byproduct in plants or microorganisms via a nonenzymatic reaction. It is one of the most important intermediates in the inorganic nitrogenous compound metabolism in nitrogen-fixing organisms and plants. In plants in the presence of an efficient reductive pathway, there is also an oxidative pathway for the production of NO. A specialized enzyme, nitric oxide synthase (NOS), catalyzes the production of NO by converting L-arginine to L-citrulline and NO, through double mono-oxygenation (Astier et al. 2018). This enzyme become functional in homodimer form and has two functional domains, the C-terminal domain, which has reducing activity, and the N-terminal domain, which has oxidative activity in the presence of cofactors like tetrahydrobiopterin, flavin mononucleotide, and flavin adenine dinucleotide. These two functional domains remain joined at the calmodulin-binding site, which initiates a conformational change in homodimers, to become functional.

10.3 The Role of NO in Nutrient Acquisition for Phytohormone Production and Regulation of Photosynthesis

The water and nutrient supply to the entire plant depends on the plant root function. The major limitation for crop productivity is less accessibility of a few important mineral nutrients to the root, which is common in the majority of agricultural systems. Macronutrients like nitrogen, sulfur, phosphorus, and carbon are components of all important biomolecules. So plants have also developed many methods for the extraction of mineral nutrients from complex compounds to aid survival under low soil nutrient conditions. Changes in root growth patterns, efficient transport systems, production of metabolites to alter the solubility of complex compounds, and altered root permeability for better nutrient uptake by the root are the main modifications in plants (Pourrut et al. 2011). For efficient function, continuous change takes place in the root architecture, including change in root hair pattern and density and position of side roots, to adapt to the surrounding environment. The availability of different types and amounts of mineral nutrients in the soil directly influences the root architecture. All these modifications meet the requirement for efficient nutrient acquisition for better growth and biomass production. A previous study showed that NO has a regulatory role in major metabolic functions as well as determining specific root architecture. Actually, NO plays a pivotal role in efficient acquisition, assimilation, and utilization of important mineral nutrients like nitrogen, potassium, phosphorus, zinc, and iron (Jha and Subramanian 2013a).

Nitrogen is one of the essential mineral nutrients for plant cell division and it is required in higher amounts than other elements as it is necessary for formation of DNA, RNA, and proteins, and all these biomolecules are important in plant cells. Ammonium and nitrate are two chemical forms of nitrogen absorbed by plants and seem to be interconnected for the formation of NO by NR activity in plants. Ammonium and nitrate are the initial forms for the assimilation of nitrogen from soil. The NO and NO reservoirs directly influence the integration of nitrate in the plant. Therefore, plant assimilation of nitrogen is directly influenced by modification of the root system architecture. NO and nitrate nutrition have inverse effects in plants via localized effects. The induction of development of lateral roots and inhibition of transverse roots take place as systemic effects of NO, which modulates the distribution of carbon among roots and shoots.

Nitrogen assimilates in the plant mostly in inorganic forms as nitrate and ammonium from the soil and low levels of nitrogen in the soil result in low assimilation of nitrogen by the plant root, which is responsible for limited plant growth. Of the two, nitrate is the most easily assimilated form of nitrogen to be readily absorbed by the root as it does not form complexes with colloids. Plant enzymes like nitrite and NRs reduce the plant root-assimilated nitrate into nitrite and then ammonium. The ammonium generated acts as a precursor via enzyme glutamine synthase for the formation of glutamate and glutamate is used by glutamate oxyglutarate aminotransferase for the formation of glutamine in the plants (van Heeswijk et al. 2013). This conversion produces important intermediates, which serve as key nitrogen/amino donors for the production of N-containing compounds and amino acids required for the continuous growth of plant under normal as well as under stress.

Phosphorus is another macronutrient necessary for healthy plant growth, as it has a significant role in utilizing the conversion of sunlight energy into essential plant compounds, acting as a catalyst for several metabolic activities in plants. Not only that, phosphorus is a major component in essential biomolecules like DNA, RNA, ATP etc., which carry information for the new progeny cells for sustainable growth of plants. The only energy currency for the living world is ATP, which is also required for all plant functions. ATP is produced in the plant via cellular respiration and photosynthesis. NO also helps enhance acquisition of phosphorus from the soil under low phosphorus conditions by inducing formation of root clusters, and release of low molecular weight organic acids and protons (Jha and Subramanian 2014). Similarly potassium also has many critical roles for plant growth and survival. It helps with carbohydrate and water intake, reduces water loss and wilting, maintains turgor in the plant tissue, and competes with sodium to minimize osmotic stress. It is also involved in activation of several plant enzymes necessary for vital plant metabolism. Potassium acquisition in plants takes place in the root via selective and nonselective pathways from the soil. In low or moderate soil potassium concentrations the plant initiates the selective pathway, whereas in high soil potassium conditions nonselective pathways are operated, which results in reduced uptake of potassium (Jha and Subramanian 2016). However, the presence of an NO donor increases the uptake of potassium for the activation of vital metabolic pathways allowing better adaptation of plants under osmotic stress. NO is also directly responsible for the modulation of the homeostasis of transition metals like Fe and Zn in the plant under sufficient or deficient conditions, for the proper growth of plants. The higher plants have multiple systems with a set of organs for diverse functions and nutritional requirements, therefore requiring well-coordinated signaling pathways, and phytohormones are the signaling biomolecules involved in these processes. Besides having a role in major nutrient acquisition NO also regulates cross-talk with phytohormone signaling pathways and photosynthesis. Takei et al. (2001) reported that the specific pattern of phytohormone production and accumulation in the root and its transportation to leaves and other shoot parts directly depend on nitrogen nutrient concentration. It has been reported that NO mediates auxin, which directly stimulates the formation of lateral and adventitious roots in plants (Jha and Subramanian 2013b). Similarly, the production of NO by NR enzyme is connected with production of indole-3-butyric acid (IBA), which is responsible for the induction of lateral root development in *Arabidopsis thaliana*. Among the phytohormones, auxin, cytokinins, and ABA are directly influenced by nitrogen signaling.

10.4 The Role of NO in Phytohormone Production and Regulation of Photosynthesis

Plants have developed composite systems to sense, react, and acclimatize to the changing surrounding environment to enable their survival in such environments. Such frequently changing environments are responsible for adverse effects on plant metabolism, which is directly related to crop productivity and results in yield loss (Pandey et al. 2016). Plants generally communicate or send information from one site to other for regulation of various functions via signal molecules known as phytohormones. The actions of phytohormones are directly related to plant metabolic activity, which is directly responsible for regulating plant reaction to stress and for maintaining plant growth and development. For the regulation of metabolic and physiological activities, plants produce different types of plant hormones in a coordinated manner to enable their survival under normal/adverse environmental conditions. There are different types of phytohormones for different metabolic activities, such as auxin for apical growth, cytokinins for lateral growth and senescence, gibberellins for intermodal growth, ABA against stress, and ethylene for stress and fruit ripening. Recently a few more growth regulators with advanced coordinated functions have been identified, including jasmonic acid, NO, brassinosteroids, and salicylic acid (Jha 2020). Additionally, phytohormones regulate various physiological activities by altering cell sensitivity as well as modulating cell concentrations.

When plants encounter a changing environment, the plant responses are efficiently regulated by secondary messengers of signal transduction mechanisms like reactive oxygen species (ROS), calcium ions, cyclic nucleotides, NO, and plant hormones to establish multifaceted coordinated signaling networks to adjust metabolic activity.

Due to its gaseous nature NO has great ability to diffuse rapidly and efficiently to reach its intracellular targets. It does not require any carrier to cross membranes and serve as an impending cellular signaling molecule. It is possible that the pathway for the production of NO is either denitrification or nitrification of nitrogenous compounds in plant cells (Thomson et al. 2011). NO has specific interactions with a range of phytohormones to modulate metabolism of all physiological processes from germination of seeds to plant senescence. NO interacts with plant growth regulators to regulate plant growth and development under stress. In addition, NO is able to interact directly with the plant for stimulation of specific hormones for regulation of plant metabolism.

NO regulates the production, allocation, degradation, and formation of complexes with the effectors involved in phytohormone stimulation, transportation, and signaling. There are clear crossovers of NO signaling within signaling cascades of phytohormones under normal as well as stress conditions, as both NO and phytohormones have common functions as signaling effectors. Phytohormones not only induce desired modifications in plant but also regulate important physiological activities like photosynthesis. Induction of phytohormone in response to NO can increase the rates of photosynthesis per unit leaf area by altering the photosynthetic CO_2 uptake, through regulating the stomatal conductance, or modulating the activity of photosynthetic enzymes. Major events like carbohydrate accumulation, photosynthesis, and fruit establishment in tomato are regulated by auxin as reported by Yuan et al. (2019). Similarly cytokinin hormone forms a complex with the stress receptor proteins and induces a specific photosynthesis protective signaling pathway for the protection of photosynthesis under stress. The phytohormone ethylene has an inverse role in photosynthesis by reducing leaf area index, light interception, and formation of carbonate from CO_2. For efficient photosynthesis under stress the maintenance of electron transport through photosystem II is necessary, which is stimulated by the presence of a NO donor. Photosynthesis is possible only in the presence of light, where synthesis of new carbon skeletons takes place for the incorporation of nitrogen, which has been assimilated by the plant root in the form of nitrate/ammonia. The activity of NR enzyme and nitrogen content in the plant are regulated by NO and ethylene (Iqbal et al. 2015). Ethylene formation in the stressed plant is directly modulated by availability of NO and it is responsible for enhanced salinity tolerance and improves photosynthesis to maintain plant growth. Under high nitrogen levels, nitrate is reduced to ammonia, which needs to be assimilated into the amino acids. However, under stress plants cannot incorporate all the reduced ammonia, which initiates stress responses in the plant and induces ethylene biosynthesis. Therefore, there is balance between reduction in nitrates and sugar biosynthesis during photosynthesis in plants. The ethylene produced under stress inhibits root growth, limits leaf expansion, and reduces photosynthesis by regulating stomatal movement. ABA hormone also efficiently regulates the distribution of the photosynthate pool between source and sink tissues for proper functioning of photosynthesis, as has been reported by Barickman et al. (2014). NO directly reduces the rate of photosynthesis and causes a marginal stimulation of the dark reaction.

The production of reactive oxygen/nitrogen species (ROS/RNS) during stress can affect the photosynthetic machinery and overall photosynthesis. NO can reversibly bind with the tyrosine residue of the oxido-reductase enzyme at several sites in PSII and inhibit electron transfer by decreasing the redox potential in thylakoid membrane (Wang et al. 2017).

Similarly, Rather et al. (2020) reported in mung bean the presence of sodium nitroprusside (SNO), the NO donor responsible for the reduced Rubisco subunit binding with G-protein β-subunit/Rubisco and enhancing the oxygen evolving complex content of PSII, which results in a reduced rate of photosynthesis. Carbonic anhydrase is one of the main chloroplast enzymes involved in C3 carbon fixation, and is responsible for the interconversion of CO_2 and HCO_3 – which are reported to be decreased in the presence of higher SNP concentrations in tomato. It is presumed to play a role in photosynthetic carbon assimilation.

10.5 The Role of NO-Mediated Phytohormone Production for Photosynthesis

Photosynthesis is a complex physiological process in the plant, where light energy is captured by the colored chlorophyll pigment, followed by assimilation of CO_2, excitation of electrons, generation of ATP, production of sugar, and liberation of oxygen, which takes place by reduction of water to $NADP^+$. This multistep process is carried out in chloroplasts, where there are two separate steps for production of carbohydrates, resulting in an accumulation of starch granules in the chloroplasts, which is distributed to the entire plant tissue through phloem. Therefore, proper growth and development of the plant require sufficient assimilation of carbon dioxide and coordinated allocation of assimilates. Any imbalance in the source and sink coordination affects all life processes of the plant and such imbalance is common in plants due to reduced expression of photosynthetic genes or accelerated expression of genes responsible for leaf senescence under abiotic stress. The chloroplast serves as a factory for the process of photosynthesis and is the site for the production of NO also (Wei et al. 2014). There are two separate pathways for the production of NO in chloroplasts, one dependent on nitrite and the other dependent on an NOS-like enzyme. The generation of ROS in the presence of high NO takes place in chloroplasts and is responsible for adverse effects on the photosynthetic machinery. NO has the potential to directly block the chloroplast electron transport system in a reversible manner, which ultimately causes reduced CO_2 assimilation and induced stomatal closure in both monocotyledonous and dicotyledonous plants. The effects of NO on stomatal movement are very often connected with the activity of the enzyme NR responsible for the production of NO in guard cells or ABA-induced stomatal closure under extreme environmental conditions (Gayatri et al. 2013). There are several reports that have shown the positive effect of NO on the chlorophyll content, photosynthetic electron transport chain, and generation of pH gradient across the thylakoid membrane. It modifies the maximal quantum efficiency and increases the rate of ATP synthesis in various economic crop plants. The effect of NO on stomatal conductance causes proportionate changes in the rates of photosynthesis and transpiration.

NO is an essential signaling chemical directly related to the modification of plant root architecture, which acts downstream of auxin in plants (Corpas and Barroso 2015). The phytohormone indole-3-acetic acid (IAA) is responsible for stomatal opening and promotes the rate of photosynthesis in plants. It also has a positive effect on CO_2 fixation, photophosphorylation in chloroplasts, and reduces the aging of chloroplasts. NO has either synergistic or antagonistic effects on cytokinin metabolism, depending on the plant cell status (Liu et al.

2013). Cytokinins are responsible for enhanced leaf chlorophyll content, modify photosynthesis components, regulate CO_2 assimilation capacity, accelerate chloroplast development, and coordinate the activity of photosynthetic enzymes to promote photosynthesis. Cytokinins induce light-induced morphogenic effects on plants, chiefly associated with pigment synthesis and de-etiolation, and have the ability to abolish the generation time of chlorophyll synthesis during successive illumination. It has also been observed that wheat seedlings allowed to grow in the absence of light and in the presence of NO have fairly high levels of chlorophyll. NO signaling depends on its specific interaction with phytohormones and on the primary reaction of photosynthesis under drought stress. Under osmotic stress NO acts as a secondary messenger to regulate the cytokinin-induced photosynthetic activity and accumulation of ions, and to maintain mineral nutrient concentration and photosynthetic activity to re-establish plant growth. To develop the stress tolerance ability of plants, the NO signaling mechanism induces antioxidant activity, regulates adverse osmotic effects, and modulates phytohormone balance and photosynthesis.

ABA-induced NO production is necessary for stomatal closure and movement of guard cells. In guard cells, ABA-induced NO inactivates K^+ influx, and drives K^+ efflux and free cytoplasmic Ca^{2+} by depolarizing the plasma membrane potential, which is responsible for the loss of guard cell turgor for stomatal closure. Early reports showed that NO acts as an activator for the transcription of gibberellin gene (Sanz et al. 2015), which codes for two main enzymes for the biosynthesis of active gibberellic acid (GA), and GAs are also promoters of photosynthesis. NO may establish a coordinated reduction of ABA under stress by closing the stomata to impose dormancy, which eventually causes decreased intracellular levels of CO_2, which is required for the dark reaction of photosynthesis. ABA also directly reduces carboxylase activity and increases oxygenase activity of Rubisco to induce photorespiration. Induction of ethylene biosynthesis has been reported by Gniazdowska et al. (2010) for the removal of NO, which induces dormancy and stimulation of germination in apple, but in the absence of ethylene it inhibits photosynthesis primarily by affecting stomatal conductance.

In summary, NO coordinates with phytohormones, which are produced in very small amounts and are responsible for the coordinated stimulation and modulation of the major photosynthetic enzymes like fructose-1,6-bisphosphatase, NADP malic enzyme, Rubisco, phosphoenolpyruvate carboxylase, pyruvate orthophosphate dikinase, and Rubisco activase involved in photosynthesis under normal as well as stress conditions (Freschi 2013).

10.6 The Role of NO in Phytohormone Production for Regulation of Programmed Cell Death for Photosynthesis

The global population is continuously increasing and requires a proportionate increase in crop productivity to meet the enhanced demand for food. The increasing population leads to urbanization, which results in reduced crop land and is responsible for global environmental change. Such gradual environmental change results in abiotic environmental stress, which is responsible for adverse impacts on worldwide crop production. Abiotic environmental stresses are considered as a major limiting factor for plant productivity and are a great threat to global food security. When plants are exposed to such environmental stress, it induces numerous physiological changes in plants and major plant physiological activities such as photosynthesis. The key physiological changes in plants in response to stress include: altered nutrient uptake and translocation ability, altered photosynthetic rates and assimilate translocation, modified antioxidant activity, altered water uptake and transpiration responses, altered gene expression and corresponding enzyme activity, and programmed cell death (Jha and Subramanian 2015).

In all living organisms, synchronized death of tissues and cell takes place as an essential activity for maintenance of the life cycle. This is known as programmed cell death (PCD) and is characterized by cytoplasmic organelle and DNA fragmentation. Abiotic stress results in induction of ROS and RNS in plants and stimulates PCD in affected cells to protect neighboring cells. NO is a diffusible, trans-cellular, multifunctional messenger associated with many pathological and numerous physiological conditions, which is endogenous synthesized in the plant by NO synthase enzyme from the precursor L-arginine. The cellular functions of the plant are affected by NO directly during posttranslational modification of proteins, i.e. nitration and nitrosylation, or indirectly by modification of functional proteins, i.e. ribosylation and methylation. The associated protein phosphorylation and activation of guanylate cyclase enzyme for the formation of cyclic guanosine monophosphate (cGMP) is the main physiological signaling pathway of NO (Gonzalez-Fontes et al. 2014). The interaction of NO with transition metals, superoxide, oxygen, and thiols and its interactions with targets decide whether the signals of NO are cytotoxic or cytostatic. Under extended stress, the production of NO and its derivatives is enhanced, which is one of the major causes of NO stress in the plant cell. Although several plants produce NO in sufficient amounts in normal condition, NO stress causes the nitrosylation of important biomolecules like proteins, lipids, and nucleic acids responsible for dysfunction of metabolic activity, degradation of cellular metabolites, and severe metabolic injury leading to PCD (Jha 2019a). Chloroplasts and mitochondria are the major sites in plants for the production of NO and have a pivotal role in induction of PCD. Among the two major organelles, the chloroplast is the major contributor of NO and ROS, particularly under stress conditions. Plant chloroplasts are the main site for the production of ROS and RNS during photosynthetic photochemical reactions but they are also produced in the chlorophyll biosynthetic pathway by tetrapyrrole intermediate molecules (Brzezowski et al. 2015). The PSI and PSII reactions in the thylakoid membranes of chloroplasts are the main sites for production of ROS and RNS during the light reactions of photosynthesis. During chloroplast formation, synthesis of plastid and chlorophyll proteins takes place, which is necessary for reduction of phototoxicity of the free tetrapyrrolic intermediates and the stability of the Chl–protein complex for chlorophyll formation. The accumulation of chlorophyll pigment takes place at the time of greening of leaves and red light responses in greening leaves are directly stimulated by NO. The photosynthetic pigments, chlorophyll, act as phenotypic markers for both senescence and chloroplast development in leaves. Under abiotic stress the protection of photosynthetic machinery is carried out by escalating the PSI and PSII processes with NO. Liu and Guo (2013) reported a positive response of NO in enhancing chlorophyll content in pea leaves under normal conditions and reduced chlorophyll degradation in *Phytophthora*-infected potato leaves. Experimentally it has been observed that induction of ABA biosynthesis under water stress triggered the exogenous NO donor SNP to maintain photosynthetic activity of the stressed plant.

The gaseous stress hormone ethylene plays an active role in plants and has an antagonistic effect with NO during maturation and senescence of the plant cell. A remarkable decrease in NO concentration has been associated with an increase in ethylene concentration during the maturation process of the plant cell. The release of an electron from cytochrome c of chloroplast and mitochondrial membrane induced directly by NO causes loss of development of membrane potential used for ATP synthesis. NO also inhibits the activity of cytochrome oxidase enzyme on the electron transport chain of plant mitochondria, which is responsible for synthesis of ATP. In this way it alters the mitochondrial activity and accumulation or release of electrons stimulating PCD in plant cells (Jha 2018). Release of mitochondrial cytochrome c and inhibition of the respiratory chain is the main mechanism for NO-induced PCD. The caspase-dependent apoptotic signal cascade for PCD is activated by released cytosolic cytochrome c, which is regulated by metacaspase enzyme. Metacaspase is responsible for regulation of PCD processes and metacaspase activity in plants is regulated by NO via *S*-nitrosylation of enzymes. As observed in *A. thaliana*, NO regulates proteolytic activity to regulate PCD (Wang et al. 2017). Cytochrome c release and mitochondrial dysfunction are induced by peroxynitrite, formed under stress due to generation of superoxide, which reacts with NO and induces apoptotic cell death in stressed plant cells.

10.7 The Role of NO in Phytohormone Production for Regulation of Photosynthate Partitioning During Photosynthesis

Nitrogen is one of the important elements in mineral nutrition of the plant, which is directly related to plant growth and yield. Plant growth correlates with net carbon/nitrogen gain and photosynthesis on a whole-plant basis (Jha 2019b). Photosynthesis rates of a plant are directly related to leaf nitrogen (N) content and light intensity, and the other major determinant of photosynthesis is plant carbon gain, which is directly dependent on leaf area and leaf number. So the C/N ratio is the chief component for the regulation of plant metabolism, and recently researchers have started to focus on the independent effect of nitrogen metabolism on plant physiology, enzyme systems, and genetic adaptability. This coordinated integration of plant physiological processes depends directly on source–sink regulation among photosynthetic and assimilating organs. During photosynthesis the pace of loading of photosynthetic products is regulated by assimilate acceptor tissues, but regulation of this mechanism of both source and sink of assimilates requires a highly coordinated mechanism. The translocation of photosynthetic carbohydrate from the sites of synthesis to different tissues for its utilization and accumulation is regulated at several points. The rate of photoassimilate translocation and pattern of partitioning is primarily based on the concentration gradient of photoassimilates between the source and sink. So, regulation of photoassimilate translocation and partitioning is an integrated process for proper photosynthesis (Poorter et al. 2012). The partitioning pattern depends on environmental conditions and is regulated by coordinated activity of a secondary messenger (NO) and phytohormone. The secondary messenger (NO) and phytohormone are produced in specific tissues and regulate whole-plant processes under normal and stress conditions, showing an indirect relationship of NO to source–sink process modulation. The signaling and sensing of the C/N balance requires high levels of tight coordination for the long-distance translocation of carbon and nitrogen. The carbon primarily assimilates and metabolizes in the leaves, while the NO_3^- and/or NH_4^+ are absorbed by the root from the soil and transported to the leaves.

The NO signaling system of the plant has the ability to reorganize physiological and biochemical processes for the survival of the plant. Generally NO has a signaling function as well as acting as the substrate for amino acid biosynthesis in the form of nitrate ion. Nitrate reduction by NR enzyme is the main source of NO production in plant cells/chloroplasts. The reduction of nitrate initially takes place in chloroplasts and it is initiated in the cytoplasm. Nitrate ions in plant shoots are able to trigger synthesis of metabolically active molecules like phytohormones. NO signaling induces plant hormones for the regulation of sink potential by modulating the differentiation and division of developing sinks. This process requires the coordinated action of a set of phytohormones such as ethylene, cytokinins, auxin, and ABA. Tolerance to heat stress is established by maintaining a critical balance between ABA and cytokinins, which is regulated by NO to determine sink potential by modulating cell division. Bianchetti et al. (2017) reported that the balance between auxins and cytokinins is regulated by NO in developing tissue at the time of cell differentiation, which is necessary for photosynthate partitioning. Cytokinins affect photosynthate partitioning by changing the ratio of source to sink tissue, which is directly influenced by NO. In plants there are two possible conditions associated with the source–sink relationship (Table 10.1):

1) Excess nitrates taken up by the plant roots are transported to the shoots, where NO inhibits the export of carbohydrate from the source through the phloem and encourage formation of apical points, resulting in progression of shoot formation with new leaves. These new leaves increase the leaf area for enhanced net photosynthesis, which causes production of an excess of sugars, which is finally transported into the roots. So change in the environmental condition triggers change in the sugar/nitrate flow ratio, which alters the metabolic pathway either in shoots or roots, by means of NO signaling.
2) An excess of sugar produced during photosynthesis in plant leaves is transported to the roots, which intensify the uptake of nitrate. This nitrate in the root is reduced by an enzyme to initiate root growth and de novo synthesis of root, which results in enhanced uptake of nitrates by the plant roots and causes intensification of nitrate transport into the shoots.

Table 10.1 Mechanisms and effects of nitric oxide on plants for phytohormone modulation.

Mechanisms	Targets	Changes
Transcriptional regulation of gene involved in phytohormone transport, metabolism, and signaling	Enzymes involved in phytohormone biosynthesis, degradation, complex formation	Plant hormone level
Direct chemical modification of proteins involved in phytohormone transport, metabolism, and signaling	Channel and transportation proteins	Plant hormone distribution
Direct chemical reaction between phytohormone receptors and NO derivatives	Signal transduction proteins and receptors	Plant hormone signaling

The stimulation of elevated concentrations of phytohormone like cytokinins, auxins, ABA, and gibberellins in developing seeds indicates direct impact of phytohormones in photosynthate partitioning as reported by Ozga et al. (2017).

10.8 Effect of NO to Regulate Gene Expression for Phytohormones Under Abiotic Stress for Photosynthesis

Plant growth always depends on the surrounding environment and any change in its environment directly affects growth of the plant. The appropriate micro and macronutrients, optimum temperature, pH, and osmolarity required for proper plant growth are provided by the surrounding environment. Any changes in any of the parameters directly affect plant cell activity. Under adverse environmental conditions, plants need the activation of suitable responses for survival and, with plants being static in nature, the effects are even more complex (Jha et al. 2011).

The specific plant hormones triggered by NO for modulation of transcription and posttranscriptional modification in the plant enable the ability to survive under stress. During NO-mediated posttranscriptional modifications, modification of target proteins and structure of proteins takes place. Such chemical modifications of target proteins cause specific changes in their activities and cause relative changes in corresponding cellular functions also, ultimately stimulating plant responses to act according to the NO message transduction signal (Kumar and Pathak 2018). The potential NO–phytohormone interaction involves specific mechanisms as: (i) NO/NO-derived reactive species directly interact with specific plant hormones to modify metabolic activity of the plant for its better survival. (ii) The genes for phytohormone production, metabolism, transport, and for signal transduction are influenced by NO at transcription level by chemical modification of transcription factors and other related proteins. (iii) NO also causes posttranslational modification of proteins for the signaling, production, or distribution of plant hormones at a particular target. Such posttranslational modification of target proteins by NO is by *S*-nitrosylation, i.e. NO covalently interacts with cysteine residues of the target protein. This interaction triggers an intermediate NO signal transduction pathway in plants for the modulation of several physiological activities like photosynthesis, respiration, cellular architecture etc. for the processing of genetic information for coordinated production of phytohormones to develop tolerance against abiotic and biotic stresses (Jha and Subramanian 2018).

The other NO-dependent modification in gene expression takes place by interaction of NO with ROS, such as superoxide for the formation of NO-derived species like peroxynitrite. This peroxynitrite covalently interacts with tyrosine residues of target proteins/transcriptional factors of the gene involved in numerous essential cellular processes like photosynthesis, nitrogen metabolism, and respiration (Tanou et al. 2012). NO also regulates the expression of defense-related genes, as reported by Gullner et al. (2018) in pathogen-challenged tobacco and soybean, where NO upregulates the expression of glutathione-S-transferase and phenylalanine ammonia-lyase to acquire resistance against pathogens. The induction of gene expression of the key enzymes of jasmonic acid biosynthesis like ferritin and peroxidases by NO after infection by necrotropic pathogen has been reported by Turrion-Gomez and Benito (2011).

NO-dependent modification in gene expression also takes place by metal nitrosylation, where NO covalently interacts with transition metals at the centers of metalloproteins for the regulation of plant processes like mitochondrial respiration, nodule functioning, root/shoot development, and defense responses. A study conducted by de Mia (2017) on *Arabidopsis* mutants showed that NO plays a significant role in the regulation of photosynthetic genes by decreasing Rubisco mRNA levels. This important physiological activity is directly influenced by both NO and plant hormones, but studies are needed to find out whether they act in parallel, independently, or share a common signaling cascade for the modulation of plant events.

The expression of the four photosynthesis-related genes (i.e. encoding the large subunit of Rubisco, the PSI/PSII core subunit, and two enzymes for light-independent chlorophyll synthesis) is necessary for carbon fixation, the photosynthetic electron transport chain, and chlorophyll synthesis, respectively, and is upregulated by NO in *Chlorella vulgaris*. Misra et al. (2011) also reported activation of the gene for chalcone synthase enzyme, alternative oxidase, glutathione peroxidase, and inactivation of thylacoidal ascorbate peroxidase of photosynthesis by NO in *A. thaliana*. Similarly, Courtois et al. (2008) reported that NO is an early signaling molecule for the activation of the specific protein kinases responsible for upregulation of ABA-responsive gene expression in plant cells challenged by osmotic stress. NO and Ca^{2+} collectively upregulate the protein kinase-induced pathways, which act on transcriptional regulation of the specific group of genes concerned with plant developmental processes and plant defense.

10.9 Concluding Remarks

NO and plant hormones were initially considered as biochemical intermediates or products with limited roles in the physiological activity of plants. However, now NO is recognized as a gaseous signal transducer in plant with wide activity in plant physiology. NO directly modulates photosynthesis, pathogen defense responses, and abiotic stress responses by interacting with phytohormones, for the growth of plants under normal as well as stress conditions. Photosynthesis is one of the most important physiological activities of plant and is responsible for the production of energy and food for the plant as well as for other organisms. It is regulated by many effector molecules at different stages. NO is one of these effectors, but the actual mechanism behind its regulation needs detailed study. Further, research needs to be carried out for the identification of the specific targets of NO, and its interactions with other regulators, specifically with phytohormones to activate the appropriate set of responses. As it is a gaseous growth regulator identification of its cross-talk with other target molecules and its regulation is critical and the need of the hour to achieve sustainable plant productivity.

References

Astier, J., Gross, I., and Durner, J. (2018). Nitric oxide production in plants: an update. *Journal of Experimental Botany*69: 3401–3411.

Barickman, T.C., Kopsell, D.A., and Sams, C.E. (2014). Abscisic acid increases carotenoid and chlorophyll concentrations in leaves and fruit of two tomato genotypes. *Journal of the American Society for Horticultural Science*139: 261–266.

Bianchetti, R., Cruz, A., and Oliveira, B. (2017). Phytochromobilin deficiency impairs sugar metabolism through the regulation of cytokinin and auxin signaling in tomato fruits. *Science Reports*7: 7822.

Brzezowski, P., Richter, A.S., and Grimm, B. (2015). Regulation and function of tetrapyrrole biosynthesis in plants and algae. *Biochimica et Biophysica Acta (BBA) – Bioenergetics*1847: 968–985.

Courtois, C., Besson, A., Dahan, J. et al. (2008). Nitric oxide signalling in plants: interplays with Ca^{2+} and protein kinases. *Journal of Experimental Botany*59: 155–163.

Corpas, F.J. and Barroso, J.B. (2015). Functions of nitric oxide (NO) in roots during development and under adverse stress conditions. *Plants (Basel)* 4 (2): 240–252. doi: 10.3390/plants4020240.

de Mia, M. (2017). The role of nitric oxide in the remodeling of the photosynthetic apparatus under abiotic stress in *Chlamydomonas reinhardtii*. Vegetal Biology. Université Paris Saclay (COmUE), English. NNT: 2017SACLS516.

Freschi, L. (2013). Nitric oxide and phytohormone interactions: current status and perspectives. *Frontiers in Plant Sciences*4: 398. doi: 10.3389/fpls.2013.00398.

Gayatri, G., Agurla, S., and Raghavendra, A.S. (2013). Nitric oxide in guard cells as an important secondary messenger during stomatal closure. *Frontiers in Plant Sciences*4: 425. doi: 10.3389/fpls.2013.00425.

Gniazdowska, A., Krasuska, U., and Czajkowska, K. (2010). Nitric oxide, hydrogen cyanide and ethylene are required in the control of germination and undisturbed development of young apple seedlings. *Plant Growth Regulation*61: 75–84.

Gonzalez-Fontes, A., Navarro-Gochicoa, M.T., Camacho-Cristobal, J.J. et al. (2014). Is Ca^{2+} involved in the signal transduction pathway of boron deficiency? New hypotheses for sensing boron deprivation. *Plant Science*217–218: 135–139. doi: 10.1016/j.plantsci.2013.

Gullner, G., Komives, T., Király, L. et al. (2018). Glutathione S-transferase enzymes in plant-pathogen interactions. *Frontiers in Plant Sciences*9: 1836. doi: 10.3389/fpls.2018.01836.

Iqbal, N., Umar, S., and Khan, N.A. (2015). Nitrogen availability regulates proline and ethylene production and alleviates salinity stress in mustard (*Brassica juncea*). *Journal of Plant Physiology*178: 84–91. doi: 10.1016/j.jplph.2015.02.006.

Jha, Y. (2018). Effects of salinity on growth physiology, accumulation of osmo-protectant and autophagy-dependent cell death of two maize variety. *Russian Agricultural Sciences*44 (2): 124–130.

Jha, Y. (2019a). Mineral mobilizing bacteria mediated regulation of secondary metabolites for proper photosynthesis in maize under stress. In: *Photosynthesis Productivity and Environmental Stress* (eds. P. Ahmad, M.A. Ahanger, M.N. Alyemeni, and P. Alam), 197–213. Chichester, UK: John Wiley & Sons Ltd.

Jha, Y. (2019b). Regulation of water status, chlorophyll content, sugar, and photosynthesis in maize under salinity by mineral mobilizing bacteria. In: *Photosynthesis Productivity and Environmental Stress* (eds. P. Ahmad, M.A. Ahanger, M.N. Alyemeni, and P. Alam), 75–93. Chichester, UK: John Wiley & Sons Ltd.

Jha, Y. (2020). Plant microbiomes with phytohormones: attribute for plant growth and adaptation under the stress conditions. In: *Advances in Plant Microbiome and Sustainable Agriculture* (ed.A.Yadav, A.Rastegari, N.Yadav, and D.Kour), 85–103. Microorganisms for Sustainability, vol 20. Singapore: Springer.

Jha, Y. and Subramanian, R.B. (2013a). Paddy physiology and enzymes level is regulated by rhizobacteria under saline stress. *Journal of Applied Botany and Food Quality*85: 168–173.

Jha, Y. and Subramanian, R.B. (2013b). Paddy inoculated with PGPR show better growth physiology and nutrient content under salinity. *Chilean Journal of Agricultural Research*73 (1): 213–219.

Jha, Y. and Subramanian, R.B. (2014). Under saline stress plant growth promoting bacteria affect growth, photosynthesis and antioxidant activities in paddy. *International Journal of Agriculture, Environment & Biotechnology*7 (2): 205–212.

Jha, Y. and Subramanian, R.B. (2015). Reduced cell death and improved cell membrane integrity in rice under salinity by root associated bacteria. *Theoretical and Experimental Plant Physiology*3: 227–235.

Jha, Y. and Subramanian, R.B. (2016). Regulation of plant physiology and antioxidant enzymes for alleviating salinity stress by potassium-mobilizing bacteria. In: *Potassium Solubilizing Microorganisms for Sustainable Agriculture* (ed. V.Meena, B.Maurya, J.Verma, and R.Meena), 149–162. New Delhi, India: Springer.

Jha, Y. and Subramanian, R.B. (2018). From interaction to gene induction: an eco-friendly mechanism of PGPR-mediated stress management in the plant. In: *Plant Microbiome: Stress Response. Microorganisms for Sustainability*, vol. 5 (ed.D.Egamberdieva and P.Ahmad), 217–232. Singapore: Springer.

Jha, Y., Subramanian, R.B., and Patel, S. (2011). Combination of endophytic and rhizospheric plant growth promoting rhizobacteria in *Oryza sativa* shows higher accumulation of osmoprotectant against saline stress. *Acta Physiologiae Plantarum*33 797–802.

Jha, Y., Subramanian, R.B., and Patel, S. (2012). Endophytic bacteria induced enzymes against *M. grisea* in *O. sativa* under biotic stress. *African Journal of Basic and Applied Sciences*3 (4): 136–146.

Kolbert, Z., Feigl, G., Freschi, L., and Poór, P. (2019). Gasotransmitters in action: nitric oxide-ethylene crosstalk during plant growth and abiotic stress responses. *Antioxidants*8: 167.

Kumar, P. and Pathak, S. (2018). Nitric oxide: a key driver of signaling in plants. *MOJ Ecology and Environmental Sciences*3 (3): 145–148. doi: 10.15406/mojes.2018.03.00079.

Liu, F. and Guo, F.Q. (2013). Nitric oxide deficiency accelerates chlorophyll breakdown and stability loss of thylakoid membranes during dark-induced leaf senescence in *Arabidopsis*. *PLoS ONE*8 (2): e56345. doi: 10.1371/journal.pone.0056345.

Liu, W.Z., Kong, D.D., and Gu, X.X. (2013). Cytokinins can act as suppressors of nitric oxide in *Arabidopsis*. *Proceedings of the National Academy of Sciences of the United States of America*110 (4): 1548–1553. doi: 10.1073/pnas.1213235110.

Misra, A.N., Misra, M., and Singh, R. (2011). Nitric oxide ameliorates stress responses in plants. *Plant Soil and Environment*57 (3): 95–100.

Nabi, R.B.S., Tayade, R., Hussain, A. et al. (2019). Nitric oxide regulates plant responses to drought, salinity, and heavy metal stress. *Environmental and Experimental Botany*161: 120–133.

Ozga, J.A., Kaur, H., Savada, R.P., and Reinecke, D.M. (2017). Hormonal regulation of reproductive growth under normal and heat-stress conditions in legume and other model crop species. *Journal of Experimental Botany*68: 1885–1894.

Pandey, M.K., Roorkiwal, M., Singh, V.K. et al. (2016). Emerging genomic tools for legume breeding: current status and future perspectives. *Frontiers in Plant Sciences*7: 455. doi10.3389/fpls.2016.00455.

Poorter, H., Niklas, K.J., Reich, P.B. et al. (2012). Biomass allocation to leaves, stems and roots: meta-analysis of interspecific variation and environmental control. *New Phytologist*193: 30–50.

Postila, P.A. and Róg, T. (2020). A perspective: active role of lipids in neurotransmitter dynamics. *Molecular Neurobiology*57: 910–925.

Pourrut, B., Shahid, M., Dumat, C. et al.(2011). Lead uptake, toxicity, and detoxification in plants. *Reviews of Environmental Contamination and Toxicology*213: 113–136.

Rather, B.A., Masood, A., Sehar, Z. et al. (2020). Mechanisms and role of nitric oxide in phytotoxicity-mitigation of copper. *Frontiers in Plant Sciences*11: 675. doi: 10.3389/fpls.2020.00675.

Raza, A., Razzaq, A., Mehmood, S.S. et al. (2019). Impact of climate change on crops adaptation and strategies to tackle its outcome: a review. *Plants (Basel)* 8 (2): 34. https://doi.org/10.3390/plants8020034

Salgado, I., Carmen Martínez, M., and Oliveira, H.C. (2013). Nitric oxide signaling and homeostasis in plants: a focus on nitrate reductase and *S*-nitrosoglutathione reductase in stress-related responses. *Brazilian Journal of Botany*36: 89–98.

Sanz, L., Albertos, P., Mateos, I. et al. (2015). Nitric oxide (NO) and phytohormones crosstalk during early plant development. *Journal of Experimental Botany*66: 2857–2868.

Sanz-Luque, E., Chamizo-Ampudia, A., Llamas, A. et al. (2015). Understanding nitrate assimilation and its regulation in microalgae. *Frontiers in Plant Sciences*6: 899.

Sharma, A., Kumar, V., and Shahzad, B. (2020). Photosynthetic response of plants under different abiotic stresses: a review. *Journal of Plant Growth Regulation*39: 509–531.

Takei, K., Sakakibara, H., Taniguchi, M. et al. (2001). Nitrogen-dependent accumulation of cytokinins in root and the translocation to leaf: implication of cytokinin species that induces gene expression of maize response regulator. *Plant and Cell Physiology*42: 85–93.

Tanou, G., Filippou, P., Belghazi, M. et al. (2012). Oxidative and nitrosative-based signaling and associated post-translational modifications orchestrate the acclimation of citrus plants to salinity stress. *Plant Journal*72: 585–599.

Tejada-Jimenez, M., Llamas, A., Galván, A. et al. (2019). Role of nitrate reductase in NO production in photosynthetic eukaryotes. *Plants (Basel)* 8 (3): 56. doi: 10.3390/plants8030056.

Thomson, A.J., Giannopoulos, G., Pretty, J. et al. (2011). Biological sources and sinks of nitrous oxide and strategies to mitigate emissions. *Philosophical Transactions of the Royal Society B: Biological Sciences*367 (1593): 1157–1168. doi: 10.1098/rstb.2011.0415.

Turrion-Gomez, J.L. and Benito, E.P. (2011). Flux of nitric oxide between the necrotrophic pathogen *Botrytis cinerea* and the host plant. *Molecular Plant Pathology*12 (6): 606–616. doi: 10.1111/j.1364-3703.2010.00695.x.

van Heeswijk, W.C., Westerhoff, H.V., and Boogerd, F.C. (2013). Nitrogen assimilation in *Escherichia coli*: putting molecular data into a systems perspective. *Microbiology and Molecular Biology Reviews*77 (4): 628–695. doi: 10.1128/MMBR.00025-13.

Wang, T., Yang, W., and Xie, Y. (2017). Effects of exogenous nitric oxide on the photosynthetic characteristics of bamboo (*Indocalamus barbatus* McClure) seedlings under acid rain stress. *Plant Growth Regulation*82: 69–78.

Wei, L., Derrien, B., Gautier, A. et al. (2014). Nitric oxide–triggered remodeling of chloroplast bioenergetics and thylakoid proteins upon nitrogen starvation in *Chlamydomonas reinhardtii*. *The Plant Cell*26: 353–372.

Yuan, Y., Xu, X., and Gong, Z. (2019). Auxin response factor 6A regulates photosynthesis, sugar accumulation, and fruit development in tomato. *Horticulture Research*6: 85.

11

Polyamines and Nitric Oxide Interaction in Abiotic Stress Regulation in Plants

Rinukshi Wimalasekera

Department of Botany, Faculty of Applied Sciences, University of Sri Jayewardenepura, Gangodawila, Nugegoda, Sri Lanka

11.1 Introduction

Plants are continually confronted by diverse abiotic and biotic factors. Abiotic stresses, such as drought, flooding, high salinity, extreme high temperatures, chilling, heavy metal toxicity, and nutrient deficiency, are considered as the major factors in global crop yield reduction (Wang et al. 2003; Hu and Sperotto 2021). Plants have developed complex and highly coordinated systems that employ a plethora of components for a range of stress responses. Recently, substantial progress has been made in understanding stress tolerance pathways including signaling cascade events (Hu and Sperotto 2021; Saddhe et al. 2021; Zia et al. 2021). In this chapter, the contributory roles of polyamines (PAs), nitric oxide (NO), and their interaction in some plant abiotic stress responses are comprehensively discussed.

PAs are aliphatic amine compounds abundant in all living organisms. In higher plants, PAs mainly occur in their free unconjugated form. Putrescine (Put), spermidine (Spd), and spermine (Spm) are the major PAs in plants, and they are involved in the regulation of diverse growth and development processes and defense responses (Wuddineh et al. 2018; Takahashi 2020; Gerlin et al. 2021). PAs are regarded as growth regulators in plants that contribute to a range of cellular and development processes including cell proliferation, seed germination, morphogenesis, flowering, and senescence (Tiburcio et al. 2014). PAs are key substances that confer and modulate stress responses or tolerance in response to abiotic stresses, such as drought, high salinity, extreme temperatures, heavy metal toxicity, and nutrient deficiency (Liu et al. 2007; Chen et al. 2019).The key role of PAs in stress tolerance is generally accomplished by elevation of endogenous PA levels, stabilizing cell membrane structure under stress conditions, and secondary messenger signaling (Alcázar et al. 2010). PAs are catabolized by enzymes – copper-containing diamine oxidases (CuAOs) and flavin adenine dinucleotide (FAD)-containing polyamine oxidases (PAOs) (Cona et al. 2006). The H_2O_2 produced by PA catabolism is known to be involved in the signal transduction process of plants during abiotic stress responses. Reactive oxygen species (ROS) generated by PAs are the main signaling molecules that act on defense responses to myriad abiotic stresses (Cona et al. 2006; Minocha et al. 2014). Recent research studies have shown that PAs are strongly linked with production of NO, which is involved in plant stress responses (Pál et al. 2015). NO is a gaseous, highly reactive lipophilic free radical of widespread biological significance. Biosynthesis of NO in plants occurs by enzymatic and nonenzymatic pathways and the most characterized route is the reduction of nitrite to NO (Fröhlich and Durner 2011; Corpas and Barroso 2017). NO plays a key role in the regulation of an array of plant developmental processes including fertilization, seed germination, hypocotyl elongation, root development, leaf senescence, and flowering (Wendehenne and Hancock 2011; Brouquisse 2019). NO is an important signal molecule implicated in both biotic and abiotic stress reactions of plants. NO as well as NO-derived molecules such as nitrogen dioxide, peroxynitrite, and nitrosonium cation are implicated in abiotic stresses such as drought, salinity, extreme temperatures, ozone, ultraviolet radiation, heavy metal toxicity, and mechanical wounding (Siddiqui et al. 2011; Fancy et al. 2017; Nabi et al. 2019). NO is extensively involved in cross-talk with hormones, especially with abscisic acid (ABA) in modulating environmental defense processes (Freschi 2013). Interactions between NO and PAs during several plant defense responses have been described. During abiotic stress responses, PAs may act as substrates, cofactors, or signals for promoting NO biosynthesis and subsequently NO biosynthesis could be mediated by H_2O_2 (Tun et al. 2006; Wimalasekera et al. 2011; Diao et al. 2017). The mechanism of action of the NO–PA relationship leading to a defense response is currently obscure.

11.2 PA Biosynthesis and Catabolism

The diamine Put is the initial product of PA biosynthesis. Put biosynthesis occurs either via the arginine decarboxylase pathway (ADC pathway) from arginine, catalyzed by three successive reactions or by the ornithine decarboxylase pathway (ODC pathway) directly from ornithine. In plants, Put biosynthesis mainly occurs through the ADC pathway (Hanfrey et al. 2001; Docimo et al. 2012). Put is the precursor for the biosynthesis of both Spd and Spm. Put is first converted into Spd and subsequently into Spm by the sequential action of spermidine synthase (SPDS) and spermine synthase (SPMS), respectively, and the transfer of an aminopropyl group released by decarboxylated *S*-adenosylmethionine (Kusano et al. 2008; De Oliveira et al. 2018). The expression of the genes of the biosynthetic pathway is regulated by stress factors thereby modulating the PA metabolism. Spm and Spd are known to be effective inducers of NO biosynthesis in *Arabidopsis thaliana*. However, Put and its biosynthetic precursor arginine are not known to be involved in NO biosynthesis (Tun et al. 2006).

Amine oxidases diamine oxidase (DAO) and PA oxidase (PAO) catabolize PAs in plants. DAO preferably catalyzes the oxidation of Put to 4-aminobutanal with concomitant production of NH_3 and H_2O_2, and the resulting aldehyde is converted to gamma-aminobutyric acid (GABA). PAO preferably catabolizes higher-order PAs like Spd and Spm to form 4-aminobutanal, 3-aminopropyl-4-aminobutanal,1,3-diaminopropane, and H_2O_2 (Cona et al. 2006; Wimalasekera et al. 2011). The synthesis of signaling molecule H_2O_2 from PA oxidation is considered as an important process of plants during abiotic stress responses (Mellidou et al. 2017). In some other instances, PAOs catalyze the reverse reaction of PA synthesis in the PA back-conversion pathway.

11.3 NO Biosynthesis

NO biosynthesis in plants occurs via two main pathways: a reductive pathway and an oxidative pathway. In the reductive pathway, nitrite is reduced to NO, while in the oxidative pathway aminated compounds are oxidized (Mur et al. 2013; Astier et al. 2018). In plants, reduction of nitrite to NO through nonenzymatic or enzymatic mechanisms occurs through the well-characterized NO biosynthesis pathway. The reduction of nitrate to nitrite is catalyzed by nitrate reductase (NR) enzyme using reduced nicotinamide adenine dinucleotide (NADH) as an electron donor in a rate-limiting step (Campbell 2001). NO is produced from nitrite by nitrite reductase using electrons provided by the activity of NR. In plants, another route of NO production is from nitrite via the action of the mitochondrial electron transport chain (Gupta and Kaiser 2010; Astier et al. 2018). Unlike in animal systems, the existence of nitric oxide synthase (NOS) pathway is questionable in plants. Foresi et al. (2010) have shown that NO can be produced from L-arginine, which acts as a substrate, and NADPH as an electron donor. A study with *Arabidopsis* has indicated that arginine or derivatives of arginine are potential sources of NO and arginase activity has resulted in a large amount of arginine conversion to NO (Flores et al. 2008). Tun et al. (2006) reported that treating *Arabidopsis* seedlings with PA induced rapid synthesis of NO. Because of the overlapping biosynthesis pathways of PAs and NO, a rather new phenomenon of PA-induced NO biosynthesis has been presented. Arginases catalyze the conversion of arginine into ornithine and mutants of *A. thaliana* that are impaired in arginase expression displayed an increased NO content that correlated with higher accumulation of Put and Spm levels (Flores et al. 2008). However, aspects of the pathway of PA-induced NO formation are not completely understood, such as whether PAs act as substrates, cofactors, or signals for promoting NO synthesis and the mechanisms involved in NO regulation. PAO and DAO-mediated catabolism of PA leads to H_2O_2 production and NO production either directly or indirectly from H_2O_2 is also another speculation presented for PA-mediated NO production (Tun et al. 2006; Astier et al. 2018)

11.4 PAs and NO in Abiotic Stress Responses in Plants

Plants are constantly exposed to a plethora of environmental stresses that interfere with vigorous growth and development. As sessile organisms, plants have developed a multifaceted network for perception, transduction of stress signals, and initiation of appropriate reactions at the molecular, cellular, and physiological level for existence under environmental stress situations (Akpinar et al. 2012; Zhu 2016). PAs and NO are signaling molecules that play essential roles in modulating response to abiotic stresses. Among several components, NO and NO-derived molecules play important roles in responding to a spectrum of abiotic stresses. Mounting evidence suggests the role of NO both as a significant signaling molecule and as a molecule that regulates plant redox homeostasis in abiotic stress resistance (Arasimowicz and Floryszak-Wieczorek 2007; Fancy et al. 2017). Several studies with different plant species have suggested a role of PAs in abiotic stress tolerance. Increase in endogenous PA levels, enhanced H_2O_2 production by oxidative reduction of

PAs, and homeostasis of ROS are established mechanisms in modulating stress tolerance (Alcázar et al. 2010, 2020; Minocha et al. 2014). PA-associated production of H_2O_2 and NO has overlapping functions in stress-mediated mechanisms (Groppa et al. 2008; Wimalasekera et al. 2011; Adamipour et al. 2020). NO acts in concert with other plant growth regulators, such as ABA and PAs. The PA and NO biosynthetic pathways are overlapping and share some common signal transduction pathways in exerting specific reactions to stresses. Further, a link between PAs and NO in responding to myriad environmental stresses is reported.

11.4.1 Water Stress

Among the environmental stresses that plants encounter, water stress is the most common type of stress that adversely impacts physiology of plants and ultimately leads to impaired plant growth and severely reduced plant productivity. Plants have developed an array of physiological and biochemical adaptive mechanisms for water stress tolerance (Taiz and Zeiger 2003). ROS scavenging by antioxidant systems, maintenance of cell membrane stability, expression of aquaporines and stress proteins, ion balance, and accumulation of osmolytes and osmoprotectants are essential mechanisms contributing to drought tolerance (Taiz and Zeiger 2003).

The function of NO in water stress signaling has been intensively investigated in an array of plant species (Siddiqui et al. 2011; Fancy et al. 2017; Nabi et al. 2019). NO and NO-derived molecules act as signaling molecules or link with other signaling pathways to aid the regulation of antioxidative systems, stomatal closure, and modulation of ABA in the cells in dealing with osmotic stress. The stomatal aperture is regulated by NO signaling and NO cross-talk with H_2O_2 (Desikan et al. 2002, 2004). Further, research studies have shown that ABA-induced NO generation results in stomatal closure (Neill et al. 2002; Bright et al. 2006). Recent investigations have demonstrated that NO plays a key functional role in activating stress defense genes and a process of posttranslational modification of proteins called *S*-nitrosylation (García-Mata and Lamattina 2001; Arasimowicz and Floryszak-Wieczorek 2007; Sidana et al. 2015).

Studies have shown increased biosynthesis of PAs in plants when exposed to water deficit conditions. Water stress-mediated PA biosynthesis is accompanied by accumulation of osmotic adjustment substances, such as free amino acids, soluble sugars, and proline (Wimalasekera et al. 2011; Mohammadi et al. 2018). Regulation of PA-biosynthesizing arginine decarboxylase and PAO during osmotic stress has been shown (Cona et al. 2006). Exogenous application of PAs as seed priming and/or to the plants directly confer enhanced tolerance when citrus, wheat, finger millet, maize, and tobacco plants are exposed to water stresses (Tiburcio and Alcáza 2018; Alcázar et al. 2020). Exogenous application of PAs is known to enhance water use efficiency, chlorophyll content, proline, amino acids, and soluble sugar contents, and prevent cell plasmolysis (Alcázar et al. 2020).

Some studies show the interplay or interdependency between PAs and NO in improving water stress tolerance. In *Arabidopsis*, exogenous application of PAs has led to rapid NO biosynthesis during osmotic stress and this observation has directed further research (Tun et al. 2006; Wimalasekera et al. 2011). Arasimowicz-Jelonek (2009) reported that pretreatment of cucumber seedlings with PAs resulted in early and transient NO production suggesting positive correlation of alleviation of water deficit-induced membrane permeability and lipid peroxidation. The study further indicated that in the signaling pathway, NO may act downstream of PAs in cucumber seedlings under water stress. A study with transgenic barley plants overexpressing the barley nonsymbiotic hemoglobin gene HvHb1 that oxidizes NO to NO_3^- reported that drought-related NO changes could involve the PA pathway (Montilla-Bascón et al. 2017). Reduced NO levels in these transgenic barley plants correlated with increased PA biosynthesis, particularly Spd, and a significant increase in arginine and methionine suggesting a potential interplay between NO and PA biosynthesis during water stress responses. Exogenous application of Spd had exhibited an enhanced drought tolerance in white clover. The NR and NOS pathways were shown to be activated resulting in NO release, which subsequently triggered activation of antioxidant enzymes (Peng et al. 2016). ABA plays a crucial role in drought stress tolerance. When applied with ABA, mutants of *Arabidopsis* lacking *COPPER AMINE OXIDASE1* (*CuAO1*) coding genes showed impaired NO production in roots and leaves, supporting the function of *CuAO1* in PA- and ABA-mediated NO production (Wimalasekera et al. 2011). NO has both positive and negative functional roles in regulating guard cell ABA signaling mediated by H_2O_2, small peptides, and PAs. A study with *Rosa canina* showed that regulation of opening and closing of stomatal apertures in response to drought stress is mediated by PAs, NOS, and H_2O_2 (Adamipour et al. 2020). The study further reported that at various drought stress regimes, oxidation of Put by CuAO was increased and subsequently resulted in an enhanced H_2O_2 production. The accumulated H_2O_2 acted as a secondary messenger leading to stimulation of NOS gene expression implying that the increase in NOS played as a signal for stomatal closure. Agurla et al. (2018) showed that in *Arabidopsis*, exogenous applications of Put, Spd, or Spm considerably increased the NO content in guard cells and stimulated stomatal closure. PA-induced stomatal closure is triggered by NO synthesized from NOS (Sun et al. 2019). Some other studies reported that PA-induced NO production in guard cell is impaired in the mutants of amine oxidase genes of *Arabidopsis* suggesting that PA-induced NO synthesis in guard cells may rely indirectly on ROS.

Evidence supports a clear link between PA signaling and production of H_2O_2 and NO in drought stress tolerance of plants. Drought tolerance is a result of highly complex mechanisms that involve a network of signaling.

11.4.2 Salt Stress

High salinity in soil is a critical global problem that impairs plant development by negatively affecting crop productivity. Salinity enforces water stress, cytotoxicity due to excessive uptake of Na^+ and Cl^-, nutrition deficiencies and imbalances, and ROS-generated oxidative stress.

Among many molecules involved in defense reactions against salt stressed plants, NO is known to play an essential role acting as an antioxidant, quenching ROS during oxidative stress and reducing lipid peroxidation. Further, NO can network with a wide array of molecules and signaling cross-talk preceding activities such as modification of protein activity, Glutathione (GSH) biosynthesis, *S*-nitrosylation, and peroxynitrite formation to sustain stress tolerance. External application of NO donors such as *S*-nitroacetylpenicillamine (SNAP) could mitigate the deleterious effects of NaCl stress in plants. The importance and mechanisms of function of NO in salt tolerance in numerous plant species have been comprehensively reviewed (Hayat et al. 2010; Khan et al. 2012; Kausar and Shahbaz 2013; Isayenkov and Maathuis 2019; Goyal et al. 2021).

The importance of PAs in plant defense responses against high salinity is described though the molecular mechanisms are not well understood. An increase in PA levels in plants is an important indicator of salt tolerance. Several studies have shown that exogenous application of Spd and Spm was able to reduce the inhibitory effect of salt stress in a number of plant species by accumulation of PAs, upregulated *S*-adenosyl-l-methionine decarboxylase (SAMDC) expression and downregulated arginine decarboxylase (ADC) and ornithine decarboxylase (ODC) expression and increased reactive oxygen metabolism (Wang and Bo 2014; Meng et al. 2015; Baniasadi et al. 2018; Napieraj et al. 2020).

A few recent studies have reported the interaction of PAs and NO in exerting salt stress tolerance. Combined foliar application of NO donor sodium nitroprusside (SNP) and Spd was effective to mitigate adverse effects of salinity stress in citrus seedlings. In these seedlings defense responses were associated with reduced electrolyte leakage, malondialdehyde, and H_2O_2 (Khoshbakht et al. 2018). When chickpea plants under salt stress were treated with SNP and Put, the toxic effects of salt stress were ameliorated by electrolyte leakage and lipid peroxidation and modulation of relative water content (Sheokand et al. 2008). Exogenous application of Put, Spd, and Spm to citrus plants exposed to high salinity showed upregulation of NO-associated genes; NR, NOS-like, and alternative oxidase (AOX) and *S*-nitrosoglutathione reductase activity (Tanou et al. 2014). Moreover, this study revealed several *S*-nitrosylated proteins targeted by salinity and PAs. PA-mediated NO production is important to enhance salt tolerance of plants via posttranslational protein modifications through *S*-nitrosylation, nitration of tyrosine residues, and expression of downstream salt stress-tolerant genes. The cross-talk between NO and PAs involves various strategies to respond to salt stress.

11.4.3 Extreme Temperature Stress

Plants are continuously exposed to high and low chilling temperatures and frequent fluctuations of temperatures. These extreme temperatures have several negative effects on physiology, biochemistry, and gene regulation pathways leading to low plant productivity (Wahid 2007; Bhandari and Nayyar 2014). Negative effects of temperature stress have been experimentally proved on various plants. Plants have developed different strategies for heat and cold stress tolerance including increased fluidity of the membranes, production of osmolytes and antioxidants, and accumulation of proline, glycine betaine, and soluble sugars (Wahid 2007).

Accumulated evidence suggests NO as a modulator of plant defense during extreme temperature stress (Parankusam et al. 2017). Rapid NO burst is observed depending on the severity and duration of heat stress. NO is a key signaling molecule in mediating defense responses such as oxidative defense, osmolyte accumulation, and protein modifications under heat and cold stress. At high temperatures, accumulation of NO is followed by increased Nitric oxide synthase and S-Nitrosoglutathione reductase activities. NO mitigates the heat-induced oxidative stress response by maintaining cellular redox homeostasis through moderating the carotenoids under heat stress (Santisree et al. 2015). NO-mediated protein *S*-nitrosylation and tyrosine nitration influence enzymes that confer heat stress tolerance (Romero-Puertas et al. 2013). Application of NO donors externally is shown to be an effective approach in protecting plants from heat and cold stress (Parankusam et al. 2017).

Experimental evidence shows that PAs play a critical role in plant defense in response to high and low temperature stress and this has been reviewed extensively. Increased endogenous PA levels are concurrent with maintaining membrane stability and modulating antioxidant systems (Liu et al. 2015). Transgenic tomato plants overexpressing PA biosynthesizing SAMDC showed higher antioxidant enzyme activity and the protection of membrane lipid peroxidation leading to enhanced tolerance to temperature stress (Li et al. 2009). A number of studies have revealed that priming with PAs increases endogenous PA contents and subsequently enhances tolerance to heat and cold stress. For example, exogenous application of arginine increases quality of cauliflower grown under heat stress (Collado-González et al. 2020).

The literature supports the overlapping mechanisms of action of NO and PA in heat and cold stress tolerance especially in modulating antioxidant systems – the common factor in thermotolerance. However, research studies on the relationship between NO and PAs in exerting heat and cold stress-tolerant mechanisms are limited. A study with tomato seedlings revealed that upon melatonin pretreatment, reprogramming of the PA and NO biosynthesis pathways occurs resulting in improved activation of antioxidant defense mechanisms. Scavenging of excess ROS and increased stability of the cellular membrane are helpful to alleviate heat-induced oxidative stress (Jahan et al. 2019).

A study with tomato leaves by Diao et al. (2016) showed that under chilling stress, Spd acts upstream of NO synthesis and induces NO production directly through enhancing NOS-like and NR activities and also inducing H_2O_2. They have also suggested that the defense response is probably via regulating antioxidant genes and enhancing the antioxidant activities. Wang et al. (2020) showed that NO, GABA, and proline can improve the cold resistance of tea plants. The study suggests NO and PAs act together indirectly to improve the cold resistance of tea plant by improving the ADC, ODC activities (Wang et al. 2020).

11.4.4 Heavy Metal Stress

Accumulation of toxic heavy metals such as As, Pb, Cd, Hg, Cr, Al, and Be in soil causes abnormality in growth and metabolic disorders that can adversely affect crop productivity. Plants have evolved mechanisms like sequestration of metal into vacuoles, binding to phytochelatins/metallothioneins, and activation of various antioxidants to withstand heavy metal stress (Shahid et al. 2015).

The importance of NO in reducing heavy metal stress in plants has been broadly studied (Xiong et al. 2010; Ahmad et al. 2018). Notably, several studies have investigated the role of NO in Cd stress tolerance. The mechanisms of NO-mediated regulation of plant tolerance to heavy metal stress are reported and the protective effect is mainly known to happen by stimulation of superoxide dismutase activity to counteract overproduction of superoxide radicals. Recently, the roles of *S*-nitrosylation as a regulator of plant responses to heavy metal stress has been investigated (Wei et al. 2020). Exogenously applied NO donors have been shown to enhance the defense responses against heavy metals, especially to Cd, by mitigating antioxidative systems to protect plants from oxidative stress.

Higher contents of PAs are detected in plants subjected to heavy metal stress concomitant with enhanced antioxidant function by protecting the tissues from the metal-induced oxidative damage (Groppa et al. 2007; Wimalasekera et al. 2011). Exogenous PA application has been shown to improve the heavy metal tolerance of plants; for example, Spd application improved the tolerance of *Typha latifolia* to Cd^{2+} stress by elevating the activities of antioxidant enzymes (Tang et al. 2005).

Polyamine and nitric oxide cross-talk are shown to be important in heavy metal stress tolerance in plants. Exogenous application of Put and SNP to mung seedlings exposed to Cd led to better Cd-toxicity tolerance with decreased Cd uptake, maintained osmotic balance, and water status suggesting possible cross-talk between NO and PAs (Nahar et al. 2016). In wheat roots, Cd and spermine treatments induce NO formation leading to an increase in the levels of the nonenzymatic antioxidant defense glutathione (Groppa et al. 2008). Exogenous treatment of rice seedlings with NO donors SNP and GSH mitigated the adverse effects of Cu toxicity. Cellular redox state and *S*-nitrosylation-dependent signaling are modulated by NO and GSH (Mostofa et al. 2015).

Several studies have revealed the interaction between PA and NO signaling mechanisms when the plants are subjected to drought, salinity, extreme temperatures, and other environmental stresses. Cross-talk among phytohormones have been uncovered, especially ABA, sugars, PA, and NO signaling pathways that coordinate the activation of appropriate mechanisms leading to mitigation of stresses.

11.5 Conclusion and Future Perspectives

An increasing amount of information identifies that NO and PAs are important compounds that mediate defense responses against a wide range of abiotic stresses via a vast arsenal of mechanisms such as cellular signaling, modulation of redox systems, activation of ABA synthesis, and posttranslational modification of proteins, especially *S*-nitrosylation. The existence of a link between NO and PAs during drought, salinity, heat and cold, and heavy metal stresses has been shown in different plant species. Further investigations on NO and PA interconnection and their possible cross-talk with phytohormones and other signaling compounds will provide a framework for elucidating mechanisms of environmental stress tolerance.

References

Adamipour, N., Khosh-Khui, M., Halehi, H. et al. (2020). Regulation of stomatal aperture in response to drought stress mediating with polyamines, nitric oxide synthase and hydrogen peroxide in *Rosa canina* L. *Plant Signaling & Behavior* 15: 9.

Agurla, S., Gayatri, G., and Raghavendra, A.S. (2018). Polyamines increase nitric oxide and reactive oxygen species in guard cells of *Arabidopsis thaliana* during stomatal closure. *Protoplasma* 255: 153–162.

Ahmad, P., Ahanger, M.A., Alyemeni, M.N. et al. (2018). Exogenous application of nitric oxide modulates osmolyte metabolism, antioxidants, enzymes of ascorbate-glutathione cycle and promotes growth under cadmium stress in tomato. *Protoplasma* 255: 79–93.

Akpinar, B.A., Avsar, B., Lucas, S.J. et al. (2012). Plant abiotic stress signaling. *Plant Signaling & Behavior* 7: 1450–1455.

Alcázar, R., Altabella, T., Marco, F. et al. (2010). Polyamines: molecules with regulatory functions in plant abiotic stress tolerance. *Planta* 231: 1237–1249.

Alcázar, R., Bueno, M., and Tiburcio, A.F. (2020). Polyamines: small amines with large effects on plant abiotic stress tolerance. *Cells* 9: 2373.

Arasimowicz, M. and Floryszak-Wieczorek, J. (2007). Nitric oxide as a bioactive signalling molecule in plant stress responses. *Plant Science*172: 876–887.

Arasimowicz-Jelonek, M. (2009). Interaction between polyamine and nitric oxide signaling in adaptive responses to drought in cucumber. *Journal of Plant Growth Regulation* 28: 177–186.

Astier, J., Gross, I., and Durner, J. (2018). Nitric oxide production in plants: an update. *Journal of Experimental Botany* 69: 3401–3411.

Baniasadi, F., Saffari, V.R., and Moud, A.A.M. (2018). Physiological and growth responses of *Calendula officinalis* L. plants to the interaction effects of polyamines and salt stress. *Scientia Horticulturae* 234: 312–317.

Bhandari, K. and Nayyar, H. (2014). Low temperature stress in plants: an overview of roles of cryoprotectants in defense. In: *Physiological Mechanisms and Adaptation Strategies in Plants Under Changing Environment* (ed. P. Ahmad and M. Wani), 193–265. New York: Springer.

Bright, J., Desikan, R., Hancock, J.T. et al. (2006). ABA-induced NO generation and stomatal closure in *Arabidopsis* are dependent on H_2O_2 synthesis. *The Plant Journal* 45: 113–122.

Brouquisse, R. (2019). Multifaceted roles of nitric oxide in plants. *Journal of Experimental Botany* 70: 4319–4322.

Campbell, W.H. (2001). Structure and function of eukaryotic NAD(P)H: nitrate reductase. *Cellular and Molecular Life Sciences* 58: 194–204.

Chen, D., Shao, Q., Yin, L. et al. (2019). Polyamine function in plants: metabolism regulation on development, and roles in abiotic stress responses. *Frontiers in Plant Science* 9: 1945.

Collado-González, J., Piñero, M.C., Otálora, G. et al. (2020). Merging heat stress tolerance and health-promoting properties: the effects of exogenous arginine in cauliflower (*Brassica oleracea* var *botrytis* L.). *Foods* 24: 30.

Cona, A., Rea, G., Angelini, R. et al. (2006). Functions of amine oxidases in plant development and defence. *Trends in Plant Science* 11: 80–88.

Corpas, F.J. and Barroso, J.B. (2017). Nitric oxide synthase-like activity in higher plants. *Nitric Oxide* 68: 5–6.

De Oliveira, L.F., Navarro, B.V., Cerruti, G. et al. (2018). Polyamines and amino acid related metabolism: the roles of arginine and ornithine are associated with the embryogenic potential. *Plant & Cell Physiology* 59: 1084–1098.

Desikan, R., Cheung, M.K., and Bright, J. (2004). ABA, hydrogen peroxide and nitric oxide signaling in stomatal guard cells. *Journal of Experimental Botany* 55: 205–212.

Desikan, R., Griffiths, R., Hancock, J. et al. (2002). A new role for an old enzyme: nitrate reductase-mediated nitric oxide generation is required for abscisic acid-induced stomatal closure in *Arabidopsis thaliana*. *Proceedings of the National Academy of Sciences of the United States of America* 99: 16314–16318.

Diao, Q., Song, Y., Shi, D. et al. (2017). Interaction of polyamines, abscisic acid, nitric oxide, and hydrogen peroxide under chilling stress in tomato (*Lycopersicon esculentum* mill.) seedlings. *Frontiers in Plant Science* 8: 203.

Diao, Q.N., Song, Y.J., Shi, D.M. et al. (2016). Nitric oxide induced by polyamines involves antioxidant systems against chilling stress in tomato (*Lycopersicon esculentum* Mill.) seedling. *Journal of Zhejiang University Science B* 17: 916–930.

Docimo, T., Reichelt, M., Schneider, B. et al. (2012). The first step in the biosynthesis of cocaine in *Erythroxylum coca*: the characterization of arginine and ornithine decarboxylases. *Plant Molecular Biology* 78: 599–615.

Fancy, N.N., Bahlmann, A.-K., and Loake, G.J. (2017). Nitric oxide function in plant abiotic stress. *Plant, Cell & Environment* 40: 462–472.

Flores, T., Todd, C.D., Tovar-Mendez, A. et al. (2008). Arginase-negative mutants of *Arabidopsis* exhibit increased nitric oxide signaling in root development. *Plant Physiology* 147: 1936–1946.

Foresi, N., Correa-Aragunde, N., Parisi, G. et al. (2010). Characterization of a nitric oxide synthase from the plant kingdom: NO generation from the green alga *Ostreococcus tauri* is light irradiance and growth phase dependent. *The Plant Cell* 22: 3816–3830.

Freschi, L. (2013). Nitric oxide and phytohormone interactions: current status and perspectives. *Frontiers in Plant Science* 4: 398.

Fröhlich, A. and Durner, J. (2011). The hunt for plant nitric oxide synthase (NOS): is one really needed? *Plant Science* 181: 401–404.

García-Mata, C. and Lamattina, L. (2001). Nitric oxide induces stomatal closure and enhances the adaptive plant responses against drought stress. *Plant Physiology* 126: 1196–1204.

Gerlin, L., Baroukh, C., and Genin, S. (2021). Polyamines: double agents in disease and plant immunity. *Trends in Plant Science* 11: S1360–1385.

Goyal, V., Jhanghel, D., and Mehrotra, S. (2021). Emerging warriors against salinity in plants: nitric oxide and hydrogen sulphide. *Physiologia Plantarum* 171: 896–908.

Groppa, M.D., Rosales, E.P., Iannone, M.F. et al. (2008). Nitric oxide, polyamines and Cd-induced phytotoxicity in wheat roots. *Phytochemistry* 69: 2609–2615.

Groppa, M.D., Tomaro, M.L., and Benavides, M.P. (2007). Polyamines and heavy metal stress: the antioxidant behavior of spermine in cadmium- and copper-treated wheat leaves. *Biometals* 20: 185–195.

Hanfrey, C., Sommer, S., Mayer, M.J. et al. (2001). Arabidopsis polyamine biosynthesis: absence of ornithine decarboxylase and the mechanism of arginine decarboxylase activity. *The Plant Journal* 27: 551–560.

Hayat, S., Mori, M., Pichtel, J. et al. (2010). *Nitric Oxide in Plant Physiology*. Weinheim, Germany: Wiley-VCH Verlag GmbH and Co. KGaA.

Hu, Y.B. and Sperotto, R.A. (2021). Regulatory hubs in plant stress adaptation. *Plant Biology* 23 (Suppl. 1): 3–6.

Isayenkov, S.V. and Maathuis, F.J.M. (2019). Plant salinity stress: many unanswered questions remain. *Frontiers in Plant Science* 10: 80.

Jahan, M.S., Shu, S., Wang, Y. et al. (2019). Melatonin alleviates heat-induced damage of tomato seedlings by balancing redox homeostasis and modulating polyamine and nitric oxide biosynthesis. *BMC Plant Biology* 19: 414.

Kausar, F. and Shahbaz, M. (2013). Interactive effect of foliar application of nitric oxide (NO) and salinity on wheat (*Triticum aestivum* L.). *Pakistan Journal of Botany* 45: 67–73.

Khan, M.N., Siddiqui, M.H., Mohammad, F. et al. (2012). Interactive role of nitric oxide and calcium chloride in enhancing tolerance to salt stress. *Nitric Oxide* 27: 210–218.

Khoshbakht, D., Asghari, M.R., and Haghighi, M. (2018). Effects of foliar applications of nitric oxide and spermidine on chlorophyll fluorescence, photosynthesis and antioxidant enzyme activities of citrus seedlings under salinity stress. *Photosynthetica* 56: 1313–1325.

Kusano, T., Berberich, T., Tateda, C. et al. (2008). Polyamines: essential factors for growth and survival. *Planta* 228: 367–381.

Liu, J.H., Kitashiba, H., Wang, J. et al. (2007). Polyamines and their ability to provide environmental stress tolerance to plants. *Plant Biotechnology* 24: 117–126.

Liu, J.H., Wang, W., Wu, H., Gong, X., and Moriguchi, T. (2015). Polyamines function in stress tolerance: from synthesis to regulation. *Frontiers in Plant Science* 6: 827.

Mellidou, I., Karamanoli, K., Beris, D. et al. (2017). Underexpression of apoplastic polyamine oxidase improves thermotolerance in *Nicotiana tabacum*. *Journal of Plant Physiology* 218: 171–174.

Meng, D., Hou, L., and Yang, S. (2015). Exogenous polyamines alleviating salt stress on peanuts (*Arachis hypogaea*) grown in pots. *Chinese Journal of Plant Ecology* 39: 1209–1215.

Minocha, R., Majumdar, R., and Minocha, S.C. (2014). Polyamines and abiotic stress in plants: a complex relationship. *Frontiers in Plant Science* 5: 175.

Mohammadi, H., Ghorbanpour, M., and Brestic, M. (2018). Exogenous putrescine changes redox regulations and essential oil constituents in field-grown *Thymus vulgaris* L. under well-watered and drought stress conditions. *Industrial Crops and Products* 122: 119–132.

Montilla-Bascón, G., Rubiales, D., Hebelstrup, K.H. et al. (2017). Reduced nitric oxide levels during drought stress promote drought tolerance in barley and is associated with elevated polyamine biosynthesis. *Scientific Reports* 7: 13311.

Mostofa, M.G., Seraj, Z.I., and Fujita, M. (2015). Interactive effects of nitric oxide and glutathione in mitigating copper toxicity of rice (*Oryza sativa* L.) seedlings. *Plant Signaling & Behavior* 10: 3.

Mur, L.A., Mandon, J., Persijn, S. et al. (2013). Nitric oxide in plants: an assessment of the current state of knowledge. *AoB Plants* 5: pls052.

Nabi, R.B.S., Tayade, R., Hussain, A. et al. (2019). Nitric oxide regulates plant responses to drought, salinity, and heavy metal stress. *Environmental and Experimental Botany* 161: 120–133.

Nahar, K., Hasanuzzaman, M., Alam, M.M. et al. (2016). Polyamine and nitric oxide crosstalk: antagonistic effects on cadmium toxicity in mung bean plants through upregulating the metal detoxification, antioxidant defense and methylglyoxal detoxification systems. *Ecotoxicology and Environmental Safety* 126: 245–255.

Napieraj, N., Reda, M.G., and Janicka, M.G. (2020). The role of NO in plant response to salt stress: interactions with polyamines. *Functional Plant Biology* 47: 865–879.

Neill, S.J., Desikan, R., Clarke, A. et al. (2002). Nitric oxide is a novel component of abscisic acid signaling in stomatal guard cells. *Plant Physiology* 128: 13–16.

Pál, M., Szalai, G., and Janda, T. (2015). Speculation: polyamines are important in abiotic stress signaling. *Plant Science* 237: 16–23.

Parankusam, S., Adimulam, S.S., Bhatnagar-Mathur, P. et al. (2017). Nitric Oxide (NO) in plant heat stress tolerance: current knowledge and perspectives. *Frontiers in Plant Science* 8: 1582.

Peng, D., Wang, X., Li, Z. et al. (2016). NO is involved in spermidine-induced drought tolerance in white clover via activation of antioxidant enzymes and genes. *Protoplasma* 253: 1243–1254.

Romero-Puertas, M.C., Rodríguez-Serrano, M., and Sandalio, L.M. (2013). Protein S-nitrosylation in plants under abiotic stress: an overview. *Frontiers in Plant Science* 4: 373.

Saddhe, A.A., Manuka, R., and Penna, S. (2021). Plant sugars: homeostasis and transport under abiotic stress in plants. *Physiologia Plantarum* 171: 739–755.

Santisree, P., Bhatnagar-Mathur, P., and Sharma, K.K. (2015). NO to drought-multifunctional role of nitric oxide in plant drought: do we have all the answers? *Plant Science* 239: 44–55.

Shahid, M., Khalid, S., Abbas, G. et al. (2015). Heavy metal stress and crop productivity. In: *Crop Production and Global Environmental Issues* (ed. K.R. Hakeem), 1–25. Cham, Switzerland: Springer.

Sheokand, S., Kumari, A., and Sawhney, V. (2008). Effect of nitric oxide and putrescine on antioxidative responses under NaCl stress in chickpea plants. *Physiology and Molecular Biology of Plants* 14: 355–362.

Sidana, S., Bose, J., Shabala, L. et al. (2015). Nitric oxide in drought stress signalling and tolerance in plants. In: *Nitric Oxide Action in Abiotic Stress Responses in Plants* (ed. M. Khan, M. Mobin, F. Mohammad, and F. Corpas), 95–114. Cham, Switzerland: Springer.

Siddiqui, M.H., Al-Whaibi, M.H., and Basalah, M.O. (2011). Role of nitric oxide in tolerance of plants to abiotic stress. *Protoplasma* 248: 447–455.

Sun, L.R., Yue, C.M., and Hao, F.S. (2019). Update on roles of nitric oxide in regulating stomatal closure. *Plant Signaling & Behavior* 14: 10.

Taiz, L. and Zeiger, E. (2003). *Plant Physiology*, 3e. Sunderland, MD: Sinauer Associates.

Takahashi, T. (2020). Plant polyamines. *Plants* 9: 511.

Tang, C.F., Liu, Y.G., Zeng, G.M. et al. (2005). Effects of exogenous spermidine on antioxidant system responses of *Typha latifolia* L. under Cd^{2+} stress. *Journal of Integrative Plant Biology* 47: 42834.

Tanou, G., Ziogas, V., Belghazi, M. et al. (2014). Polyamines reprogram oxidative and nitrosative status and the proteome of citrus plants exposed to salinity stress. *Plant, Cell & Environment* 37: 864–885.

Tiburcio, A.F. and Alcáza, R. (2018). Potential applications of polyamines in agriculture and plant biotechnology. *Methods in Molecular Biology* 1694: 489–508.

Tiburcio, A.F., Altabella, T., Bitrián, M. et al. (2014). The roles of polyamines during the lifespan of plants: from development to stress. *Planta* 240: 1–18.

Tun, N.N., Santa-Catarina, C., Begum, T. et al. (2006). Polyamines induce rapid biosynthesis of nitric oxide (NO) in *Arabidopsis thaliana* seedlings. *Plant & Cell Physiology* 47: 346–354.

Wahid, A. (2007). Physiological implications of metabolite biosynthesis for net assimilation and heat-stress tolerance of sugarcane (*Saccharum officinarum*) sprouts. *Journal of Plant Research* 120: 219–228.

Wang, Q. and Bo, Y. (2014). Alleviative effects of different kinds of exogenous polyamines on salt injury of soybean seedlings. *Journal of Henan Agricultural Sciences* 43: 48–50.

Wang, W., Vinocur, B., and Altman, A. (2003). Plant responses to drought, salinity and extreme temperatures: towards genetic engineering for stress tolerance. *Planta* 218: 1–14.

Wang, Y., Xiong, F., Nong, S. et al. (2020). Effects of nitric oxide on the GABA, polyamines, and proline in tea (*Camellia sinensis*) roots under cold stress. *Scientific Reports* 10: 12240.

Wei, L., Zhang, M., Wei, S. et al. (2020). Roles of nitric oxide in heavy metal stress in plants: cross-talk with phytohormones and protein *S*-nitrosylation. *Environmental Pollution* 259: 113943.

Wendehenne, D. and Hancock, J.T. (2011). New frontiers in nitric oxide biology in plant. *Plant Science* 181: 507–508.

Wimalasekera, R., Villar, C., Begum, T. et al. (2011). *COPPER AMINE OXIDASE1* (*CuAO1*) of *Arabidopsis thaliana* contributes to abscisic acid- and polyamine-induced nitric oxide biosynthesis and abscisic acid signal transduction. *Molecular Plant* 4: 663–678.

Wimalasekera, R., Tebartz, F., and Scherer, G.F. (2011). Polyamines, polyamine oxidases and nitric oxide in development, abiotic and biotic stresses. *Plant Science* 181: 593–603.

Wuddineh, W., Minocha, R., and Minocha, S.C. (2018). Polyamines in the context of metabolic networks. *Methods in Molecular Biology* 1694: 1–23.

Xiong, J., Fu, G., Tao, L. et al. (2010). Roles of nitric oxide in alleviating heavy metal toxicity in plants. *Archives of Biochemistry and Biophysics* 497: 13–20.

Zhu, J.-K. (2016). Abiotic stress signaling and responses in plants. *Cell* 167: P313–324.

Zia, R., Nawaz, M.S., Siddique, M.J. et al. (2021). Plant survival under drought stress: implications, adaptive responses, and integrated rhizosphere management strategy for stress mitigation. *Microbiological Research* 242: 126626.

12

Nitric Oxide: Interaction with Auxins, Brassinosteroids, and Abscisic Acid

Shreya Gupta[1], *Loitongbam Lorinda Devi*[1†], *and Amar Pal Singh*[1†]

[1]*National Institute of Plant Genome Research, New Delhi, India*

†*Equal contribution*

12.1 Introduction

Plants are sessile; they undergo adaptation to overcome adverse environmental conditions for their survival. Stresses like heat, salinity, drought, and nutrient deficiency limit plant growth and productivity. To adapt under unfavorable environmental conditions, plants reprogram their metabolic, physiological, and developmental signaling cascades as part of a survival strategy. Recently, nitric oxide (NO) has gained much attention as a messenger molecule that regulates various processes such as senescence, flowering, photomorphogenesis, fertilization, and defense by modulating phytohormone activity and responses (Wilson et al. 2008). NO is a gaseous, biologically active molecule, which is abundantly generated throughout the plant's life cycle (Liu et al. 2009). NO plays a key role in the plant life cycle, not only during normal developmental processes but also under environmental stress conditions, such as by inducing stomatal closure during dehydration to provide tolerance (Zhang et al. 2019). NO synthesis and removal are stringently balanced in plants and NO pools are strictly maintained inside the plant cell, both spatially and temporally. There are two well-established mechanisms of NO production. One is oxidative and the other is a reductive pathway (Gupta et al. 2011). NO generation via nitrate reductase (NR) is categorized as a reductive reaction, whereas NO generation by polyamines or arginine comes under the oxidative pathway. Widely studied processes involving NO production include nitric oxide synthase (NOS) and NR-mediated enzymatic reactions (Mur et al. 2012).

Earlier it was shown that *NO Associated 1 (NOA1)* significantly controls NO production in *Arabidopsis* (Zemojtel et al. 2006). The triple *nia1nia2noa1* mutants in *Arabidopsis* showed negligible amounts of NO and compromised developmental and physiological processes including altered meristematic activity and reduced root growth, suggesting NO is necessary for optimal growth (Sanz et al. 2014). Not only its cellular levels, but NO scavenging and its removal are also important to maintain optimum NO levels inside plants. For example, one of the NO scavenging mechanisms involves NO reacting with glutathione (GSH), which forms *S*-nitrosoglutathione (GSNO) and is further converted to glutathione disulfide and ammonium (NH_4^+) by GSNO reductase (Freschi 2013).

NO has been shown to modulate many biological processes through posttranslational modification (PTM) of target proteins mainly via (i) *S*-nitrosylation, (ii) tyrosine nitration, and (iii) metal nitrosylation (Begara-Morales et al. 2016). *S*-nitrosylation of proteins is a major NO-dependent PTM where an NO group binds reversibly to the cysteine at the thiol group of the target protein. The binding of NO with the cysteine residue in the thiol group of the protein forms *S*-nitrosothiol, which modulates the target protein functioning thereby, implicating the role of NO as a signaling molecule (Astier et al. 2011). For example, nitrosylation of GST to GSNO is considered a major bioactive and mobile NO reservoir that has a role in defense-related processes (Leitner et al. 2009). Further, *S*-nitrosoglutathione reductase (GSNOR) regulates the level of *S*-nitrosoglutathione inside the cells (Feechan et al. 2005). The nitration at the tyrosine residue involves the formation of 3-nitrotyrosine where a nitro group (NO_2^-) is added to the target protein at the tyrosine residue (Corpas et al. 2015). The nitro group mainly arises from peroxynitrite ($ONOO^{\bullet}$) as a result of a reaction between NO and a superoxide anion. The PTMs cause conformational changes in the target protein to either activate or deactivate the protein (Radi 2013). The third mechanism, metal nitrosylation, consists of metal-nitrosyl complex formation where NO binds with metals such as zinc, iron, or copper of the metalloproteins that are critical for the activity of these proteins (Bellota-Antón et al. 2011). Plant proteins like hemoglobin, aconitase, ascorbate peroxidase, cytochrome c oxidase, and catalases are also regulated by NO through metal nitrosylation (Gupta et al. 2011).

NO orchestrates various developmental and growth responses of the plant through its interaction with the growth regulators (hormones) such as auxins, brassinosteroid (BR), and abscisic acid (ABA) (Zou et al. 2018). The phytohormone auxin mediates almost all the morpho-physiological and growth-related processes of the plants. These processes include cell differentiation, elongation, and division, which are necessary for overall growth (Piacentini et al. 2020). ABA is a well-known stress hormone that modulates many stress responses such as under high-salinity conditions and plays a critical role during the germination of the seeds (Finkelstein et al. 2008). The steroid hormone BR is necessary for optimal and synchronized plant growth (Pandey et al. 2020). The complex interplay of auxins, ABA, and BR has been shown to regulate several growth features of the plant such as root and shoot growth and seed germination to ensure overall plant productivity. Recently the role of their cross-talk with NO in mediating these growth processes has been explored.

12.2 Insights from NO and Auxin Interaction Studies

NO has been reported to regulate various growth features of plants including lateral root organogenesis, primary root elongation, and formation of root hairs. To regulate these developmental processes more efficiently, multilevel interaction of NO with the auxins at the level of biosynthesis, signaling, and transport has been demonstrated (Fernández-Marcos et al. 2011).

Studies from *Arabidopsis* research show that auxin is mainly synthesized by two major pathways: (i) the de novo pathway; and (ii) auxin release from its conjugates (Zhao 2014). De novo auxin biosynthesis involves the tryptophan (Trp) dependent and independent pathways. In the Trp-dependent pathway, the amino group of Trp is removed by the members of transaminase family of enzymes, *tryptophan aminotransferase* (*TAA*) to form indole 3-pyruvate (IPA). IPA is converted through an oxidative decarboxylation reaction into indole 3-acetic acid (IAA) by flavin monooxygenases of the *YUCCA* (*YUC*) family of proteins (Wang et al. 2018; Piacentini et al. 2020). Auxin can also be synthesized through the Trp independent pathway by indole synthase (INS), which is localized in the cytoplasm (Wang et al. 2015). Plants contain both conjugated and free forms of auxin. The conjugated auxin serves as a reservoir of bioactive auxin, IAA. Conjugated auxins are mainly present as IAA-esters, IAA-sugars, and IAA-amino acids, which are degraded through hydrolysis (Zhao 2014). All the auxin biosynthesis pathways function in coordination for synchronized growth (Wang et al. 2015).

IAA binding to its receptor complex, the transport inhibitor response1/auxin signaling F-box (TIR1/AFB) proteins initiates downstream signaling events. This triggers the interaction of TIR1/AFB and AUX/IAAs (auxin/indole-3-acetic acid), which results in the degradation of AUX/IAA by E3-ubiquitination ligase complex, $SCF^{TIR1/AFB}$. Degradation of AUX/IAAs induces auxin activity, which activates auxin response factors (ARFs) that modulate the expression of auxin-responsive genes (Schepetilnikov and Ryabova 2017).

The auxin gradient and homeostasis are also maintained through the activity of its transporter proteins (Leyser 2018). The cellular auxin gradient through plasma membrane-localized influx and efflux transporters defines overall plant growth including the gravitropic response and lateral root formation (Shi et al. 2015). These transporters include the auxin efflux PIN-formed (PINs) and influx carrier proteins Auxin1/Like-AUX1 (AUX1/LAX) and the ATP-binding cassette transporter B-type (ABCB) proteins (Shi et al. 2015). Altered activity of auxin biosynthesis, signaling, or transport leads to its compromised response.

12.2.1 NO at the Level of Auxin Biosynthesis

The endogenous pool of active auxins is regulated via the modulation of its storage form and precursor of auxins such as indole-3-butyric acid (IBA) (Zolman et al. 2000). Recently it has been shown that auxin acts in a NO-dependent manner at the IBA level (Yadav et al. 2013). The IBA is converted to IAA in the peroxisome via β-oxidation, which is essential for NO generation (Zolman et al. 2000). It was shown that IBA treatment induces NO that regulates lateral roots and adventitious root formation (Schlicht et al. 2013). Moreover, it was demonstrated that the lateral root formation (LRF) via IBA depends on NO levels, which are generated from a reaction of IBA to IAA conversion. In contrast, IBA was not able to initiate LRF in roots treated with NO inhibitors such as 2-4-carboxyphenyl-4,4,5,5-tetramethylimidazoline-1-oxyl-3-oxide (cPTIO). The *NO synthase 1* mutant (*nos1*), which is defective in NO production from L-arginine, showed a similar effect (Guo and Crawford 2005). The maize *lateral rootless1* mutant (*lrt1*) is insensitive to different auxin derivatives, IAA, 2,4-dichlorophenoxyacetic acid (2,4-D) and 1-napthaleneacetic acid (NAA) in LRF. But IBA and NO treatment induce LR formation in the *lrt1* mutant. However, the *lrt1* mutant was unresponsive to IBA in the presence of cPTIO and failed to promote LR growth. These findings show that IBA to IAA conversion produces NO in the peroxisomes, which is necessary for LRF. Moreover, auxins have been shown to regulate the production of NO via NR and NOS (Chen et al. 2010). The apical dominance and axillary bud outgrowth in tomato is controlled by NO homeostasis by regulating auxin–cytokinin levels. Increased NO accumulation in the *gsnor* mutant leads to reduced auxin biosynthesis by inhibiting *FLOOZY* (*FZY*) transcription in the apical bud. This results in a reduced IAA:CK ratio, leading to induction of axillary bud outgrowth by inhibiting apical dominance. Moreover, GSNOR activity is impaired by NR-mediated NO through *S*-nitrosylation at the Cys10 residue (Yan et al. 2021).

12.2.2 NO Cross-talk at the Level of Auxin Signaling

Auxin and NO interaction in regulating root system architecture (RSA) is tightly linked (Correa-Aragunde et al. 2006). Treatment with an NO donor and IAA promoted LRF. Furthermore, the *noa1* mutant is unable to initiate LR formation. This indicates a cross-talk of NO at the level of auxin signaling. NO accumulates in response to auxin treatment, which enhances LR, root hair, and axillary root (AR) formation (Lombardo et al. 2006). Moreover, the auxin receptor mutant (*tir1-1*) shows impaired auxin signaling and is unable to induce LR initiation (Terrile et al. 2012).

NO has been shown to regulate the protein function through PTM by *S*-nitrosylation where NO binds to cysteine residue reversibly, which results in the formation of *S*-nitrosothiol. It has also been demonstrated that *S*-nitrosylation of TIR1 by NO regulates auxin signaling (Terrile et al. 2012). IAA treatment enhances NO accumulation whereas NO donor, sodium nitroprusside (SNP) treatment increases auxin reporter, *DR5-GUS* and auxin-responsive genes such as *indole-3-acetic acid inducible1* (*IAA1*) and *indole-3-acetic acid inducible5* (*IAA5*). By performing an abiotin switch assay, it was demonstrated that NO-mediated *S*-nitrosylation at C-140 of TIR1 increases the interaction of TIR1-AUX/IAAs. The interaction of TIR1-AUX/IAAs results in AUX/IAA degradation to enhance the auxin output. Additionally, the *gsnor1-3* mutant showed compromised auxin signaling (Shi et al. 2015). The *gsnor1-3* mutant leads to an increased production of *S*-nitrosylated proteins in the cells and does not show LR organogenesis and development that resembles the phenotype of the auxin signaling mutant. Nonetheless, the excess NO and impaired reactive oxygen species (ROS) homeostasis in the *gsnor1-3* mutant leads to S-nitrosylation of auxin signaling component TIR1 thereby altering auxin signaling and response.

12.2.3 NO at the Level of Auxin Transport

Treatment with exogenous NO and mutants with altered NO levels such as *chlorophyll a/b binding protein underexpressed1/nitrous oxide overexpressor1* (*cue1/nox1*) exhibit reduced primary root elongation, reduced meristematic cell activity, and early cell differentiation (Fernández-Marcos et al. 2011). These phenotypic changes resemble the auxin efflux carrier mutant, *pin1*, indicating that NO might be hampering auxin transport, thus impairing root meristem activity. The supraoptimal levels of NO reduced the number of actively dividing cells in the meristematic region as seen by using the $CycB1;1_{pro}$:*GUS* line while it promoted early cell differentiation. Using *cue1/nox1* mutants it is also shown that supraoptimal levels of endogenous NO retard the primary root elongation by reducing auxin transport protein PIN1, thus leading to impaired auxin responses (Fernández-Marcos et al. 2011).

In agreement, it has further been shown that mutants such as *noa1* and *nia1nia2*, which are defective in NO production and accumulate low levels of NO, showed enhanced acropetal auxin transport (Sanz et al. 2014). The polarity of the PIN proteins at the plasma membrane is crucial in directing the auxin flow, which in turn impacts growth processes

(Wisniewska et al. 2006). The actin cytoskeleton regulates PIN polarity through the trafficking of PIN proteins between plasma membrane and endosomes. A high level of endogenous NO due to exogenous application of 2,4-D causes PTM (*S*-nitrosylation) of the actin cytoskeleton. These perturbed actin cytoskeletal dynamics, in turn, alter PIN proteins (Rodríguez-Serrano et al. 2014). Additionally, the *gsnor1-3* mutant showed reduced levels of PIN1, PIN2, PIN3, and PIN7-GFP, and NO compromised the endocytosis of plasma membrane-localized PIN2 protein (Ni et al. 2017). The molecular mechanism of PIN2 protein internalization by NO is still vague and needs further research. Moreover, the axillary bud's outgrowth is positively controlled by NO through polar auxin transport (PAT). Supraoptimal levels of NO inhibit PAT in the apical bud, whereas they promote PAT in the axillary bud by enhancing *AUX1* and *PIN1* transcription. This enhanced PAT in the axillary bud promotes its outgrowth in tomato (Yan et al. 2021). Taken together, NO interacts with auxin at the point of biosynthesis, signaling, and transport, which in turn regulates multiple plant developmental features. The findings of NO–auxin interaction studies have been summarized in Figure 12.1.

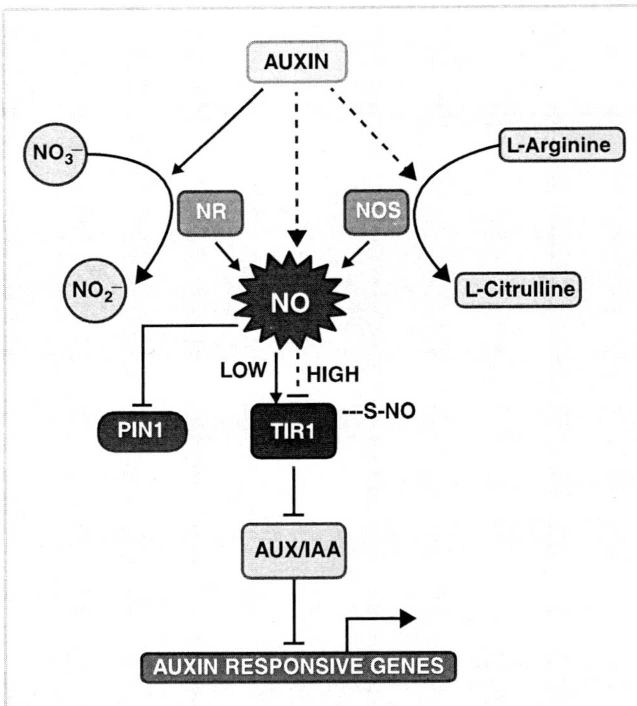

Figure 12.1 Nitric oxide (NO)–auxin interaction in regulating plant growth. Auxin promotes NO production through nitrate reductase (NR) and nitric oxide synthase (NOS) (Schlicht et al. 2013). This NO regulates plant growth by modulating auxin homeostasis. High NO leads to the inhibition of auxin receptor TIR1 via S-nitrosylation (Shi et al. 2015), while a moderate level of NO promotes TIR1 activity through S-nitrosylation at C-140 residue leading to degradation (proteasomal) of the negative regulator's auxin/indole 3-acetic acids (IAAs), thus modulating the auxin response (Terrile et al. 2012). Further NO alters auxin transport by inhibiting PIN1 activity (Fernández-Marcos et al. 2011; Ni et al. 2017).

12.2.4 NO and Auxin Interaction During Stress Responses

The constantly changing environmental cues affect overall plant growth. RSA modification is one such adaptive strategy developed by plants to overcome the changing environmental conditions. Several signaling molecules are involved in regulating these adaptive responses and NO-mediated signaling is one of them. Several studies have shown the role of NO in overcoming abiotic and biotic stresses in plants. NO regulates plant development in response to drought, salt, heavy metal stress, and nutrient stresses (Molassiotis et al. 2010; Siddiqui et al. 2011; Piacentini et al. 2020). Further, NO can also function as an antioxidant via activating ROS scavengers (Siddiqui et al. 2011). Auxin regulates various survival strategies that plants have evolved during stress conditions. Modulating RSA is a crucial survival strategy that involves modification of primary roots, lateral roots, and root hair. These changes in RSA during stress conditions are also developmentally controlled by auxin as well as NO levels (Jin et al. 2011). Recent studies have revealed the interconnected activity of NO and auxin in mitigating environmental stress in plants.

Nitrogen (N) is an essential nutrient for the plant. The major N sources, NH_4^+ and NO_3^-, are the key requirement for plant growth. NO_3^- has been reported to regulate root growth plasticity in many plants including rice and *Arabidopsis* (Sun et al. 2017a). Further, the modulation of root morphology by NO_3^- involves the phytohormone auxin. Moreover, NO has also been reported to regulate root growth under NO_3^- supply. NO is generated through NR activity; therefore, the level of nitrate determines NO generation, which in turn regulates root growth. Additionally, it has been revealed that NO_3^- treatment enhances seminal root (SR) length and induces the formation of lateral roots in a NO–auxin-dependent manner (Sun et al. 2018). NO_3^- is known to promote NO production through NR activity as compared to NH_4^+ supply. Treatment with SNP in NH_4^+-grown plants enhanced auxin transporter PIN1 while cPTIO treatment in the NO_3^- condition reduces it. Additionally, IAA treatment could not enhance NO levels in NH_4^+-grown roots indicating auxin may be acting downstream of NO (Sun et al. 2018). In conclusion, root elongation in response to N nutrition is regulated by NO–auxin cross-talk.

NO along with ethylene regulates root hair formation under magnesium (Mg) deficient conditions (Liu et al. 2018). NO and ethylene promote the expression of *PIN1*, *PIN2*, and *AUX1* genes, which are necessary for the auxin transport response. The high level of auxin in a positive feedback loop enhances NO production and ethylene by increasing the expression of *NR/nitric oxide synthase like* (*NR/NOS-L*) and *1-aminocyclopropane-1-carboxylate* (*ACC*)*/1-aminocyclopropane-1-carboxylate* (*ACC*) *oxidase* (*ACO*)*/ACC synthase* (*ACS*), respectively. The NO/ethylene–auxin loop has been shown to regulate the formation of root hair during Mg deficiency and auxin seems to act downstream of NO and ethylene signaling pathways (Liu et al. 2018).

Copper (Cu) and iron (Fe) are micronutrients that are necessary for plant survival and growth. However, supraoptimal levels of these are toxic for the plant. High levels of Cu inhibit cotyledon development and primary root growth. Cu toxicity has been demonstrated to be regulated by NO–auxin interaction. In cotyledon, Cu toxicity enhanced NO levels thus hampering growth, while in primary root NO mitigates stress response by reducing auxin levels (Peto et al. 2011). The deficiency of iron induces root branching in *Solanum lycocarpum* and *Trifolium pratense* L. (Jin et al. 2008). Auxin and NO positively regulate the branching of roots. This root branching promotes ferric chelate reductase activity to enhance iron uptake. It has been reported that root branching under Fe deficiency is reduced by NO scavenger, cPTIO, and auxin transport inhibitors, 2,3,5-triiodobenzoic acid (TIBA)/naphthylphthalamic acid (NPA) (Jin et al. 2011). Additionally, root branching inhibition by NPA is known to be recovered by NO donor NONOate treatment but NAA treatment failed to recover cPTIO-inhibited root branching. This indicates that during Fe deficiency, NO and auxin finetune root branching (Jin et al. 2011). NO–auxin activity in regulating RSA modification of rice plants under Fe deficiency has been reported (Sun et al. 2017b). Fe deficiency increases LR density and inhibits SR length, which is correlated with increased NO production. Further, using the *ospin1b* mutant, it has been observed that NO regulates SR growth by reducing PAT activity. Moreover, SR growth inhibition under Fe deficiency is due to reduced root meristematic activity as shown by reduced *pCYCB1;1::GUS* expression. Taken together, studies from different plant species demonstrate that auxin and NO together play a pivotal role to modulate plant development under adverse environmental conditions.

12.3 NO and ABA Interactions

The role of ABA has been demonstrated in maintaining stomatal physiology and preventing oxidative damage in plants. Synergistic or antagonistic activity of ABA and NO has been demonstrated in regulating plant development and yield (Delledonne 2005; Hancock et al. 2011). In the past few years, a lot of work has been done to understand the complex regulatory network between NO and ABA.

ABA binds to pyrabactin resistance1 (PYR1)/PYR1-like (PYL)/regulatory components of ABA receptors (RCAR) and initiates the signaling cascade and forms a ternary complex to inactivate type 2C protein phosphatases (PP2Cs). This activates sucrose nonfermenting1 (SNF1)-related protein kinase2 (SnRK2) via autophosphorylation (González-Guzman et al. 2012). Further, these protein kinases mediate ABA responses by targeting and phosphorylating its downstream proteins. For example, in guard cells, ion channels like S-type anion channels (SLAC1) and inward rectifying K^+ channels (KAT1) are phosphorylated by SnRK2, which leads to stomatal closure during water scarcity (Chen et al. 2020). In seeds, SnRK2 phosphorylates the ABA signaling transcription factor, ABSCISIC ACID INSENSITIVE5 (ABI5), and inhibits seed germination. SnRK2 activity is further inhibited in a negative feedback manner by an increase in the transcription of group A PP2C (TYPE 2 C PROTEIN PHOSPHATASES) in the ABA response pathway (Ng et al. 2014).

12.3.1 Role of NO in the ABA Signaling Pathway

ABI5 is considered a central hub for growth repression and maintains seed dormancy (Hilhorst and Karssen 1992). It is reported that transcriptional regulation of ABI5 by ethylene response factors of group VII (VII ERFs) depends on NO levels (Gibbs et al. 2014). ABI5 *S*-nitrosylation at Cys-153 leads to its degradation, which eventually assists the germination processes (Albertos et al. 2015). Thus, NO promotes the germination of seeds by targeting ABI5 both at transcriptional (degradation of VII ERFs by NO) and posttranslational level. Additionally, NO also targets the ABA receptors (PYR/PYL/RCAR) and leads to their PTM. NO-mediated tyrosine nitration of the ABA receptors, PYR/PYL/RCAR, eventually inactivates the ABA-mediated responses (Castillo et al. 2015).

ABA is a major inhibitor of seed germination and plays a role in maintaining seed dormancy (Rodríguez-Gacio et al. 2009). Contrarily, NO promotes seed germination and breaks seed dormancy (Penfield 2017). NO is accumulated in the aleurone layer of the imbibed seeds that suppresses ABA levels by ABA catabolism, thus promoting the seed germination rate (Bethke et al. 2006). Additionally, the negative relation of NO–ABA signaling during seed germination and breakage of dormancy are correlated with the higher protein levels of ABA catabolic enzyme, CYP707A2 (Liu et al. 2009). Overexpression of *CYP707A2* in *Arabidopsis* rapidly decreases ABA levels and releases seed dormancy (Arc et al. 2013) (Figure 12.2). This is further confirmed in *nia1nia2* and *nia1nia2noa1-2* mutants, which displayed delayed seed germination and were more dormant as compared with wild-type plants (Lozano-Juste and León 2010). The role of NO-dependent ABA during seed germination appears conserved across various species such as wheat, lettuce, barley, and *Arabidopsis* (Beligni and Lamattina 2000; Bethke et al. 2006; Liu et al. 2009; Jacobsen et al. 2013).

12.3.2 NO–ABA Interaction During Abiotic Stresses

NO seems to act downstream of ABA during abiotic stress conditions (Prakash et al. 2019). It is also known that during rapid dehydration, ABA levels are induced by NO in the hydrogen peroxide (H_2O_2)-dependent pathway (Desikan et al. 2004). There is an undeniable specific interaction between NO and ABA under different physiological conditions. For example, the closure of stomata during drought condition is mediated by NO, which acts as an intermediate secondary messenger in the ABA signaling cascade to provide tolerance against antioxidants (Murata et al. 2015; Shabbir et al. 2016). However, this NO-dependent stomatal closure is still unclear and additional experimental evidence is required. Some findings reveal that NO is not a crucial molecule in stomatal closure mediated by ABA, but instead, there could be an alternative pathway that is involved in stomatal closure (Van Meeteren et al. 2020).

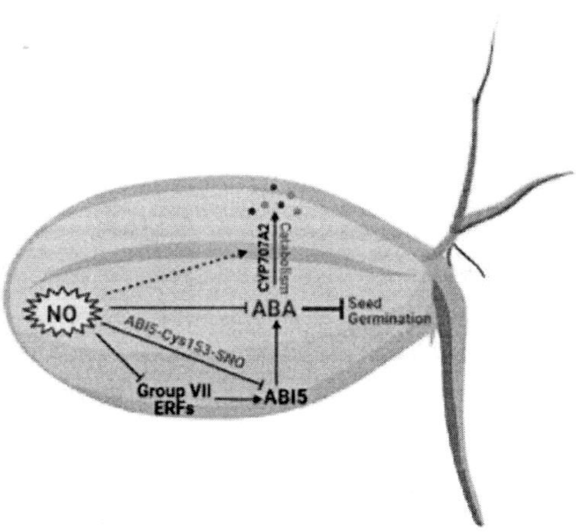

Figure 12.2 Nitric oxide (NO)–abscisic acid (ABA) interaction during seed germination. NO inhibits ABA by promoting its catabolism via CYP707A2. Lower ABA levels are required for breaking seed dormancy and promoting seed germination. NO suppresses transcription factor ABI5 either by degrading it via S-nitrosylation or by inhibiting VII ethylene response factors, which results in lower ABA (Based on Liu et al. 2009).

During heavy metal stresses such as As and Cd toxicity, NO plays an important role in ABA responses (Bücker-Neto et al. 2017). In *Vicia faba*, NO represses As toxicity by reducing ABA levels (Mohamed et al. 2016). Similarly, in *Vigna unguiculata*, the exogenous supply of NO reduces ABA levels, which confers lead (Pb) stress tolerance (Sadeghipour 2017). These studies indicate the role of NO in ABA signaling through posttranslationally modifying its targeted proteins like SnRKs under heavy metal stress (Zhu et al. 2017).

Under high-salinity stress, NO–ABA interactions have been proposed to modulate auxin levels to modify RSA in tomato. NO–ABA-mediated altered auxin levels inhibit primary root growth and lateral root numbers as a survival strategy (Santos et al. 2020). In *Arabidopsis*, *NO ASSOCIATED 1* mutant (*atnoa1*), which has reduced NO levels and accumulates high sodium, is associated with higher oxidative damage. However, the exogenous supply of NO has been shown to reduce the salinity stress damage (Zhao et al. 2007; Manai et al. 2014).

Earlier studies have shown the synergistic role of NO–ABA in providing cold/thermal tolerance in plants (Song et al. 2006). The callus of *Phragmites communis* when subjected to heat stress displayed electrolytic leakage. However, exogenous applications of both NO and ABA alleviated the ion leakage in the callus (Song et al. 2008). *Medicago sativa* plants under cold treatment displayed a higher accumulation of S-adenosylmethionine synthetase, one of the precursors of the phytohormone ethylene and polyamines that eventually generate H_2O_2 and prevent the plants from sustaining oxidative and chilling stress damage (Guo et al. 2014).

Under ultraviolet-B (UV-B) (280–320 nm), there is an increase in ABA, which in turn increases the NO levels to prevent cellular damage and maintain cell homeostasis in leaves of *Zea mays* (Tossi et al. 2012). These studies demonstrate that the NO–ABA module is essential during abiotic stress conditions encountered by plants. However, mechanistic understanding of NO–ABA interactions in regulating heavy metal stress is still cryptic and needs further investigation.

12.4 NO and BR Interactions

BRs are the plant steroidal hormones that play an extensive role in the growth and development of plants; however, their roles in stress responses are still emerging (Nolan et al. 2020; Pandey et al. 2020). BR levels and their spatiotemporal distribution in different plant tissues are still not very well studied. BR ligand binding to the receptor, Brassinosteroid insensitive1 (BRI1) leads to the activation of the plasma membrane-associated signaling complex. As a result, the BRI1 KINASE INHIBITOR1 (BKI1) dissociates from the receptor complex (Li and Chory 1997). This dissociation is followed by the formation of the BRASSINOSTEROID INSENSITIVE1 and BRI1-ASSOCIATED RECEPTOR KINASE (BRI1-BAK1) complex, which further initiates a series of downstream phosphorylation or dephosphorylation signaling events. This eventually causes dephosphorylation of BR INSENSITIVE2 (BIN2) kinase, which negatively regulates the BR signaling pathway (Russinova et al. 2004). The dephosphorylation inactivates BIN2 and promotes its degradation via the proteasomal degradation pathway, thus activating the BR transcriptional effectors, BRASSINAZOLE RESISTANT1/BRI1-EMS-SUPPRESSOR 1(BZR1/BES1) to initiate BR responses (Yin et al. 2002). Several reports talk about the key roles played by NO under various abiotic stress conditions including water deficiency, heavy metal lethality, and salinity. However, the interactions and cross-talk between NO and BRs are still not well explored.

Studies in *Z. mays* have been conducted under drought stress, where BR-dependent NO levels provide stress tolerance. This BR-mediated increase in NO levels further activates ABA biosynthesis and provides drought stress tolerance (Zhang et al. 2010). Similarly, in a different study, under drought conditions, it was shown that epi-brassinolide (biologically active BR) application induces endogenous NO levels that activate antioxidant mechanisms to provide drought tolerance in *Capsicum annuum* (Kaya et al. 2019). Therefore, it appears that NO–BR signaling pathways integrate to regulate plant physiology and development; however, little is known about the NO–BR interaction for improving plant growth during stress conditions.

In *Arabidopsis*, it has been shown that the BR-dependent RSA modification is regulated by the downstream induction of NO levels. This was shown by depleting NO in the media using its scavenger c-PTIO, which caused a reduction in BR-dependent root growth responses (Tossi et al. 2013). As mentioned earlier, NO mediates auxin responses during the formation of adventitious roots (Pagnussat et al. 2002); similarly, BRs also play a major part in root development. In a recent study, the relationship of NO and BR in the adventitious root formation in cucumber was explored (Li et al. 2020). This relationship was further confirmed by using a NO scavenger and NO inhibitors, which reduced adventitious root formation mediated by BR (Li et al. 2020). In tomato, synergistic activity of NO–BR has been demonstrated, which protected the plants from oxidative damage and prevented electrolytic leakage (Bajguz and Hayat 2009). BRs have recently been investigated for their roles in flavonoid biosynthesis in many plant species such as *Camellia sinensis* (Li et al. 2017). Optimum NO levels are required for BR to induce flavonoid biosynthesis for enhancing the quality of tea. This is further verified when the NO scavenger decreased the transcription of phenylalanine ammonia-lyase (PAL) in leaves, an important enzyme in the flavonoid biosynthesis pathway (Li et al. 2017).

It is very well established that ROS and NO together regulate numerous signaling pathways. During chilling and photo-oxidative stress conditions, the levels of antioxidant enzymes such as superoxide dismutase (SOD), ascorbate peroxidase (APX), and catalase (CAT) increase to protect the plant against these stresses (Zhang et al. 2007). One such study shows their role in BR-mediated stress responses in cucumber (Cui et al. 2011). The study revealed the role of BR in mediating tolerance against oxidative damage along with an increase in the antioxidant enzymes, which depends on the induction of NO by BR in a H_2O_2-dependent manner. Thus, these findings suggested the potential interaction between NO, H_2O_2, and BR signaling cascades to alleviate damage during abiotic stresses. BR has been shown to protect plants during biotic stress caused by organisms such as viruses, fungi, and bacteria (Wu et al. 2017). The application of low levels of BR provides tolerance against these pathogens in various plants including tomato, tobacco, and cucumber, and thus increases yield and improves plant growth (Bajguz and Hayat 2009). In *Arabidopsis*, BR-induced NO accumulation via the *nitrate reductase1* (*NIA1*)-dependent pathway resulted in the elimination of Cucumber mosaic virus (Zou et al. 2018). Thus, not only during abiotic stress but also under biotic stress conditions, there is an interaction between the NO and BR pathways that improves plant immunity responses. In summary, NO–BR interactions play a crucial role in mitigating the effects of various abiotic stresses in plants. So far, these studies have been done at the physiological level. However, understanding of the molecular mechanisms underlying the NO–BR interaction is still lacking, especially under extreme environmental challenges.

12.5 Role of NO During Stress Tolerance: Regulatory Interactions with Hormones

NO has become an enigma with regard to its roles in abiotic stress conditions. This signaling molecule becomes more complex in context with its interactions with various plant hormones such as auxin, ABA, and BR. With the recent advancements and extensive research in this area, the complexity of such interactions has been simplified; however, some aspects are still unclear. The interaction of NO and auxin has been explored at multiple levels. NO regulates auxin responses by regulating its biosynthesis, signaling, and transport, and is correlated with stress responses. For example, the toxicity associated with heavy metals such as As and Cd leads to oxidative stress that causes lipid peroxidation and alters auxin signaling and response. The As stress induces auxin biosynthesis gene *osanthranilate synthase alpha subunit2* (*OsASA2*) in rice, while Cd toxicity induces *OsYUCCA2*, leading to an increased endogenous IAA level and disrupting auxin balance (Piacentini et al. 2020). Exogenous application of NO mitigates the high auxin level through oxidation of IAA by ROS. Further, NO enhances the expression of *AUX1*, which is involved in auxin influx in the presence of Cd and As, thus establishing auxin equilibrium and promoting LR induction. In *Medicago truncculata*, NO improves Cd stress tolerance by enhancing IAA levels via reducing IAA oxidase activity (Freschi 2013). Similarly, NO–ABA interactions play a major role in mitigating heavy metal toxicity where NO reduces ABA levels to provide tolerance (Bücker-Neto et al. 2017). These instances indicate a plausible tri-interaction between NO, ABA, and auxin. Such evidence is still vague and needs further research.

Even during nutrient deficiency in the soil, especially nitrogen stress, there are reports where NO positively regulates root growth by increasing auxin levels (Sun et al. 2018). Similarly, NO–BR interactions play a major role in regulating the formation of adventitious roots (Li et al. 2020). It is well established that the interdependent relationship between auxin and BR activity defines root growth and cell elongation (Chaiwanon and Wang 2015). For example, BR and auxin regulate the development of different root zones in a concentration-dependent manner. This differential distribution of BR–auxin gradient in the root tissues regulates a plethora of genes that are involved in meristem maintenance, cell elongation, and differentiation (Chaiwanon and Wang 2015). However, there is no information about the role of NO with respect to these genes, so there might be inter-regulatory cross-talk between NO, BR, and auxin to control root development and morphogenesis under stress conditions. However, this needs to be explored and investigated. Thus, a mixture of positive and negative genetic interactions occur between NO and the phytohormones during different abiotic stress conditions as a way to overcome the negative effects of these stresses.

12.6 Conclusions and Future Perspectives

NO plays a key role in the hormone synthesis, transport, and signaling pathways by modulating various signaling elements like receptors or transcription factors. In recent years, research on the interaction between NO and phytohormones, mainly auxin, has revealed the significance of various instances of cross-talk between these two signaling molecules in regulating a plethora of morpho-physiological processes in plants. Recent findings have described the involvement of NO in auxin biosynthesis, signaling, and transport. It has been shown that the formation of IAA from IBA results in the production of NO, which is responsible for regulating LRF. Moreover, auxin is also regulated by NO at multiple levels including the posttranslational modification of its receptor TIR1 by *S*-nitrosylation at the C-140 residue. Auxin regulates RSA through modulation of auxin transport and NO alters the auxin transporters that define an auxin gradient to mediate the growth. Several studies have reported that NO perturbed auxin distribution by interfering with PIN activity. However, the roles of NO in regulating their import and trafficking need further investigation.

Although the interaction of NO with plant hormones has gained much attention recently, there are still some missing pieces in the puzzle that need to be revealed. For example, there is a lack of information on genetic and molecular components involved in the interactions and the cross-talk between various plant hormones, especially ABA and BR with NO. The most discussed study concerning NO and phytohormone interaction focuses on individual stress conditions or a single parameter. However, the combinatorial effect of abiotic and biotic stresses on NO and phytohormone interaction is yet to be explored, thus leaving us far behind in understanding the holistic mechanism of NO action with the plant phytohormones. Additionally, dealing with the complexity of different stresses is a real challenge as some stresses are often coupled together, for example, salt stress and nutrient deficiency (Kawa et al. 2016). Future work in this area of study could be the construction of a multidimensional interactome between different hormone networks and NO and their corresponding regulatory genes under combined stress conditions. This will help to understand the multilevel complexity of NO-mediated effects on plant growth. Furthermore, the concentration-dependent control of hormones and dosage dependency of NO supplied either exogenously and its internal dynamics and how they influence different plant developmental aspects are yet to be explored. Unraveling the knowledge of genetics and the complex hormonal symphony of NO homeostasis at the cellular level can be manipulated to ameliorate stress resistance for improving yield and productivity in crop plants.

References

Albertos, P., Romero-Puertas, M.C., Tatematsu, K. et al. (2015). S-nitrosylation triggers ABI5 degradation to promote seed germination and seedling growth. *Nature Communications*. doi: 10.1038/ncomms9669.

Arc, E., Sechet, J., Corbineau, F. et al. (2013). ABA crosstalk with ethylene and nitric oxide in seed dormancy and germination. *Frontiers in Plant Science*. doi: 10.3389/fpls.2013.00063.

Astier, J., Rasul, S., Koen, E. et al. (2011). S-nitrosylation: an emerging post-translational protein modification in plants. *Plant Science*. doi: 10.1016/j.plantsci.2011.02.011.

Bajguz, A. and Hayat, S. (2009). Effects of brassinosteroids on the plant responses to environmental stresses. *Plant Physiology and Biochemistry*. doi: 10.1016/j.plaphy.2008.10.002.

Begara-Morales, J.C., Sánchez-Calvo, B., Chaki, M. et al. (2016). Antioxidant systems are regulated by nitric oxide-mediated post-translational modifications (NO-PTMs). *Frontiers in Plant Science*. doi: 10.3389/fpls.2016.00152.

Beligni, M.V. and Lamattina, L. (2000). Nitric oxide stimulates seed germination and de-etiolation, and inhibits hypocotyl elongation, three light-inducible responses in plants. *Planta*. doi: 10.1007/PL00008128.

Bellota-Antón, C., Munnoch, J., Robb, K. et al. (2011). Spectroscopic analysis of protein Fe-NO complexes. *Biochemical Society Transactions*. doi: 10.1042/BST0391293.

Bethke, P.C., Libourel, I.G.L., and Jones, R.L. (2006). Nitric oxide reduces seed dormancy in *Arabidopsis*. *Journal of Experimental Botany*. doi: 10.1093/jxb/erj060.

Bücker-Neto, L., Paiva, A.L.S., Machado, R.D. et al. (2017). Interactions between plant hormones and heavy metals responses. *Genetics and Molecular Biology*. doi: 10.1590/1678-4685-gmb-2016-0087.

Castillo, M.C., Lozano-Juste, J., González-Guzmán, M. et al. (2015). Inactivation of PYR/PYL/RCAR ABA receptors by tyrosine nitration may enable rapid inhibition of ABA signaling by nitric oxide in plants. *Science Signaling*. doi: 10.1126/scisignal.aaa7981.

Chaiwanon, J. and Wang, Z.Y. (2015). Spatiotemporal brassinosteroid signaling and antagonism with auxin pattern stem cell dynamics in *Arabidopsis* roots. *Current Biology*. doi: 10.1016/j.cub.2015.02.046.

Chen, K., Li, G.J., Bressan, R.A. et al. (2020). Abscisic acid dynamics, signaling, and functions in plants. *Journal of Integrative Plant Biology*. doi: 10.1111/jipb.12899.

Chen, W.W., Yang, J.L., Qin, C. et al. (2010). Nitric oxide acts downstream of auxin to trigger root ferric-chelate reductase activity in response to iron deficiency in Arabidopsis. *Plant Physiology*. doi: 10.1104/pp.110.161109.

Corpas, F.J., Begara-Morales, J.C., Sánchez-Calvo, B. et al. (2015). Nitration and S-nitrosylation: two post-translational modifications (PTMs) mediated by reactive nitrogen species (RNS) and their role in signalling processes of plant cells. doi: 10.1007/978-3-319-10079-1_13.

Correa-Aragunde, N., Lanteri, M.L., García-Mata, C. et al. (2006). Nitric oxide functions as intermediate in auxin, abscisic acid, and lipid signaling pathways. doi: 10.1007/7089_2006_087.

Cui, J.X., Zhou, Y.H., Ding, J.G. et al. (2011). Role of nitric oxide in hydrogen peroxide- dependent induction of abiotic stress tolerance by brassinosteroids in cucumber. *Plant, Cell & Environment*. doi: 10.1111/j.1365-3040.2010.02248.x.

Delledonne, M. (2005). NO news is good news for plants. *Current Opinion in Plant Biology*. doi: 10.1016/j.pbi.2005.05.002.

Desikan, R., Cheung, M.K., Bright, J. et al. (2004). ABA, hydrogen peroxide and nitric oxide signalling in stomatal guard cells. *Journal of Experimental Botany*. doi: 10.1093/jxb/erh033.

Feechan, A., Kwon, E., Yun, B.W. et al. (2005). A central role for S-nitrosothiols in plant disease resistance. *Proceedings of the National Academy of Sciences of the United States of America*. doi: 10.1073/pnas.0501456102.

Fernández-Marcos, M., Sanz, L., Lewis, D.R. et al. (2011). Nitric oxide causes root apical meristem defects and growth inhibition while reducing PIN-FORMED 1 (PIN1)-dependent acropetal auxin transport. *Proceedings of the National Academy of Sciences of the United States of America*. doi: 10.1073/pnas.1108644108.

Finkelstein, R., Reeves, W., Ariizumi, T. et al. (2008). Molecular aspects of seed dormancy. *Annual Review of Plant Biology*. doi: 10.1146/annurev.arplant.59.032607.092740.

Freschi, L. (2013). Nitric oxide and phytohormone interactions: current status and perspectives. *Frontiers in Plant Science*. doi: 10.3389/fpls.2013.00398.

Gibbs, D.J., MdIsa, N., Movahedi, M. et al. (2014). Nitric oxide sensing in plants is mediated by proteolytic control of group VII ERF transcription factors. *Molecular Cell*. doi: 10.1016/j.molcel.2013.12.020.

González-Guzmán, M., Pizzio, G.A., Antoni, R. et al. (2012). Arabidopsis PYR/PYL/RCAR receptors play a major role in quantitative regulation of stomatal aperture and transcriptional response to abscisic acid. *The Plant Cell*. doi: 10.1105/tpc.112.098574.

Guo, F.Q. and Crawford, N.M. (2005). Arabidopsis nitric oxide synthase1 is targeted to mitochondria and protects against oxidative damage and dark-induced senescence. *The Plant Cell*. doi: 10.1105/tpc.105.037770.

Guo, Z., Tan, J., Zhuo, C. et al. (2014). Abscisic acid, H_2O_2 and nitric oxide interactions mediated cold-induced S-adenosylmethionine synthetase in *Medicago sativa* subsp. *Falcata* that confers cold tolerance through up-regulating polyamine oxidation. *Plant Biotechnology Journal* doi: 10.1111/pbi.12166.

Gupta, K.J., Fernie, A.R., Kaiser, W.M. et al. (2011). On the origins of nitric oxide. *Trends in Plant Science*. doi: 10.1016/j.tplants.2010.11.007.

Hancock, J.T., Neill, S.J., and Wilson, I.D. (2011). Nitric oxide and ABA in the control of plant function. *Plant Science*. doi: 10.1016/j.plantsci.2011.03.017.

Hilhorst, H.W.M. and Karssen, C.M. (1992). Seed dormancy and germination: the role of abscisic acid and gibberellins and the importance of hormone mutants. *Plant Growth Regulation*. doi: 10.1007/BF00024561.

Jacobsen, J.V., Barrero, J.M., Hughes, T. et al. (2013). Roles for blue light, jasmonate and nitric oxide in the regulation of dormancy and germination in wheat grain (*Triticum aestivum* L.). *Planta*. doi: 10.1007/s00425-013-1878-0.

Jin, C.W., Chen, W.W., Meng, Z.B. et al. (2008). Iron deficiency-induced increase of root branching contributes to the enhanced root ferric chelate reductase activity. *Journal of Integrative Plant Biology*. doi: 10.1111/j.1744-7909.2008.00654.x.

Jin, C.W., Du, S.T., Shamsi, I.H. et al. (2011). NO synthase-generated NO acts downstream of auxin in regulating Fe-deficiency-induced root branching that enhances Fe-deficiency tolerance in tomato plants. *Journal of Experimental Botany*. doi: 10.1093/jxb/err078.

Kawa, D., Julkowska, M.M., Sommerfeld, H.M. et al. (2016). Phosphate-dependent root system architecture responses to salt stress. *Plant Physiology*. doi: 10.1104/pp.16.00712.

Kaya, C., Akram, N.A., Sürücü, A. et al. (2019). Alleviating effect of nitric oxide on oxidative stress and antioxidant defence system in pepper (*Capsicum annuum* L.) plants exposed to cadmium and lead toxicity applied separately or in combination. *Scientia Horticulturae*. doi: 10.1016/j.scienta.2019.05.029.

Leitner, M., Vandelle, E., Gaupels, F. et al. (2009). NO signals in the haze. Nitric oxide signalling in plant defence. *Current Opinion in Plant Biology* doi: 10.1016/j.pbi.2009.05.012.

Leyser, O. (2018). Auxin signaling. *Plant Physiology*. doi: 10.1104/pp.17.00765.

Li, J. and Chory, J. (1997). A putative leucine-rich repeat receptor kinase involved in brassinosteroid signal transduction. *Cell*. doi: 10.1016/s0092-8674(00)80357-8.

Li, X., Zhang, L., Ahammed, G.J. et al. (2017). Nitric oxide mediates brassinosteroid-induced flavonoid biosynthesis in *Camellia sinensis* L. *Journal of Plant Physiology*. doi: 10.1016/j.jplph.2017.04.005.

Li, Y., Wu, Y., Liao, W. et al. (2020). Nitric oxide is involved in the brassinolide-induced adventitious root development in cucumber. *BMC Plant Biology*. doi: 10.1186/s12870-020-2320-y.

Liu, M., Zhang, H., Fang, X. et al. (2018). Auxin acts downstream of ethylene and nitric oxide to regulate magnesium deficiency-induced root hair development in *Arabidopsis thaliana*. *Plant & Cell Physiology*. doi: 10.1093/pcp/pcy078.

Liu, Y., Shi, L., Ye, N. et al. (2009). Nitric oxide-induced rapid decrease of abscisic acid concentration is required in breaking seed dormancy in Arabidopsis. *The New Phytologist*. doi: 10.1111/j.1469-8137.2009.02899.x.

Lombardo, M.C., Graziano, M., Polacco, J.C. et al. (2006). Nitric oxide functions as a positive regulator of root hair development. *Plant Signaling & Behavior*. doi: 10.4161/psb.1.1.2398.

Lozano-Juste, J. and León, J. (2010). Enhanced abscisic acid-mediated responses in nia1nia2noa1-2 triple mutant impaired in NIA/NR- and AtNOA1-dependent nitric oxide biosynthesis in Arabidopsis. *Plant Physiology*. doi: 10.1104/pp.109.148023.

Manai, J., Gouia, H., and Corpas, F.J. (2014). Redox and nitric oxide homeostasis are affected in tomato (*Solanum lycopersicum*) roots under salinity-induced oxidative stress. *Journal of Plant Physiology*. doi: 10.1016/j.jplph.2014.03.012.

Mohamed H.I., Latif H.H., and Hanafy R.S. (2016). Influence of nitric oxide application on some biochemical aspects, endogenous hormones, minerals and phenolic compounds of *Vicia faba* plant grown under arsenic stress. *Gesunde Pflanzen*. doi: 10.1007/s10343-016-0363-7.

Molassiotis, A., Tanou, G., and Diamantidis, G. (2010). NO says more than "YES" to salt tolerance. Salt priming and systemic nitric oxide signaling in plants. *Plant Signaling & Behavior*. doi: 10.4161/psb.5.3.10738.

Mur, L.A.J., Sivakumaran, A., Mandon, J. et al. (2012). Haemoglobin modulates salicylate and jasmonate/ethylene mediated resistance mechanisms against pathogens. *Journal of Experimental Botany*. doi: 10.1093/jxb/ers116.

Murata, Y., Mori, I.C., and Munemasa, S. (2015). Diverse stomatal signaling and the signal integration mechanism. *Annual Review of Plant Biology*. doi: 10.1146/annurev-arplant-043014-114707.

Ng, L.M., Melcher, K., Teh, B.T. et al. (2014). Abscisic acid perception and signaling: structural mechanisms and applications. *Acta Pharmacologica Sinica*. doi: 10.1038/aps.2014.5.

Ni, M., Zhang, L., Shi, Y.F. et al. (2017). Excessive cellular S-nitrosothiol impairs endocytosis of auxin efflux transporter PIN2. *Frontiers in Plant Science*. doi: 10.3389/fpls.2017.01988.

Nolan, T.M., Vukasinović, N., Liu, D. et al. (2020). Brassinosteroids: multidimensional regulators of plant growth, development, and stress responses. *The Plant Cell* 32, 295–318.

Pagnussat, G.C., Simontacchi, M., Puntarulo, S. et al. (2002). Nitric oxide is required for root organogenesis. *Plant Physiology*. doi: 10.1104/pp.004036.

Pandey, A., Devi, L.L., and Singh, A.P. (2020). Review: emerging roles of brassinosteroid in nutrient foraging. *Plant Science*. doi: 10.1016/j.plantsci.2020.110474.

Penfield, S. (2017). Seed dormancy and germination. *Current Biology*. doi: 10.1016/j.cub.2017.05.050.

Peto, A., Lehotai, N., Lozano-Juste, J. et al. (2011). Involvement of nitric oxide and auxin in signal transduction of copper-induced morphological responses in *Arabidopsis* seedlings. *Annals of Botany*. doi: 10.1093/aob/mcr176.

Piacentini, D., Della Rovere, F., Sofo, A. et al. (2020). Nitric oxide cooperates with auxin to mitigate the alterations in the root system caused by cadmium and arsenic. *Frontiers in Plant Science*. doi: 10.3389/fpls.2020.01182.

Prakash, V., Singh, V.P., Tripathi, D.K. et al. (2019). Crosstalk between nitric oxide (NO) and abscisic acid (ABA) signalling molecules in higher plants. *Environmental and Experimental Botany*. doi: 10.1016/j.envexpbot.2018.10.033.

Radi, R. (2013). Protein tyrosine nitration: biochemical mechanisms and structural basis of functional effects. *Accounts of Chemical Research*. doi: 10.1021/ar300234c.

Rodríguez-Gacio, M.D.C., Matilla-Vázquez, M.A., and Matilla, A.J. (2009). Seed dormancy and ABA signaling: the breakthrough goes on. *Plant Signaling & Behavior* 4: 1035–1048. doi: 10.4161/psb.4.11.9902.

Rodríguez-Serrano, M., Pazmiño, D.M., Sparkes, I. et al. (2014). 2,4-Dichlorophenoxyacetic acid promotes S-nitrosylation and oxidation of actin affecting cytoskeleton and peroxisomal dynamics. *Journal of Experimental Botany*. doi: 10.1093/jxb/eru237.

Russinova, E., Borst, J.W., Kwaaitaal, M. et al. (2004). Heterodimerization and endocytosis of Arabidopsis brassinosteroid receptors BRI1 and AtSERK3 (BAK1). *The Plant Cell*. doi: 10.1105/tpc.104.025387.

Sadeghipour, O. (2017). Amelioration of salinity tolerance in cowpea plants by seed treatment with methyl jasmonate. *Legume Research*. doi: 10.18805/lr.v0i0.8394.

Santos, M.P., Zandonadi, D.B., de Sá, A.F.L. et al. (2020). Abscisic acid-nitric oxide and auxin interaction modulates salt stress response in tomato roots. *Theoretical and Experimental Plant Physiology*. doi: 10.1007/s40626-020-00187-6.

Sanz, L., Fernández-Marcos, M., Modrego, A. et al. (2014). Nitric oxide plays a role in stem cell niche homeostasis through its interaction with auxin. *Plant Physiology*. doi: 10.1104/pp.114.247445.

Schepetilnikov, M. and Ryabova, L.A. (2017). Auxin signaling in regulation of plant translation reinitiation. *Frontiers in Plant Science*. doi: 10.3389/fpls.2017.01014.

Schlicht, M., Ludwig-Müller, J., Burbach, C. et al. (2013). Indole-3-butyric acid induces lateral root formation via peroxisome-derived indole-3-acetic acid and nitric oxide. *The New Phytologist*. doi: 10.1111/nph.12377.

Shabbir, R.N., Waraich, E.A., Ali, H. et al. (2016). Supplemental exogenous NPK application alters biochemical processes to improve yield and drought tolerance in wheat (*Triticum aestivum* L.). *Environmental Science and Pollution Research* doi: 10.1007/s11356-015-5452-0.

Shi, Y.F., Wang, D.L., Wang, C. et al. (2015). Loss of GSNOR1 function leads to compromised auxin signaling and polar auxin transport. *Molecular Plant*. doi: 10.1016/j.molp.2015.04.008.

Siddiqui, M.H., Al-Whaibi, M.H., and Basalah, M.O. (2011). Role of nitric oxide in tolerance of plants to abiotic stress. *Protoplasma*. doi: 10.1007/s00709-010-0206-9.

Song, L., Ding, W., Shen, J. et al. (2008). Nitric oxide mediates abscisic acid induced thermotolerance in the calluses from two ecotypes of reed under heat stress. *Plant Science*. doi: 10.1016/j.plantsci.2008.08.005.

Song, L., Ding, W., Zhao, M. et al. (2006). Nitric oxide protects against oxidative stress under heat stress in the calluses from two ecotypes of reed. *Plant Science* 171: 449–458.

Sun, C.H., Yu, J.Q., and Hu, D.G. (2017a). Nitrate: a crucial signal during lateral roots development. *Frontiers in Plant Science* 8: 485.

Sun, H., Feng, F., Liu, J. et al. (2018). Nitric oxide affects rice root growth by regulating auxin transport under nitrate supply. *Frontiers in Plant Science*. doi: 10.3389/fpls.2018.00659.

Sun, H., Feng, F., Liu, J. et al. (2017b). The interaction between auxin and nitric oxide regulates root growth in response to iron deficiency in rice. *Frontiers in Plant Science*. doi: 10.3389/fpls.2017.02169.

Terrile, M.C., París, R., Calderõn-Villalobos, L.I.A. et al. (2012). Nitric oxide influences auxin signaling through S-nitrosylation of the Arabidopsis TRANSPORT INHIBITOR RESPONSE 1 auxin receptor. *The Plant Journal*. doi: 10.1111/j.1365-313X.2011.04885.x.

Tossi, V., Cassia, R., Bruzzone, S. et al. (2012). ABA says NO to UV-B: a universal response? *Trends in Plant Science* doi: 10.1016/j.tplants.2012.05.007.

Tossi, V., Lamattina, L., and Cassia, R. (2013). Pharmacological and genetical evidence supporting nitric oxide requirement for 2,4-epibrassinolide regulation of root architecture in *Arabidopsis thaliana*. *Plant Signaling & Behavior*. doi: 10.4161/psb.24712.

Van Meeteren, U., Kaiser, E., Malcolm Matamoros, P. et al. (2020). Is nitric oxide a critical key factor in ABA-induced stomatal closure? *Journal of Experimental Botany*. doi: 10.1093/jxb/erz437.

Wang, B., Chu, J., Yu, T. et al. (2015). Tryptophan-independent auxin biosynthesis contributes to early embryogenesis in *Arabidopsis*. *Proceedings of the National Academy of Sciences of the United States of America*. doi: 10.1073/pnas.1503998112.

Wang, Y., Zhang, T., Wang, R. et al. (2018). Recent advances in auxin research in rice and their implications for crop improvement. *Journal of Experimental Botany*. doi: 10.1093/jxb/erx228.

Wilson, I.D., Neill, S.J., and Hancock, J.T. (2008). Nitric oxide synthesis and signalling in plants. *Plant, Cell & Environment*. doi: 10.1111/j.1365-3040.2007.01761.x.

Wisniewska, J., Xu, J., Seifartová, D. et al. (2006). Polar PIN localization directs auxin flow in plants. *Science* 80 (312): 883.

Wu, W., Zhang, Q., Ervin, E.H. et al. (2017). Physiological mechanism of enhancing salt stress tolerance of perennial ryegrass by 2,4-epibrassinolide. *Frontiers in Plant Science*. doi: 10.3389/fpls.2017.01017.

Yadav, S., David, A., Baluška, F. et al. (2013). Rapid auxin-induced nitric oxide accumulation and subsequent tyrosine nitration of proteins during adventitious root formation in sunflower hypocotyls. *Plant Signaling & Behavior*. doi: 10.4161/psb.23196.

Yan, Y., Shi, Q., and Gong, B. (2021). S-nitrosoglutathione reductase-mediated nitric oxide affects axillary buds outgrowth of *Solanum lycopersicum* L. by regulating auxin and cytokinin signaling. *Plant & Cell Physiology* doi: 10.1093/pcp/pcab002.

Yin, Y., Wang, Z.Y., Mora-Garcia, S. et al. (2002). BES1 accumulates in the nucleus in response to brassinosteroids to regulate gene expression and promote stem elongation. *Cell*. doi: 10.1016/S0092-8674(02)00721-3.

Zemojtel, T., Fröhlich, A., Palmieri, M.C. et al. (2006). Plant nitric oxide synthase: a never-ending story? *Trends in Plant Science* doi: 10.1016/j.tplants.2006.09.008.

Zhang, A., Jiang, M., Zhang, J. et al. (2007). Nitric oxide induced by hydrogen peroxide mediates abscisic acid-induced activation of the mitogen-activated protein kinase cascade involved in antioxidant defense in maize leaves. *The New Phytologist*. doi: 10.1111/j.1469-8137.2007.02071.x.

Zhang, A., Zhang, J., Ye, N. et al. (2010). ZmMPK5 is required for the NADPH oxidase-mediated self-propagation of apoplastic H_2O_2 in brassinosteroid-induced antioxidant defence in leaves of maize. *Journal of Experimental Botany*. doi: 10.1093/jxb/erq243.

Zhang, L., Shi, X., Zhang, Y. et al. (2019). CLE9 peptide-induced stomatal closure is mediated by abscisic acid, hydrogen peroxide, and nitric oxide in *Arabidopsis thaliana*. *Plant, Cell & Environment*. doi: 10.1111/pce.13475.

Zhao, M.G., Tian, Q.Y., and Zhang, W.H. (2007). Nitric oxide synthase-dependent nitric oxide production is associated with salt tolerance in *Arabidopsis*. *Plant Physiology*. doi: 10.1104/pp.107.096842.

Zhao, Y. (2014). Auxin biosynthesis. *The Arabidopsis Book*. doi: 10.1199/tab.0173.

Zhu, M., Zhang, T., Ji, W. et al. (2017). Redox regulation of a guard cell SNF1-related protein kinase in *Brassica napus*, an oilseed crop. *The Biochemical Journal*. doi: 10.1042/BCJ20170070.

Zolman, B.K., Yoder, A., and Bartel, B. (2000). Genetic analysis of indole-3-butyric acid responses in *Arabidopsis thaliana* reveals four mutant classes. *Genetics*. doi: 10.1093/genetics/156.3.1323.

Zou, L.J., Deng, X.G., Zhang, L.E. et al. (2018). Nitric oxide as a signaling molecule in brassinosteroid-mediated virus resistance to Cucumber mosaic virus in *Arabidopsis thaliana*. *Physiologia Plantarum*. doi: 10.1111/ppl.12677.

13

Nitric Oxide: Interaction with Salicylic Acid, Jasmonic Acid, and Ethylene

Harsimran Kaur[1], Sukhmeen Kaur Kohli[1,2], and Renu Bhardwaj

[1]*Plant Protection Division, PG Department of Agriculture, Khalsa College, Amritsar, Punjab, India*
[2]*Plant Stress Physiology Lab, Department of Botanical and Environment Sciences, Guru Nanak Dev University, Amritsar, Punjab, india*

13.1 Introduction

A universal plant signaling molecule, nitric oxide (NO), is involved in pathways that execute specific cellular functions in plants (Ávila et al. 2019) including growth and development, stomatal closure, seed germination, and photomorphogenesis (Hasanuzzaman et al. 2017). In addition, it acts as a scavenger of reactive radicals and upregulates several genes implicated in signal transduction, nutrient uptake, cell death, and stress acclimation (Mostofa et al. 2019; Siddiqui et al. 2020). NO is also associated with plant adaptive behavior against environmental stress. Several authors have described NO-mediated plant protection under salt, heavy metal, water deficit, osmotic, ozone, and pathogen stress (Singh et al. 2017; Ávila et al. 2019; Campos et al. 2019; Wei et al. 2020). NO can both act as an antioxidant and mitigate oxidative stress or as a reactive nitrogen species source exhibiting a higher oxidizing capacity, thus participating in several biological processes (Domingos et al. 2015; Sahay and Gupta 2017). NO also reduces free radical-induced cytotoxicity like cell death, DNA fragmentation, and ion leakage (Zhang et al. 2016).

Plants, being sessile, rely on extremely sensitive signaling networks and control their growth and metabolism with persistent alterations in the surrounding environment. Plant hormones incorporate external and internal cues into well-coordinated developmental reactions, which maximize its adjustment under disturbing environmental conditions. Several phytohormones are critical in regulating resistance mechanisms against biotic and abiotic stress factors. These phytohormones stimulate signaling cascades terminating in novel transcriptional programs that aim at synthesizing defense-related proteins (Wei et al. 2020). To carry out these critical functions effectively, phytohormones interact not only among themselves but also with endogenous signaling molecules (Freschi 2013). Thus, NO participation in a plethora of plant growth and developmental processes is attributed to cross-talk between the hormone-stimulated signaling pathways (Ávila et al. 2019; Wei et al. 2020). Remarkably, NO aids as a secondary messenger in plant hormone signaling pathways such as that of ethylene (ET), jasmonic acid (JA), and salicylic acid (SA), which are involved in stomatal regulation under environmental cues (Shah et al. 2019). Taking into account the common role displayed by plant regulatory mechanisms in maintaining normal plant growth and augmenting plant forbearance toward abiotic and biotic stressors, it is postulated that plant hormone-facilitated stress defense is concomitant with NO production. Thus, the existing knowledge of NO and plant hormone signaling pathways in plants is primarily focused on in this chapter.

13.2 NO Synthesis

The research on NO strongly signifies a strong signaling network coordinated by NO in several aspects of the plant life cycle. Initially, NO was thought to be a lethal gaseous molecule. The studies conducted in late 2000 described it as an integral constituent of nitrogen metabolism in plants (Rai et al. 2019). However, the source of NO generation is a matter of concern. Though enzymatic and nonenzymatic routes have been suggested for NO production, our present knowledge about the chief source of NO and its regulation in plants is lacking. The nitrification and denitrification processes generate a gaseous byproduct, i.e. NO, upon oxidation of nitrous oxide and release it into the atmosphere. It has been suggested that metabolism of inorganic nitrogen compounds produces NO as a vital intermediate in plants. Nonenzymatically, NO is endogenously synthesized by carotenoids during the day by transformation of nitrogen dioxide (Domingos et al. 2015). In addition, NO is also synthesized in the apoplast in reaction to abscisic acid and gibberellins, where nitrite is converted into NO under acidic conditions (Rai et al. 2019). Hydroxylamine and L-arginine also release NO into the plant tissue upon oxidation. Seven prospective enzymatic routes of NO synthesis have been discovered in plants to date (Gupta, Fernie et al. 2011). The most likely candidates for NO production presently include NO synthase (NOS-like) and nitrate reductase (NR) (Freschi 2013). NR is a nicotinamide adenine dinucleotide phosphate (NADPH)-dependent cytosolic enzyme linked to nitrogen assimilation and reduces nitrate to nitrite (Kolbert et al. 2019). It can further reduce nitrite to NO (Ferreira and Cataneo 2010). Campbell 1999 discovered two NR genes, *NIA1* and *NIA2*, in *Arabidopsis*. A substantial fall in NO production and dissimilar influence on its generation in varied tissues was observed in a comparative study of single and double mutants, *nia1/nia2* (Modolo et al. 2006).

13.3 NO as a Potent Antioxidant

Abiotic stress beyond the plant's tolerance limit triggers reactive oxygen species (ROS) generation and accumulation in the cells by interfering with biochemical mechanisms such as photorespiration-induced metabolic turbulences, alteration of radical scavenging pathways, respiration, carbon dioxide incorporation, and by replacing redox-active iron from proteins (Mostafa et al. 2019). Since ROS generation during harsh conditions is spontaneous and an inevitable cellular process, plants have developed several detoxification mechanisms for elimination and reduction of these cytotoxic molecules (Chandra et al. 2015; Yadu et al. 2017). The enzymatic constituents of the detoxification pathways comprise enzymes like catalase (CAT), superoxide dismutase (SOD), ascorbate peroxidase (APX), and guaiacol peroxidase (POX), whereas the nonenzymatic part comprises ascorbic acid, sugar, carotenoids, glutathione, α-tocopherol etc. The ROS detoxification process is initiated by SOD, a metalloenzyme that converts two superoxide ($O_2^{-\bullet}$) molecules into hydrogen peroxide (H_2O_2) and oxygen (O_2) followed by restraint in H_2O_2 accumulation that involves the ascorbate–glutathione cycle members, CAT, POX, and APX, which dissociates H_2O_2 into O_2 and water (Yadu et al. 2017).

Owing to its free radical behavior, NO interacts with antioxidants and pro-oxidants by adopting an energetically conducive electron arrangement by losing or gaining an electron, thus existing in three interchangeable forms: NO (radical); NO^- (nitroxyl radical); and NO^+ (nitrosonium cation). NO is somewhat water soluble and its solubility increases when ferrous salts are present. Thus, it can freely diffuse in the aqueous cell parts like the cytoplasm and also pass readily through the lipid bilayer. Therefore, it moves easily not only within the cell but also between different cells (Wei et al. 2020). Being a highly reactive radical, NO can scavenge other reactive molecular intermediates, thus putting an end to chain propagating reactions. Two mechanisms have been postulated by which NO is believed to abate stress. First, NO directly acts as an antioxidant and scavenges ROS like $O_2^{-\bullet}$, which is converted into peroxynitrite ($ONOO^\bullet$). Second, NO behaves as a signal molecule, which results in altered gene expression (Ferreira and Cataneo 2010). NO plays a cytoprotective role by rapidly reducing the toxicity and levels of free radicals. As an essential signaling molecule, NO is implicated in the mitochondrial electron transport network and controls ROS production by tempering the antioxidant status of plants under stress. Exogenous NO application in plants confer them resistance toward abiotic stress by augmenting the potential of antioxidant enzymes, thus curtailing ROS (Yadu et al. 2017). The cross-talk amid ROS scavengers and NO boosts plant stress forbearance (Wei et al. 2020).

NO has been reported to overcome ROS-triggered toxicity under stressed conditions. Upon stimulation of ROS formation during stressful conditions, NO eases oxidative stress by foraging the ROS and encouraging the antioxidative reaction (Sharma et al. 2019). In addition, the cross-talk amid free radical scavenging enzymes and NO augments plant stress forbearance (Arora et al. 2016; Wei et al. 2020). NO plays an important role in moderating several plant hormone-mediated stress responses. NO accumulation induces expression of genes implicated in amassing defense-related proteins through stress events and plant recovery. Escalating evidence suggests the importance of NO in conserving cellular homeostasis and reducing the potency of oxidative damage instigated by several stresses (Parankusam et al. 2017). Table 13.1 lists the reports establishing the antioxidative nature of NO in different plants exposed to abiotic stress.

13.4 NO as a Signaling Molecule

NO has arisen as a potent endogenic plant signaling molecule that intercedes in several physiological and developmental processes such as stomatal closure, flowering, programmed cell death, xylogenesis, pathogen defense, etc. At low concentrations NO is a growth promoter, whereas its high concentrations are toxic to the plant. Being a small diatomic molecule that lacks any charge and possesses a short half-life, NO is an ideal messenger in plants involved in several signaling pathways (Yu et al. 2014; Parankusam et al. 2017). Even though plants receive and respond to NO a specific NO receptor in plants has not yet been identified. Plant cells may perceive NO stimulus in several ways

Table 13.1 Influence of NO on antioxidative defense reactions in plants under the influence of abiotic stressors.

Plant name	Stress	NO dose (SNP)	Plant response to exogenous NO treatment	Mode of action	Reference
Oryza sativa	High temperature	100 µM	• Higher photosynthetic NUE, photosynthetic SUE, nitrogen, and sulfur content • Better photosynthetic rate, stomatal frequency, plant growth, and plant dry mass • Enhanced activity of PSII and Rubisco and NR • Amended high temperature stress tolerance	• Reduced amount of TBARS and H_2O_2 • Amplified accretion of proline • Upregulated ascorbate–glutathione cycle activity • Augmented APOX, GR, and SOD activity • Improved GSH content	Gautam et al. 2021
Glycine max	Drought	100 µM	• Promoted plant growth • Improved osmotic adjustment • Upregulated antioxidant system • Cell membrane protection	• Reduced H_2O_2 and MDA levels • SOD, POX, APOX, CAT, PPO • High proline and glycine betaine levels • Elevated PAL and TAL activity • Higher total phenol content, flavanol, and tocopherol levels • Reduced electrolyte leakage and LOX activity	Rezayian et al. 2020
Oryza sativa	Heavy metal (Se)	100 µM	• Improved plant growth and root architecture (length, volume, tip number, and average diameter) • Reduced Se uptake • Enhanced photosynthetic rate • Upregulation of photosynthetic parameters (P_n, C_i and T_r)	• Augmented SOD, POD, and CAT activity • Improved concentrations of GSH and AsA • Reduced levels of MDA, H_2O_2, and proline • Downregulated *OsPT2*, *OsSAMS1*, and *OsCS* gene expression (Se transport and metabolism-related genes)	Dai et al. 2020
Solanum lycopersicum	Nutrient stress (sulfur)	100 µM	• Improved plant growth (length, fresh weight, and dry weight of both root and shoot) • Increased photosynthetic parameters (P_n, G_s, and C_i) and chlorophyll content • Enhanced sulfur uptake and cysteine biosynthesis (sulfur-containing compound)	• Suppressed accretion of MDA, H_2O_2, and $O_2^{\bullet -}$ • Elevated AsA and GSH levels and GSH/GSSG ratio • Enhanced Rubisco, δ-ALAD, and CA and Rubisco activity • Augmented proline content • Upregulation of sulfur assimilation enzymes (ATP-R, ATP-S, OAST-L, and SiR)	Siddiqui et al. 2020

(*Continued*)

Table 13.1 (Continued)

Plant name	Stress	NO dose (SNP)	Plant response to exogenous NO treatment	Mode of action	Reference
Solanum melongena	Heavy metal stress (Ni)	100 μM	• Improved photosynthesis, chlorophyll content, stomatal conductance, and gas exchange parameters • Augmented RWC and antioxidant system • Enhanced nitrogen, potassium, and calcium uptake, and decremented Ni uptake	• Upregulated activity of SOD, CAT, GST, APOX, and GR • Decremented H_2O_2, lipid peroxidation, and lipoxygenase enzyme activity • Increased ascorbate, GSH, proline, and glycine betaine content • Improved electron transport and maintained NADP levels • Escalated GLY-I activity • Reduced methylglyoxal content	Soliman et al. 2019
Lolium perenne	Heavy metal (Cd)	500 μM	• Increased growth parameters, root/shoot ratio, and root volume • Enhanced chlorophyll content • Increased soluble protein • Raised iron, copper, and zinc contents • Reduced Cd content	• Diminished contents of $O_2^{-\bullet}$, H_2O_2, and MDA • Augmented SOD, POD, and CAT activity • Reestablished H^+-ATPase activity	Chen et al. 2018
Brassica napus	Osmotic stress	500 μM	• Improved plant growth • Enhanced antioxidant defense	• Diminished H_2O_2 and MDA levels • Enhanced AsA, GSH, and proline content • High GSH/GSSG ratio • Upregulated GR, GST, DHAR, MDHAR, GLY-I, and GLY-II	Hasanuzzaman et al. 2017
Brassica juncea	Salt stress	100 μM	• Well-developed rearranged chloroplast thylakoids and appropriately stacked grana • Improved photosynthetic parameters (P_n, G_s, C_i, PSII, and Rubisco activity) • Enhanced sulfur assimilation • Increased plant dry mass	• Reduced accumulation of $O_2^{-\bullet}$, H_2O_2, and TBARS • Upregulated CAT, APOX, GR, and ATP-S activity • Improved cysteine, sulfur, and GSH content, and GSH/GSSG • Reduced amount of Na^+ and Cl^-	Fatma et al. 2016

Gossypium hirsutum	Salt stress	1.00 mM	• Enhanced plant growth, stem, and root length • Upregulated photosynthetic rate • Improved stress tolerance	• Reduced $O_2^{-\bullet}$ generation rate • Improved CAT and SOD activity • Enhanced proline content • Improved cellular osmotic pressure and cytoplasmic viscosity • Increased Ca^{2+} absorption and cell permeability	Dong et al. 2014
Triticum aestivum	Chilling stress	0.1 mM	• Improved tolerance of seedlings toward cold stress	• Upregulated SOD, POX, and CAT activity • Reduced H_2O_2, $O_2^{-\bullet}$, and MDA levels	Esim et al. 2014
Triticum aestivum	High temperature stress	0.5 mM	• Improved chlorophyll content • Upregulated antioxidant guard system • Augmented methylglyoxalde toxification mechanism	• Reduced lipid peroxidation and H_2O_2 content • Augmented levels of AsA, GSH, and GSH/GSSG • Upregulated APOX, DHAR, MDHAR, GR, GST, CAT, and Gly-I activity	Hasanuzzaman et al. 2012
Cucumis sativus	Chilling stress	1.0 mmol/1	• Increased chl-a and chl-b levels • Upregulated levels of soluble sugars and proteins	• Enhanced SOD, POD, GR, and CAT activity • Lowered MDA levels	Liu et al. 2011

NO, nitric oxide; SNP, sodium nitroprusside; NUE, nitrogen use efficiency; SUE, sulfur use efficiency; PSII, photosystem II; Rubisco, ribulose-1,5-bisphosphate carboxylase/oxygenase; NR, nitrate reductase, H_2O_2, hydrogen peroxide; TBARS, thiobarbituric acid reactive substances; APOX, ascorbate peroxidase; GR, glutathione reductase; SOD, superoxide dismutase; GSH, glutathione, MDA, malondialdehyde; POX, peroxidase; CAT, catalase; PPO, polyphenol oxidase; PAL, phenylalanine ammonia lyase; TAL, tyrosine ammonia lyase; LOX, lipoxygenase, Se, selenium; P_n, net photosynthetic rate; T_r, transpiration rate, C_i, intercellular CO_2; G_s, stomatal conductance; AsA, ascorbic acid; $O_2^{-\bullet}$, superoxide anion; GSSG, glutathione disulfide; δ-ALAD, δ-aminolevulinic acid dehydratase; CA, carbonic anhydrase; APS-R, adenosine 5-phosphosulfate reductase, ATP-S, ATP sulfurylase; OAST-L, O-acetylserine (thiol) lyase; SiR, sulfide reductase, Ni, nickel; RWC, relative water content; GST, glutathione-S-transferase; NADP, nicotinamide adenine dinucleotide phosphate; GLY-I, glyoxylase-I; Cd, cadmium; DHAR, dehydroascorbate reductase; MDHAR, monodehydroascorbate reductase; GLY-II, glyoxalase-II; Na^+, sodium ion; Cl^-, chlorine ion, Ca^{2+}, calcium ion; chl-a, chlorophyll-a; chl-b, chlorophyll-b.

depending upon environmental stimuli, cell type, biochemical microenvironment, and intracellular location (Misra et al. 2010). Initial studies revealed that NO can promote defense gene expression in the aleurone layer of barley via cyclic guanosine monophosphate (cGMP) and cyclic adenosine diphosphate ribose (cADPR) in tobacco. Although NO seems to affect mRNA stability, functioning of several transcription factors, upstream signaling networks, and translation processes, there is still no substantiation of the DNA promoter sequences and elements from the eukaryotic genetic material that react or bind to NO (Domingos et al. 2015).

The physiological functions exercised by NO are attributed to discrete posttranslational modifications (PTMs) of the target proteins. *S*-nitrosylation, is a principal modification among them that has recently emerged as an indispensable mechanism in response to abiotic stress (Kolbert et al. 2019). NO initiates *S*-nitrosylation by inducing covalent reversible reactions with cystenine thiol-containing proteins leading to *S*-nitrosothiol formation. This alters protein activity, its localization, and interaction (Ferreira and Cataneo 2010; Kolbert et al. 2019). *S*-nitrosylation has been implicated in controlling several plant processes including cellular architecture, processing of genetic information, protection from oxidative stress, photosynthesis, biotic and abiotic stress defense responses, and hormonal signaling (Astier et al. 2012; Astier and Lindermayr 2012). At present, the best-described example of *S*-nitrosylation includes alteration of phytohormone biosynthetic enzymes, receptors, and signal transduction proteins in plants (Freschi 2013), and the enzymes involved in the ascorbate–glutathione cycle (Singh et al. 2017).

Another NO influenced PTM is tyrosine nitration, which depends upon the NO and ROS reaction that leads to generation of NO-derived reactive species like peroxynitrite ($ONOO^-$), leading to covalent modification of the tyrosine residues (Astier and Lindermayr 2012). Tyrosine nitration often leads to transient alterations in the occurrence of nitrated proteins. Initially tyrosine nitration was thought to be an irreversible process, but now it is believed to occur in both enzymatic and nonenzymatic ways (Astier and Lindermayr 2012; Kolbert et al. 2019). The third NO-related PTM involves conjugation of NO with metalloproteins, i.e. metal nitrosylation. Although cGMP in plants is a key intermediate in various NO-mediated responses like root growth, nodule functioning, defense mechanisms, and mitochondrial respiration, further experimental data are essential to elucidate regulation of plant-soluble guanylate cyclase by metal nitrosylation (Freschi 2013).

Regardless of the type of PTM initiated by NO, the chemical modifications brought about by NO represent the core mechanisms by which it impacts the signaling cascades that control growth and metabolism in plants. In responses controlled by the plant hormones, PTMs expedite the impact of NO on hormone synthesis and their responses via distinctive nonexclusive processes. The first such mechanism involves NO-triggered PTM of the proteins directly related to plant hormone production, conjugation, transport, degradation, perception, and signal transduction. Another indirect mechanism involves NO-dependent chemical modifications of transcription factors, regulatory proteins, and channels, which may impact the occurrence of several proteins that are more closely involved in these specific roles. The third possibility described in literature lately comprises a chemical reaction amid the NO derivatives such as $ONOO^-$ and plant hormones like zeatin, which generates products with transformed biological activity (Freschi 2013).

The developmental and physiological effects regulated by NO imply alteration in gene expression upon exposure to NO. A number of genes are reported to be initiated by NO. Akram et al. (2020) advocated that NO may trigger stress-allied genes thus increasing antioxidant enzyme activity in plants dealing with stress. Exogenous NO downregulates *OsPT2*, *OsSAMS1*, and *OsCS* gene expression, the genes involved in selenium (Se) transport and metabolism, to prevent Se toxicity in rice (Dai et al. 2020). Similarly, NO modulates metal transporters, stress-associated genes, transcription factors, amino acids, hormone signaling, calcium signaling, and secondary metabolism genes involved in detoxification of arsenic (Singh et al. 2017). *S*-nitrosoglutathione (GSNO) formed after a reaction between glutathione and NO acts as a NO reservoir in plant cells as well as regulating the expression of certain transcription factors including WRKY and MYB, implicated in defense responses to pathogenic and abiotic stressors (Baudouin 2011).

13.5 Interaction Between NO and Phytohormones

Owing to its exceedingly reactive behavior NO is involved in communications with a plethora of signaling cascades in order to accomplish specific functions associated with cells, tissues, or organelles in plants (Simontacchi et al. 2013). Lately, several research groups have affirmed after a sequence of experimental studies the role of NO in phytohormone-reconciled responses (Freschi 2013). In recent years, wide-ranging investigations have surveyed various mechanisms underlying plant hormones and NO cross-talk (Simontacchi et al. 2013). NO and phytohormones are known to interact at diversified, multiple levels. NO is confirmed to function both upstream and downstream of phytohormones depending upon the signaling network (Freschi 2013). Although, additional knowledge is necessary to elaborate the role of NO and its concurrent interaction with plant hormone-related proteins at transcriptional levels, a number of experimental studies have clearly revealed the effect of NO on synthesis, catabolism, transportation, conjugation, perception, and transduction of several phytohormone signals including ET, SA, and JA. Although still not fully elaborated, the underlying mechanism of NO and phytohormone interaction has lately been examined in several plant species displaying the regulatory role of these interactions on metabolic and developmental plant events (Figure 13.1). Taking into account the general function of the plant hormones in augmenting plant tolerance to varied environmental cues, it can be conjectured that plant hormone-mediated stress responses are co-related to NO biosynthesis (Shah et al. 2019). The following sections focus on current knowledge on cross-talk between signaling cascades of NO and SA, NO and JA, and NO and ET under normal conditions and varied stresses.

13.5.1 Interaction of NO with SA

Experimental evidence on mechanisms employed by NO to interact with SA signaling cascades has been particularly advanced in recent years (Mur et al. 2013). A widespread bioinformatics investigation of NO-receptive promoters revealed that cis-elements that are associated with SA signaling elements such as octopine synthase element-like sequences (OCSEs) and W-boxes were prominent in *Arabidopsis* (Palmieri

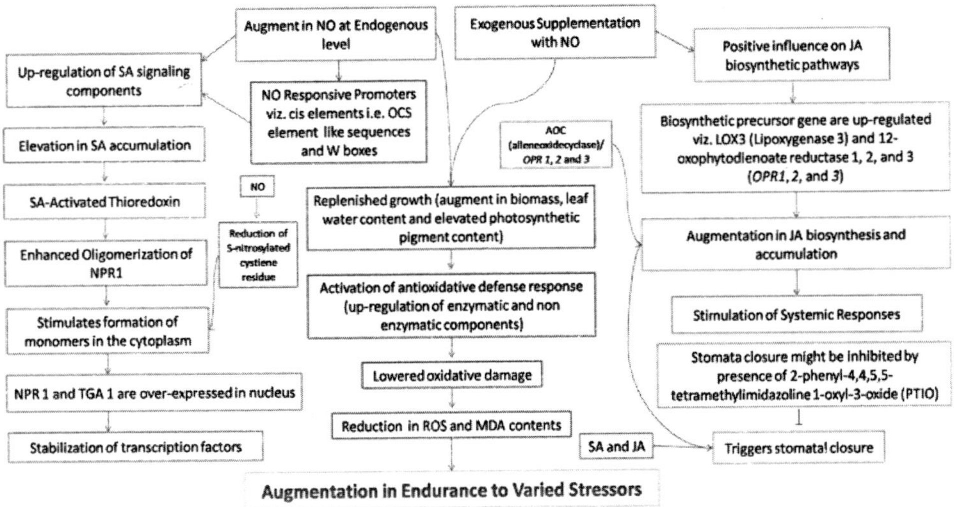

Figure 13.1 Influence of augmentation in endogenous and exogenous NO levels on SA and JA pathways. SA, salicylic acid; NPR 1, NONEXPRESSOR OF PATHOGENESIS-RELATED PROTEINS1; TGA 1, TGA Class Transcription Factor 1; NO, nitric oxide; ROS, reactive oxygen species; MDA, malondialdehyde; AOC, allene oxide cyclase; OPR1, 2, and 3, 12-oxophytodienoatereductase 1, 2, and 3; JA, jasmonic acid; LOX 3, lipooxygenase 3; PTIO, 2-phenyl-4,4,5,5-tetramethylimidazoline 1-oxyl-3-oxide.

et al. 2008). Transgenic exploitation of overexpressing and silencing approaches for endogenous NO production in *Arabidopsis* in retaliation to infestation by *Pseudomonas syringae* have been investigated lately. Upregulation in NO production was accompanied by augmentation of SA content in stressed plants. The above observation affirmed that SA is present downstream to NO and could have a key role in enhancing NO generation (Gémes et al. 2011). Furthermore, it has been revealed to trigger initiation of stomatal closure (Sun et al. 2010). Oligomeric NPR1 (NONEXPRESSOR OF PATHOGENESIS-RELATED PROTEINS1) is vital for its activity, and another component, i.e. *S*-nitrosylation of cysteine-156, is reported to assist oligomerization (Tada 2009). *S*-nitrosylated cysteine residue reduction is assisted by SA-induced thioredoxin, which triggers the production of monomers (Tada 2009). NO is thereby suggested to have a contradictory role as it enhances SA accumulation and triggers the NPR1 monomer formation and translocates them to the nucleus, and also initiates oligomerization by inducing nitrosylation (Lindermayr et al. 2010). The additive impact of NO on SA accumulation is elevated in TGACG-BINDING FACTOR 1 (TGA1) *S*-nitrosylation, which leads to stabilization of the transcription factors and its binding ability to the similar promoter sequence.

Recent observations made by Mostofa et al. (2019) revealed that *Oryza sativa* plants exposed to toxic cadmium (Cd) levels when exogenously supplemented with SA and sodium nitroprusside (SNP) showed replenished growth as indicated by increased biomass, leaf water content, and raised photosynthetic pigment levels. Both the signaling components were effectual in hampering Cd uptake and sequestration by lowering oxidative damage, which was accomplished by reducing ROS levels and malondialdehyde (MDA) generation. This was accompanied by stimulation of enzymatic and nonenzymatic constituents of the antioxidant defense pathways. The results further affirmed a synergistic effect between NO and SA in alleviating Cd- triggered adverse impacts on rice and other crop plants. Similarly, Gondor et al. (2016) and Per et al. (2017) also affirmed positive modulation of Cd toxicity resistance in maize and mustard plants, respectively. It was demonstrated that growth regulators might act as a feasible alternative to combat the toxic effect of Cd in important edible plants like mustard and maize. Also, a similar observation was made by Naser Alavi et al. (2014), who revealed SA and SNP upregulate growth in *Triticum aestivum* exposed to osmotic stress.

13.5.2 Interaction of NO with JA

An earlier study by Liu et al. (2005) was an attempt to understand the impact of interplay between JA and NO in modulating stomatal movements. Direct involvement of NO was affirmed in inducing the signaling cascade of JA, which triggers stomatal closure via enhanced NO biosynthesis in the guard cells. Also, JA-mediated stomata closing was mainly inhibited in the presence of 2-phenyl-4,4,5,5-tetramethylimidazoline1-oxyl-3-oxide (PTIO). NO positively induces synthesis of jasmonates (Mur et al. 2012) by augmenting JA synthesis gene expression (Palmieri et al. 2008). They further suggested the upregulation in expression of *LOX3* (At1g17420), 12-oxophytodienoate reductase 1, 2, and 3 (*OPR1*, *2*, and *3*) in response to exogenous NO supply. Strikingly, the expression of an important intermediate enzyme of JA biosynthesis, namely allene oxide cyclase (AOC), was revealed to be repressed upon NO application (Palmieri et al. 2008). Natural defense of *Solanum quitoense* was elicited by employing JA and SNP (a NO donor), protecting the plant from future infestation with *Fusarium oxysporum* f. sp. *quitoense* in response to triggered defense signaling networks (Ávila et al. 2019). *Fusarium oxysporum* is a hemibiotrophic pathogen, i.e. it begins life as a biotrophic pathogen and alters to a necrotrophic nature in later stages of the life cycle (Dean et al. 2012). JA foliar spray treatment as well as NO supplementation resulted in providing protection against the pathogen infestation by inducing systemic responses (Espunya et al. 2012). Similarly, the *def1* mutant in tomato plants, which has faulty JA production, also exhibited heightened vulnerability to *F. oxysporum* f. sp. *lycopersici*, which was reportedly curbed by exogenously supplied JA. Furthermore, a study by Sun et al. (2013) revealed that exogenous foliar spray application of methyl-JA reduced disease frequency in banana leaves affected by *F. oxysporum* f. sp. *cubense* race 4. As indicated by transcriptional profiling of watermelon roots, the majority of the genes involved in JA biosynthesis were overexpressed and had an irreconcilable interaction with *F. oxysporum* f. sp. *niveum* in comparison with amicable interaction with the same pathogen, which further indicated the positive influence of JA application in activation of defense mechanisms against pathogen infestation (Lü et al. 2011). In addition, in strawberry roots infested with *F. oxysporum* f. sp. *fragariae* the proteome analysis revealed that JA bestows early defense in plants (Fang et al. 2013).

13.5.3 Interaction of NO with ET

NO and ET interact during the course of plant development. Along with other regulatory signals, they both influence an array of plant developmental processes. These two gaseous signal molecules have been established to interact closely and regulate the key biological processes during the early developmental stages of plants, vegetative growth, fruit ripening, and leaf senescence (Kolbert et al. 2019). A substantial number of available reports pertaining to NO and ET interaction reveal an antagonistic relationship between the two (Manjunatha et al. 2010). Exogenous NO application postpones senescence of floral and vegetative organs in plants by adversely altering an array of elements implicated in ET production (Freschi 2013). During light-induced chloroplast differentiation, NO applied either exogenously or produced endogenously promoted de-etiolation-associated developments by suppressing ACO activity, thereby inhibiting ET production in tomato cotyledons. Furthermore, genetic and pharmacological approaches showed that augmented ET levels repressed formation of NO and NR activity, thus revealing antagonistic communication between NO and ET signals (Melo et al. 2016). Further evidence indicated that ET strongly influences sexual reproduction in plants by modulating multiple mechanisms including growth regulation of pollen tube, lifespan of the ovule, and the fruit set, but the probable NO–ET cross-talk in the course of these responses remains undescribed (Kolbert et al. 2019).

NO extend the post-harvest shelf life in fruits, acting as an antagonist of ET in most of these plants (Mukherjee 2019). Ripening-related processes including chlorophyll degradation, cell wall softening, and production of new pigments chiefly stimulated by ET are delayed or inhibited by NO supplementation, thus resulting in extended post-harvest fruit shelf life (Corpas et al. 2018). NO–ET interplay strictly regulates the biochemical pathways resulting in the synthesis of some essential nutritional fruit compounds like flavonoids, carotenoids, and ascorbate (Cruz et al. 2018). However, a synergistic relationship exists at the time of seed germination between NO and ET. NO counteracts seed dormancy imposed by abscisic acid and facilitates gibberellin-induced germination. It overcomes embryonic dormancy by encouraging ethylene emission accompanied by stimulation of ACO and ACS activities in apple (Gniazdowska et al. 2010; Arc et al. 2013). Exogenous NO-induced seed germination in *Amaranthus retroflexus*, which was followed by augmented ET synthesis, demonstrates that NO-mediated seed dormancy breaking depends upon ET as well. In addition, seed germination induced by ET also requires the presence of NO. Furthermore, ET- and NO-induced germination of seed is reported to be linked with cell cycle upregulation prior to radicle emergence (Kepczynski et al. 2017).

13.6 Jamming between NO, SA, and JA/ET

The SA- and ET/JA-assisted signaling cascades are the spine of plant immune responses against biotic aggressors. There exists increasing evidence of SA- and ET/JA-assisted defense pathways, which might be antagonistic (Li et al. 2019). The interplay between the hormones has been found to occur at multifaceted layers of modulation, which involves alteration of phytohormone metabolism, gene expression, and protein variations (Pieterse et al. 2009). Infection with hemibiotrophic pathogen NO bestows stimulation of SA biosynthesis by alleviating EIN3 (ET receptors) inhibition of *isochorismate synthase1* (*ICS1*) transcription (Chen et al. 2009). This alleviation of inhibition of EIN3 occurs in response to the restraint of ET biosynthesis via MAT1 (methionine adenosyltransferase 1) nitrosylation resulting in degradation of EIN3 to proteasome. As suggested earlier, SA also triggers thioredoxin levels leading to a reduction in NPR1 protein content resulting in its translocation to the nucleus from the cytoplasm (Lindermayr et al. 2010). Here the NPR1 binds to TGA transcription factors. Nitrosylation of these transcription factors eventually augments the binding affinity for the promoter, which is similar. A few TGA factors might also bind to promoters such as *ORA59* and hinder JA and ET biosynthesis-initiating genes (Mur et al. 2013). NPR1 has also been affirmed to provoke the expression of WKRY70 (transcription factor), which further stifles the regulation of several JA and ET genes (Li et al. 2004).

The transcript levels of marker genes involved in the signaling cascade of SA, NO, and JA were measured in *Solanum quitoense* (sensitive plants) and *Solanum hirtum* (resistant plants) shortly after infestation with *F. oxysporum* f. sp. *quitoense* (Ávila et al. 2019). Employing bioinformatic tools, oligonucleotide pairs were designed corresponding to the conserved regions of the marker *GSNOR*, *PR1*, and *PTI* genes for NO, SA, and JA NO signaling cascades, respectively. The results affirmed that the expression of *PTI*, a marker of JA biosynthesis, was augmented in *S. hirtum* plants but not *S. quitoense*, revealing significant participation of the JA signaling cascade in the early stages of infestation. Also the *S. quitoense* plants that were pretreated with JA demonstrated strapping and swift stimulation of JA signaling networks in response to *F. oxysporum* infestation, which asserted that immunity to pathogen was stimulated by JA supplementation. Another recent study on *Trichoderma* strain revealed it replenished plant growth and lowering of the gray mold disease index in cucumber plants infected with *Botrytis cinerea*. The novel *Trichoderma* strain was named *T. longibrachiatum* H9 and was inoculated in cucumber roots, elevating the endurance of the plant in response to biotic cues. Varied omic approaches, namely transcriptomic and proteomic analysis, suggested a wide array of differentially expressed genes (DEGs) and differentially expressed proteins (DEPs) were expressed in response to fungal inoculation. This augmentation in DEGs and DEPs was directly linked to stress retaliation, triggered secondary metabolism, and signaling phytohormones, which included JA, ET, and SA. Also it was revealed by the study that the endogenous JA and SA levels were elevated in *T. longibrachiatum* H9-inoculated plants (Yuan et al. 2019).

13.7 Conclusions and Future Prospects

Regardless of the conceptual complexity and methodological difficulties involved in elucidation of precise and meticulous mechanisms accountable for interrelating different phytohormones and NO signaling cascades for synchronization of plant development and metabolism, certain progressive understandings about NO–phytohormone cross-talk have been achieved recently. Even though our knowledge about NO interactions with phytohormones has radically improved in recent times, still several pieces of this puzzle need to be discovered. It is now recognized that NO synchronizes with the phytohormones and controls not only the expression of genes but also the functioning of several antioxidative enzymes during hostile environmental conditions. Interestingly, the studies associated with NO–phytohormone communications include trials conducted in organized and controlled research laboratories, thus limiting our knowledge related to cross-talk among these signal molecules at the time of flowering or during development of the grains. Our understanding related to the physiological importance of NO–phytohormone interaction is chiefly associated with NO-induced PTMs of the transcription factors and of the biosynthetic enzymes, but the same has been poorly studied in plants, offering a rich and vast area for future exploration. As well as this, we need to elucidate the relationships among NO, ROS, phytohormones, protein kinases, and the cytoskeletal proteins to recognize the convoluted networks of NO signaling during different environmental conditions.

References

Akram, N.A., Hafeez, N., Farid-ul-Haq, M. et al. (2020). Foliage application and seed priming with nitric oxide causes mitigation of salinity-induced metabolic adversaries in broccoli (*Brassica oleracea* L.) plants. *Acta Physiologiae Plantarum* 42: 1–9.

Arc, E., Sechet, J., Corbineau, F. et al. (2013). ABA crosstalk with ethylene and nitric oxide in seed dormancy and germination. *Frontiers in Plant Science* 4: 63.

Arora, D., Jain, P., Singh, N. et al. (2016). Mechanisms of nitric oxide crosstalk with reactive oxygen species scavenging enzymes during abiotic stress tolerance in plants. *Free Radical Research* 50: 291e303.

Astier, J., Kulik, A., Koen, E. et al. (2012). Protein S-nitrosylation: what's going on in plants? *Free Radical Biology and Medicine* 53: 1101–1110.

Astier, J. and Lindermayr, C. (2012). Nitric oxide-dependent posttranslational modification in plants: an update. *International Journal of Molecular Sciences* 13: 15193–15208.

Ávila, A.C., Ochoa, J., Proaño, K. et al. (2019). Jasmonic acid and nitric oxide protects naranjilla (*Solanum quitoense*) against infection by *Fusarium oxysporum* f. sp. *quitoense* by eliciting plant defense responses. *Physiological and Molecular Plant Pathology* 106: 129–136.

Baudouin, E. (2011). The language of nitric oxide signaling. *Plant Biology* 13: 233–242.

Campbell, W.H. (1999). Nitrate reductase structure, function and regulation: bridging the gap between biochemistry and physiology. *Annual Review of Plant Physiology and Plant Molecular Biology* 50: 277–303.

Campos, F.V., Oliveira, J.A., Pereira, M.G. et al. (2019). Nitric oxide and phytohormone interactions in the response of *Lactuca sativa* to salinity stress. *Planta* 250 (5): 1475–1489.

Chandra, J., Tandon, M., and Keshavkant, S. (2015). Increased rate of drying reduces metabolic inequity and critical water content in radicles of *Cicer arietinum* L. *Physiology and Molecular Biology of Plants* 21: 215–223.

Chen, H., Xue, L., Chintamanani, S. et al. (2009). ETHYLENE INSENSITIVE3 and ETHYLENE INSENSITIVE3-LIKE1 repress SALICYLIC ACID INDUCTION DEFICIENT2 expression to negatively regulate plant innate immunity in *Arabidopsis*. *The Plant Cell* 21 (8): 2527–2540.

Chen, W., Dong, Y., Hu, G. et al. (2018). Effects of exogenous nitric oxide on cadmium toxicity and antioxidative system in perennial ryegrass. *Journal of Soil Science and Plant Nutrition* 18 (1): 129–143.

Corpas, F.J., Freschi, L., Rodriguez-Ruiz, M. et al. (2018). Nitro-oxidative metabolism during fruit ripening. *Journal of Experimental Botany* 69: 3449–3463.

Cruz, A.B., Bianchetti, R.E., Alves, F.R.R. et al. (2018). Light, ethylene and auxin signaling interaction regulates carotenoid biosynthesis during tomato fruit ripening. *Frontiers in Plant Science* 9: 1370.

Dai, Z., Rizwan, M., Gao, F. et al. (2020). Nitric oxide alleviates selenium toxicity in rice by regulating antioxidation, selenium uptake, speciation and gene expression. *Environmental Pollution* 1: 257.

Dean, R., Van Kan, J.A., Pretorius, Z.A. et al. (2012). The top 10 fungal pathogens in molecular plant pathology. *Molecular Plant Pathology* 13 (4): 414–430.

Domingos, P., Prado, A.M., Wong, A. et al. (2015). Nitric oxide: a multitasked signaling gas in plants. *Molecular Plant* 8 (4): 506–520.

Dong, Y.J., Jinc, S.S., Liu, S. et al. (2014). Effects of exogenous nitric oxide on growth of cotton seedlings under NaCl stress. *Journal of Soil Science and Plant Nutrition* 14 (1): 1–3.

Esim, N., Atici, O., and Mutlu, S. (2014). Effects of exogenous nitric oxide in wheat seedlings under chilling stress. *Toxicology and Industrial Health* 30 (3): 268–274.

Espunya, M.C., De Michele, R., Gómez-Cadenas, A. et al. (2012). S-Nitrosoglutathione is a component of wound-and salicylic acid-induced systemic responses in *Arabidopsis thaliana*. *Journal of Experimental Botany* 63 (8): 3219–3227.

Fang, X., Jost, R., Finnegan, P.M. et al. (2013). Comparative proteome analysis of the strawberry–*Fusarium oxysporum* f. sp. *fragariae* pathosystem reveals early activation of defense responses as a crucial determinant of host resistance. *Journal of Proteome Research* 12 (4): 1772–1788.

Fatma, M., Masood, A., Per, T.S. et al. (2016). Nitric oxide alleviates salt stress inhibited photosynthetic performance by interacting with sulfur assimilation in mustard. *Frontiers in Plant Science* 7: 521.

Ferreira, L.C. and Cataneo, A.C. (2010). Nitric oxide in plants: a brief discussion on this multifunctional molecule. *Scientia Agricola* 67 (2): 236–243.

Freschi, L. (2013). Nitric oxide and phytohormone interactions: current status and perspectives. *Frontiers in Plant Science* 4: 398.

Gautam, H., Sehar, Z., Rehman, M.T. et al. (2021). Nitric oxide enhances photosynthetic nitrogen and sulfur-use efficiency and activity of ascorbate-glutathione cycle to reduce high temperature stress-induced oxidative stress in rice (*Oryza sativa* L.) plants. *Biomolecules* 11 (2): 305.

Gémes, K., Poór, P., Horváth, E. et al. (2011). Cross-talk between salicylic acid and NaCl-generated reactive oxygen species and nitric oxide in tomato during acclimation to high salinity. *Physiologia Plantarum* 142 (2): 179–192.

Gniazdowska, A., Krasuska, U., and Bogatek, R. (2010). Dormancy removal in apple embryos by nitric oxide or cyanide involves modifications in ethylene biosynthetic pathway. *Planta* 232: 1397–1407.

Gondor, O.K., Pál, M., Darkó, É. et al. (2016). Salicylic acid and sodium salicylate alleviate cadmium toxicity to different extents in maize (*Zea mays* L.). *PLoS ONE* 11 (8): e0160157.

Gupta, K.J., Fernie, A.R., Kaiser, W.M. et al. (2011). On the origins of nitric oxide. *Trends in Plant Science* 16: 160–168.

Gupta, K.J., Igamberdiev, A.U., Manjunatha, G. et al. (2011). The emerging roles of nitric oxide (NO) in plant mitochondria. *Plant Science* 181: 520–526.

Hasanuzzaman, M., Nahar, K., Alam, M.M. et al. (2012). Exogenous nitric oxide alleviates high temperature induced oxidative stress in wheat (*Triticum aestivum* L.) seedlings by modulating the antioxidant defense and glyoxalase system. *Australian Journal of Crop Science* 6 (8): 1314–1323.

Hasanuzzaman, M., Nahar, K., Hossain, M.S. et al. (2017). Nitric oxide pretreatment enhances antioxidant defense and glyoxalase systems to confer PEG-induced oxidative stress in rapeseed. *Journal of Plant Interactions* 12 (1): 323–331.

Kepczynski, J., Cembrowska-Lech, D., and Sznigir, P. (2017). Interplay between nitric oxide, ethylene, and gibberellic acid regulating the release of *Amaranthus retroflexus* seed dormancy. *Acta Physiologiae Plantarum* 39: 254.

Kolbert, Z., Feigl, G., Freschi, L. et al. (2019). Gasotransmitters in action: nitric oxide-ethylene crosstalk during plant growth and abiotic stress responses. *Antioxidants* 8 (6): 167.

Li, J., Brader, G., and Palva, E.T. (2004). The WRKY70 transcription factor: a node of convergence for jasmonate-mediated and salicylate-mediated signals in plant defense. *The Plant Cell* 16 (2): 319–331.

Li, N., Han, X., Feng, D. et al. (2019). Signaling crosstalk between salicylic acid and ethylene/jasmonate in plant defense: do we understand what they are whispering? *International Journal of Molecular Sciences* 20 (3): 671.

Lindermayr, C., Sell, S., Müller, B. et al. (2010). Redox regulation of the NPR1-TGA1 system of *Arabidopsis thaliana* by nitric oxide. *The Plant Cell* 22 (8): 2894–2907.

Liu, X., Shi, W., Zhang, S. et al. (2005). Nitric oxide involved in signal transduction of jasmonic acid-induced stomatal closure of *Vicia faba* L. *Chinese Science Bulletin* 50 (6): 520–525.

Liu, X., Wang, L., Liu, L. et al. (2011). Alleviating effect of exogenous nitric oxide in cucumber seedling against chilling stress. *African Journal of Biotechnology* 10 (21): 4380–4386.

Lü, G., Guo, S., Zhang, H. et al. (2011). Transcriptional profiling of watermelon during its incompatible interaction with *Fusarium oxysporum* f. sp. *niveum*. *European Journal of Plant Pathology* 131 (4): 585–601.

Manjunatha, G., Lokesh, V., and Neelwarne, B. (2010). Nitric oxide in fruit ripening: trends and opportunities. *Biotechnology Advances* 28: 489–499.

Melo, N.K., Bianchetti, R.E., Lira, B.S. et al. (2016). Nitric oxide, ethylene, and auxin cross talk mediates greening and plastid development in deetiolating tomato seedlings. *Plant Physiology* 170: 2278–2294.

Misra, A.N., Misra, M., and Singh, R. (2010). Nitric oxide biochemistry, mode of action and signaling in plants. *Journal of Medicinal Plants Research* 25: 2729–2739.

Modolo, L.V., Augusto, O., Almeida, I.M.G. et al. (2006). Decreased arginine and nitrite levels in nitrate reductase-deficient *Arabidopsis thaliana* plants impair nitric oxide synthesis and the hypersensitive response to *Pseudomonas syringae*. *Plant Science* 171: 34–40.

Mostofa, M.G., Rahman, M., Ansary, M. et al. (2019). Interactive effects of salicylic acid and nitric oxide in enhancing rice tolerance to cadmium stress. *International Journal of Molecular Sciences* 20 (22): 5798.

Mukherjee, S. (2019). Recent advancements in the mechanism of nitric oxide signaling associated with hydrogen sulfide and melatonin crosstalk during ethylene-induced fruit ripening in plants. *Nitric Oxide* 82: 25–34.

Mur, L.A., Prats, E., Pierre, S. et al. (2013). Integrating nitric oxide into salicylic acid and jasmonic acid/ethylene plant defense pathways. *Frontiers in Plant Science* 4: 215.

Mur, L.A., Sivakumaran, A., Mandon, J. et al. (2012). Haemoglobin modulates salicylate and jasmonate/ethylene-mediated resistance mechanisms against pathogens. *Journal of Experimental Botany* 63 (12): 4375–4387.

Naser Alavi, S.M., Arvin, M.J., and Manoochehri Kalantari, K. (2014). Salicylic acid and nitric oxide alleviate osmotic stress in wheat (*Triticum aestivum* L.) seedlings. *Journal of Plant Interactions* 9 (1): 683–688.

Palmieri, M.C., Sell, S., Huang, X. et al. (2008). Nitric oxide-responsive genes and promoters in *Arabidopsis thaliana*: a bioinformatics approach. *Journal of Experimental Botany* 59 (2): 177–186.

Parankusam, S., Adimulam, S.S., Bhatnagar-Mathur, P. et al. (2017). Nitric oxide (NO) in plant heat stress tolerance: current knowledge and perspectives. *Frontiers in Plant Science* 8: 1582.

Per, T.S., Masood, A., and Khan, N.A. (2017). Nitric oxide improves S-assimilation and GSH production to prevent inhibitory effects of cadmium stress on photosynthesis in mustard (*Brassica juncea* L.). *Nitric Oxide* 68: 111–124.

Pieterse, C.M., Leon-Reyes, A., Van der Ent, S. et al. (2009). Networking by small-molecule hormones in plant immunity. *Nature Chemical Biology* 5 (5): 308–316.

Rai, K.K., Pandey, N., and Rai, S.P. (2019). Salicylic acid and nitric oxide signaling in plant heat stress. *Physiologia Plantarum* 168 (2): 241–255.

Rezayian, M., Ebrahimzadeh, H., and Niknam, V. (2020). Nitric oxide stimulates antioxidant system and osmotic adjustment in soybean under drought stress. *Journal of Soil Science and Plant Nutrition* 16: 1–11.

Sahay, S. and Gupta, M. (2017). An update on nitric oxide and its benign role in plant responses under metal stress. *Nitric Oxide* 67: 39–52.

Shah, T., Wahid, S., Ilyas, M. et al. (2019). Nitric oxide and phytohormones cross-talk during abiotic stresses responses in plants. In: *Reactive Oxygen, Nitrogen and Sulfur Species in Plants: Production, Metabolism, Signaling and Defense Mechanisms* (eds. M. Hasanuzzaman, V. Vasileios, K. Nahar, and M. Fujita), 533–554. Hoboken, NJ: John Wiley & Sons Ltd.

Sharma, A., Soares, C., Sousa, B. et al. (2019). Nitric oxide-mediated regulation of oxidative stress in plants under metal stress: a review on molecular and biochemical aspects. *Physiologia Plantarum* 168 (2): 318–344.

Siddiqui, M.H., Alamri, S., Alsubaie, Q.D. et al. (2020). Exogenous nitric oxide alleviates sulfur deficiency-induced oxidative damage in tomato seedlings. *Nitric Oxide* 94: 95–107.

Simontacchi, M., García-Mata, C., Bartoli, C.G. et al. (2013). Nitric oxide as a key component in hormone-regulated processes. *Plant Cell Reports* 32 (6): 853–866.

Singh, P.K., Indoliya, Y., Chauhan, A.S. et al. (2017). Nitric oxide mediated transcriptional modulation enhances plant adaptive responses to arsenic stress. *Scientific Reports* 7 (1): 1–3.

Soliman, M., Alhaithloul, H.A., Hakeem, K.R. et al. (2019). Exogenous nitric oxide mitigates nickel-induced oxidative damage in eggplant by upregulating antioxidants, osmolyte metabolism, and glyoxalase systems. *Plants* 8 (12): 562.

Sun, D., Lu, X., Hu, Y. et al. (2013). Methyl jasmonate induced defense responses increase resistance to *Fusarium oxysporum* f. sp. *cubense* race 4 in banana. *Scientia Horticulturae* 164: 484–491.

Sun, L.R., Hao, F.S., Lu, B.S. et al. (2010). AtNOA1 modulates nitric oxide accumulation and stomatal closure induced by salicylic acid in *Arabidopsis*. *Plant Signaling & Behavior* 5 (8): 1022–1024.

Tada, Y. (2009). Plant immunity requires conformational changes of NPR1 via S-nitrosylation and thioredoxins. *Science* 325 (5944): 1072.

Wei, L., Zhang, M., Wei, S. et al. (2020). Roles of nitric oxide in heavy metal stress in plants: cross-talk with phytohormones and protein S-nitrosylation. *Environmental Pollution* 259: 113943.

Yadu, S., Dewangan, T.L., Chandrakar, V. et al. (2017). Imperative roles of salicylic acid and nitric oxide in improving salinity tolerance in *Pisum sativum* L. *Physiology and Molecular Biology of Plants* 23 (1): 43–58.

Yu, M., Lamattina, L., Spoel, S.H. et al. (2014). Nitric oxide function in plant biology: a redox cue in deconvolution. *The New Phytologist* 202: 1142–1156.

Yuan, M., Huang, Y., Ge, W. et al. (2019). Involvement of jasmonic acid, ethylene and salicylic acid signaling pathways behind the systemic resistance induced by *Trichoderma longibrachiatum* H9 in cucumber. *BMC Genomics* 20 (1): 1–13.

Zhang, L., Li, X., Li, X. et al. (2016). Exogenous nitric oxide protects against drought-induced oxidative stress in *Malus* rootstocks. *Turkish Journal of Botany* 40: 17–27.

Index